Baoding Liu

Uncertainty Theory

Studies in Computational Intelligence, Volume 300

Editor-in-Chief

Prof. Janusz Kacprzyk
Systems Research Institute
Polish Academy of Sciences
ul. Newelska 6
01-447 Warsaw
Poland
E-mail: kacprzyk@ibspan.waw.pl

Baoding Liu

Uncertainty Theory

A Branch of Mathematics for Modeling Human Uncertainty

 Springer

Baoding Liu
Department of Mathematical Sciences
Tsinghua University
Beijing 100084
China
E-mail: liu@tsinghua.edu.cn

ISBN 978-3-642-42248-5   ISBN 978-3-642-13959-8 (eBook)

DOI 10.1007/978-3-642-13959-8

Studies in Computational Intelligence   ISSN 1860-949X

© 2010 Springer-Verlag Berlin Heidelberg

Typeset & Cover Design: Scientific Publishing Services Pvt. Ltd., Chennai, India.

Printed on acid-free paper

9 8 7 6 5 4 3 2 1

springer.com

Contents

Preface

Some information and knowledge are usually represented by human language like "about 100km", "approximately 39 °C", "roughly 80kg", "low speed", "middle age", and "big size". Perhaps some people think that they are subjective probability or they are fuzziness. However, a lot of surveys showed that those imprecise quantities behave neither like randomness nor like fuzziness. How do we understand them? How do we model them? These questions provide a motivation to invent another mathematical tool to model those imprecise quantities. In order to do so, an uncertainty theory was founded and became a branch of axiomatic mathematics. Since then, uncertainty theory has been developed steadily and applied widely.

Chapter 1 is devoted to the uncertainty theory. The first fundamental concept in uncertainty theory is uncertain measure that is used to measure the truth degree of an uncertain event. The second one is uncertain variable that is used to represent imprecise quantities. The third one is uncertainty distribution that is used to describe uncertain variables in an incomplete but easy-to-use way. Uncertainty theory is thus deduced from those three foundation stones, and plays the role of mathematical model to deal with uncertain phenomena.

Uncertain programming is a type of mathematical programming involving uncertain variables. A key problem in uncertain programming is how to rank uncertain variables, and different ranking criteria produce different classes of uncertain programming. Chapter 2 will introduce four ranking criteria and then provide a spectrum of uncertain programming with applications to project scheduling problem, vehicle routing problem and machine scheduling problem.

The term risk has been used in different ways in literature. In this book the risk is defined as the accidental loss plus the uncertain measure of such loss, and a risk index is defined as the uncertain measure that some specified loss occurs. Chapter 3 will introduce uncertain risk analysis that is a tool to quantify risk via uncertainty theory.

Reliability index is defined as the uncertain measure that some system is working. Thus reliability and risk have the same root in mathematics. They are separately treated for application convenience in practice rather than theoretical demand. Chapter 4 will introduce uncertain reliability analysis that is a tool to deal with system reliability via uncertainty theory.

An uncertain process is essentially a sequence of uncertain variables indexed by time or space. Thus an uncertain process is usually used to model uncertain phenomena that vary with time or space. Some basic concepts of uncertain process will be presented in Chapter 5.

Uncertain calculus is a branch of mathematics that deals with differentiation and integration of function of uncertain processes. As the very core of uncertain calculus, canonical process is a Lipschitz continuous uncertain process that has stationary and independent increments and every increment is a normal uncertain variable. Chapter 6 will introduce the uncertain calculus including canonical process, uncertain integral and chain rule.

Uncertain differential equation is a type of differential equation driven by canonical process. Chapter 7 will discuss the existence, uniqueness and stability of solutions of uncertain differential equations, and will design a numerical method for solving monotone uncertain differential equations. This chapter will also present an application of uncertain differential equation in finance.

Uncertain logic is a generalization of mathematical logic for dealing with uncertain knowledge via uncertainty theory. A key point in uncertain logic is that the truth value of an uncertain proposition is defined as the uncertain measure that the proposition is true. One advantage of uncertain logic is the well consistency with classical logic. In other words, uncertain logic obeys the law of truth conservation, and is consistent with the law of excluded middle and the law of contradiction. Chapter 8 will discuss uncertain propositional logic and uncertain predicate logic.

Uncertain entailment is a methodology for calculating the truth value of an uncertain formula via the maximum uncertainty principle when the truth values of other uncertain formulas are given. That is, we will assign an uncertain formula a truth value as close to 0.5 as possible. Chapter 9 will introduce an entailment model from which modus ponens, modus tollens and hypothetical syllogism are deduced.

Uncertain set is a measurable function from an uncertainty space to a collection of sets. In other words, uncertain set is a set-valued function on an uncertainty space. Thus the main difference between uncertain set and uncertain variable is that the former takes values of set and the latter takes values of point. The concepts of membership function and uncertainty distribution are two basic tools to describe uncertain sets, where membership function is intuitionistic for us but frangible for arithmetic operations, and uncertainty distribution is hard-to-understand for us but easy-to-use for arithmetic operations. Fortunately, an uncertainty distribution may be uniquely determined by a membership function. In practice, we first determine membership functions for uncertain sets, and convert membership functions to uncertainty distributions. Then we perform arithmetic operations on uncertain sets via uncertainty distributions rather than membership functions. Chapter 10 will provide an uncertain set theory that is a generalization of uncertainty theory to the domain of uncertain sets.

Some knowledge and evidence in human brain are actually uncertain sets rather than fuzzy sets or random sets. This fact encourages us to propose a theory of uncertain inference that is a process of deriving consequences from uncertain knowledge or evidence via the tool of conditional uncertain set. Chapter 11 will present an inference rule with applications to uncertain system and inference control.

The book is suitable for mathematicians, researchers, engineers, designers, and students in the field of mathematics, information science, operations research, system science, industrial engineering, computer science, artificial intelligence, finance, control, and management science. The readers will learn the axiomatic approach of uncertainty theory, and find this work a stimulating and useful reference.

Lecture Slides

If you need lecture slides for uncertainty theory, please download them from the website at http://orsc.edu.cn/liu/resources.htm.

A Guide for the Reader

The readers are not required to read the book from cover to cover. The logic dependence of chapters is illustrated by the figure below.

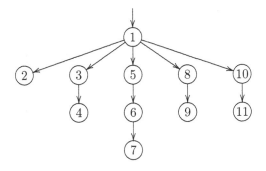

Acknowledgment

I am indebted to a series of grants from from National Natural Science Foundation, Ministry of Education, and Ministry of Science and Technology of China. I also express my deep gratitude to Professor Janusz Kacprzyk for the invitation to publish this book in his series, and Dr. Thomas Ditzinger of Springer for his wonderful cooperation and helpful comments.

February 2010   Baoding Liu
                                                                            Tsinghua University
                                                                         http://orsc.edu.cn/liu

To My Wife Jinlan

Chapter 1

Uncertainty Theory

Some information and knowledge are usually represented by human language like "about 100km", "approximately 39 °C", "roughly 80kg", "low speed", "middle age", and "big size". How do we understand them? Perhaps some people think that they are subjective probability or they are fuzzy concepts. However, a lot of surveys showed that those imprecise quantities behave neither like randomness nor like fuzziness. This fact provides a motivation to invent another mathematical tool, namely uncertainty theory.

Uncertainty theory was founded by Liu [120] in 2007. Nowadays uncertainty theory has become a branch of mathematics based on normality, monotonicity, self-duality, countable subadditivity, and product measure axioms. The first fundamental concept in uncertainty theory is uncertain measure that is used to measure the belief degree of an uncertain event. The second one is uncertain variable that is used to represent imprecise quantities. The third one is uncertainty distribution that is used to describe uncertain variables in an incomplete but easy-to-use way. Uncertainty theory is thus deduced from those three foundation stones, and provides a mathematical model to deal with uncertain phenomena.

The emphasis in this chapter is mainly on uncertain measure, uncertain variable, uncertainty distribution, independence, operational law, expected value, variance, moments, critical values, entropy, distance, convergence almost surely, convergence in measure, convergence in mean, convergence in distribution, and conditional uncertainty.

1.1 Uncertain Measure

Let Γ be a nonempty set. A collection $\mathcal{L}$ of subsets of Γ is called a σ-algebra if (a) $\Gamma \in \mathcal{L}$; (b) if $\Lambda \in \mathcal{L}$, then $\Lambda^c \in \mathcal{L}$; and (c) if $\Lambda_1, \Lambda_2, \cdots \in \mathcal{L}$, then $\Lambda_1 \cup \Lambda_2 \cup \cdots \in \mathcal{L}$. Each element Λ in the σ-algebra $\mathcal{L}$ is called an event. Uncertain measure is a function from $\mathcal{L}$ to $[0,1]$. In order to present an axiomatic definition of uncertain measure, it is necessary to assign to each event Λ a number $\mathcal{M}\{\Lambda\}$ which indicates the belief degree that Λ will occur. In order to ensure that the number $\mathcal{M}\{\Lambda\}$ has certain mathematical properties, Liu [120] proposed the following four axioms:

Axiom 1. *(Normality Axiom)* $\mathcal{M}\{\Gamma\} = 1$ *for the universal set* Γ.

Axiom 2. *(Monotonicity Axiom)* $\mathcal{M}\{\Lambda_1\} \leq \mathcal{M}\{\Lambda_2\}$ *whenever* $\Lambda_1 \subset \Lambda_2$.

B. Liu: Uncertainty Theory: A Branch of Mathematics, SCI 300, pp. 1–79.
springerlink.com   © Springer-Verlag Berlin Heidelberg 2010

Axiom 3. *(Self-Duality Axiom)* $\mathcal{M}\{\Lambda\} + \mathcal{M}\{\Lambda^c\} = 1$ *for any event* Λ.

Axiom 4. *(Countable Subadditivity Axiom) For every countable sequence of events* $\{\Lambda_i\}$*, we have*

$$\mathcal{M}\left\{\bigcup_{i=1}^{\infty} \Lambda_i\right\} \leq \sum_{i=1}^{\infty} \mathcal{M}\{\Lambda_i\}. \tag{1.1}$$

Remark 1.1: The law of contradiction tells us that a proposition cannot be both true and false at the same time, and the law of excluded middle tells us that a proposition is either true or false. The law of truth conservation is a generalization of the law of contradiction and the law of excluded middle, and says that the sum of truth values of a proposition and its negative proposition is identical to 1. Self-duality is in fact an application of the law of truth conservation in uncertainty theory. This is the main reason why self-duality axiom is assumed.

Remark 1.2: Pathology occurs if subadditivity is not assumed. For example, suppose that a universal set contains 3 elements. We define a set function that takes value 0 for each singleton, and 1 for each set with at least 2 elements. Then such a set function satisfies all axioms but subadditivity. Is it not strange if such a set function serves as a measure?

Remark 1.3: Pathology occurs if countable subadditivity axiom is replaced with finite subadditivity axiom. For example, assume the universal set consists of all real numbers. We define a set function that takes value 0 if the set is bounded, 0.5 if both the set and complement are unbounded, and 1 if the complement of the set is bounded. Then such a set function is finitely subadditive but not countably subadditive. Is it not strange if such a set function serves as a measure? This is the main reason why we accept the countable subadditivity axiom.

Remark 1.4: Although probability measure satisfies the above four axioms, probability theory is not a special case of uncertainty theory because the product probability measure does not satisfy the fifth axiom, namely product measure axiom on Page 7.

Definition 1.1 *(Liu [120]). The set function* $\mathcal{M}$ *is called an uncertain measure if it satisfies the normality, monotonicity, self-duality, and countable subadditivity axioms.*

Example 1.1: Let $\Gamma = \{\gamma_1, \gamma_2, \gamma_3\}$. For this case, there are only 8 events. Define

$$\mathcal{M}\{\gamma_1\} = 0.6, \quad \mathcal{M}\{\gamma_2\} = 0.3, \quad \mathcal{M}\{\gamma_3\} = 0.2,$$

$$\mathcal{M}\{\gamma_1, \gamma_2\} = 0.8, \quad \mathcal{M}\{\gamma_1, \gamma_3\} = 0.7, \quad \mathcal{M}\{\gamma_2, \gamma_3\} = 0.4,$$

$$\mathcal{M}\{\emptyset\} = 0, \quad \mathcal{M}\{\Gamma\} = 1.$$

Then $\mathcal{M}$ is an uncertain measure because it satisfies the four axioms.

Example 1.2: Suppose that $\lambda(x)$ is a nonnegative function on $\Re$ satisfying

$$\sup_{x \neq y} \left(\lambda(x) + \lambda(y) \right) = 1. \tag{1.2}$$

Then for any set Λ of real numbers, the set function

$$\mathcal{M}\{\Lambda\} = \begin{cases} \sup_{x \in \Lambda} \lambda(x), & \text{if } \sup_{x \in \Lambda} \lambda(x) < 0.5 \\ 1 - \sup_{x \in \Lambda^c} \lambda(x), & \text{if } \sup_{x \in \Lambda} \lambda(x) \geq 0.5 \end{cases} \tag{1.3}$$

is an uncertain measure on $\Re$.

Example 1.3: Suppose $\rho(x)$ is a nonnegative and integrable function on $\Re$ such that

$$\int_{\Re} \rho(x)\mathrm{d}x \geq 1. \tag{1.4}$$

Then for any Borel set Λ of real numbers, the set function

$$\mathcal{M}\{\Lambda\} = \begin{cases} \displaystyle\int_{\Lambda} \rho(x)\mathrm{d}x, & \text{if } \displaystyle\int_{\Lambda} \rho(x)\mathrm{d}x < 0.5 \\ 1 - \displaystyle\int_{\Lambda^c} \rho(x)\mathrm{d}x, & \text{if } \displaystyle\int_{\Lambda^c} \rho(x)\mathrm{d}x < 0.5 \\ 0.5, & \text{otherwise} \end{cases} \tag{1.5}$$

is an uncertain measure on $\Re$.

Example 1.4: Suppose $\lambda(x)$ is a nonnegative function and $\rho(x)$ is a nonnegative and integrable function on $\Re$ such that

$$\sup_{x \in \Lambda} \lambda(x) + \int_{\Lambda} \rho(x)\mathrm{d}x \geq 0.5 \quad \text{and/or} \quad \sup_{x \in \Lambda^c} \lambda(x) + \int_{\Lambda^c} \rho(x)\mathrm{d}x \geq 0.5 \tag{1.6}$$

for any Borel set Λ of real numbers. Then the set function

$$\mathcal{M}\{\Lambda\} = \begin{cases} \sup_{x \in \Lambda} \lambda(x) + \displaystyle\int_{\Lambda} \rho(x)\mathrm{d}x, & \text{if } \sup_{x \in \Lambda} \lambda(x) + \displaystyle\int_{\Lambda} \rho(x)\mathrm{d}x < 0.5 \\ 1 - \sup_{x \in \Lambda^c} \lambda(x) - \displaystyle\int_{\Lambda^c} \rho(x)\mathrm{d}x, & \text{if } \sup_{x \in \Lambda^c} \lambda(x) + \displaystyle\int_{\Lambda^c} \rho(x)\mathrm{d}x < 0.5 \\ 0.5, & \text{otherwise} \end{cases}$$

is an uncertain measure on $\Re$.

Theorem 1.1. *Suppose that* $\mathcal{M}$ *is an uncertain measure. Then the empty set* $\emptyset$ *has an uncertain measure zero, i.e.,*

$$\mathcal{M}\{\emptyset\} = 0. \tag{1.7}$$

Proof: It follows from the normality that $\mathcal{M}\{\Gamma\} = 1$. Since $\emptyset = \Gamma^c$, the self-duality axioms yields $\mathcal{M}\{\emptyset\} = 1 - \mathcal{M}\{\Gamma\} = 1 - 1 = 0$.

Theorem 1.2. *Suppose that* $\mathcal{M}$ *is an uncertain measure. Then we have*

$$0 \leq \mathcal{M}\{\Lambda\} \leq 1 \tag{1.8}$$

for any event Λ.

Proof: It follows from the monotonicity axiom that $0 \leq \mathcal{M}\{\Lambda\} \leq 1$ because $\emptyset \subset \Lambda \subset \Gamma$ and $\mathcal{M}\{\emptyset\} = 0$, $\mathcal{M}\{\Gamma\} = 1$.

Theorem 1.3. *Suppose that* $\mathcal{M}$ *is an uncertain measure. Then for any events* Λ_1 *and* Λ_2, *we have*

$$\mathcal{M}\{\Lambda_1\} \vee \mathcal{M}\{\Lambda_2\} \leq \mathcal{M}\{\Lambda_1 \cup \Lambda_2\} \leq \mathcal{M}\{\Lambda_1\} + \mathcal{M}\{\Lambda_2\}. \tag{1.9}$$

Proof: The left-hand inequality follows from the monotonicity axiom and the right-hand inequality follows from the countable subadditivity axiom immediately.

Theorem 1.4. *Suppose that* $\mathcal{M}$ *is an uncertain measure. Then for any events* Λ_1 *and* Λ_2, *we have*

$$\mathcal{M}\{\Lambda_1\} + \mathcal{M}\{\Lambda_2\} - 1 \leq \mathcal{M}\{\Lambda_1 \cap \Lambda_2\} \leq \mathcal{M}\{\Lambda_1\} \wedge \mathcal{M}\{\Lambda_2\}. \tag{1.10}$$

Proof: The right-hand inequality follows from the monotonicity axiom and the left-hand inequality follows from the self-duality and countable subadditivity axioms, i.e.,

$$\mathcal{M}\{\Lambda_1 \cap \Lambda_2\} = 1 - \mathcal{M}\{(\Lambda_1 \cap \Lambda_2)^c\} = 1 - \mathcal{M}\{\Lambda_1^c \cup \Lambda_2^c\}$$
$$\geq 1 - (\mathcal{M}\{\Lambda_1^c\} + \mathcal{M}\{\Lambda_2^c\})$$
$$= 1 - (1 - \mathcal{M}\{\Lambda_1\}) - (1 - \mathcal{M}\{\Lambda_2\})$$
$$= \mathcal{M}\{\Lambda_1\} + \mathcal{M}\{\Lambda_2\} - 1.$$

The inequalities are verified.

Null-Additivity Theorem

Null-additivity is a direct deduction from subadditivity. We first prove a more general theorem.

Theorem 1.5. *Let $\{\Lambda_i\}$ be a sequence of events with $\mathcal{M}\{\Lambda_i\} \to 0$ as $i \to \infty$. Then for any event Λ, we have*

$$\lim_{i\to\infty} \mathcal{M}\{\Lambda \cup \Lambda_i\} = \lim_{i\to\infty} \mathcal{M}\{\Lambda\backslash\Lambda_i\} = \mathcal{M}\{\Lambda\}. \qquad (1.11)$$

Proof: It follows from the monotonicity and countable subadditivity axioms that

$$\mathcal{M}\{\Lambda\} \le \mathcal{M}\{\Lambda \cup \Lambda_i\} \le \mathcal{M}\{\Lambda\} + \mathcal{M}\{\Lambda_i\}$$

for each i. Thus we get $\mathcal{M}\{\Lambda \cup \Lambda_i\} \to \mathcal{M}\{\Lambda\}$ by using $\mathcal{M}\{\Lambda_i\} \to 0$. Since $(\Lambda\backslash\Lambda_i) \subset \Lambda \subset ((\Lambda\backslash\Lambda_i) \cup \Lambda_i)$, we have

$$\mathcal{M}\{\Lambda\backslash\Lambda_i\} \le \mathcal{M}\{\Lambda\} \le \mathcal{M}\{\Lambda\backslash\Lambda_i\} + \mathcal{M}\{\Lambda_i\}.$$

Hence $\mathcal{M}\{\Lambda\backslash\Lambda_i\} \to \mathcal{M}\{\Lambda\}$ by using $\mathcal{M}\{\Lambda_i\} \to 0$.

Remark 1.5: It follows from the above theorem that the uncertain measure is null-additive, i.e., $\mathcal{M}\{\Lambda_1 \cup \Lambda_2\} = \mathcal{M}\{\Lambda_1\} + \mathcal{M}\{\Lambda_2\}$ if either $\mathcal{M}\{\Lambda_1\} = 0$ or $\mathcal{M}\{\Lambda_2\} = 0$. In other words, the uncertain measure remains unchanged if the event is enlarged or reduced by an event with uncertain measure zero.

Asymptotic Theorem

Theorem 1.6 *(Asymptotic Theorem). For any events $\Lambda_1, \Lambda_2, \cdots$, we have*

$$\lim_{i\to\infty} \mathcal{M}\{\Lambda_i\} > 0, \quad \text{if } \Lambda_i \uparrow \Gamma, \qquad (1.12)$$

$$\lim_{i\to\infty} \mathcal{M}\{\Lambda_i\} < 1, \quad \text{if } \Lambda_i \downarrow \emptyset. \qquad (1.13)$$

Proof: Assume $\Lambda_i \uparrow \Gamma$. Since $\Gamma = \cup_i \Lambda_i$, it follows from the countable subadditivity axiom that

$$1 = \mathcal{M}\{\Gamma\} \le \sum_{i=1}^{\infty} \mathcal{M}\{\Lambda_i\}.$$

Since $\mathcal{M}\{\Lambda_i\}$ is increasing with respect to i, we have $\lim_{i\to\infty} \mathcal{M}\{\Lambda_i\} > 0$. If $\Lambda_i \downarrow \emptyset$, then $\Lambda_i^c \uparrow \Gamma$. It follows from the first inequality and self-duality axiom that

$$\lim_{i\to\infty} \mathcal{M}\{\Lambda_i\} = 1 - \lim_{i\to\infty} \mathcal{M}\{\Lambda_i^c\} < 1.$$

The theorem is proved.

Example 1.5: Assume Γ is the set of real numbers. Let α be a number with $0 < \alpha \le 0.5$. Define a set function as follows,

$$\mathcal{M}\{\Lambda\} = \begin{cases} 0, & \text{if } \Lambda = \emptyset \\ \alpha, & \text{if } \Lambda \text{ is upper bounded} \\ 0.5, & \text{if both } \Lambda \text{ and } \Lambda^c \text{ are upper unbounded} \\ 1 - \alpha, & \text{if } \Lambda^c \text{ is upper bounded} \\ 1, & \text{if } \Lambda = \Gamma. \end{cases} \qquad (1.14)$$

It is easy to verify that $\mathcal{M}$ is an uncertain measure. Write $\Lambda_i = (-\infty, i]$ for $i = 1, 2, \cdots$ Then $\Lambda_i \uparrow \Gamma$ and $\lim_{i \to \infty} \mathcal{M}\{\Lambda_i\} = \alpha$. Furthermore, we have $\Lambda_i^c \downarrow \emptyset$ and $\lim_{i \to \infty} \mathcal{M}\{\Lambda_i^c\} = 1 - \alpha$.

Independence of Events

Definition 1.2. *The events* $\Lambda_1, \Lambda_2, \cdots, \Lambda_n$ *are said to be independent if*

$$\mathcal{M}\left\{\bigcap_{i=1}^{n} \Lambda_i^*\right\} = \min_{1 \le i \le n} \mathcal{M}\{\Lambda_i^*\} \tag{1.15}$$

where Λ_i^* *are arbitrarily chosen from* $\{\Lambda_i, \Lambda_i^c\}$, $i = 1, 2, \cdots, n$, *respectively.*

Note that (1.15) represents 2^n equations. For example, when $n = 2$, the four equations are

$$\begin{aligned}
\mathcal{M}\{\Lambda_1 \cap \Lambda_2\} &= \mathcal{M}\{\Lambda_1\} \wedge \mathcal{M}\{\Lambda_2\}, \\
\mathcal{M}\{\Lambda_1^c \cap \Lambda_2\} &= \mathcal{M}\{\Lambda_1^c\} \wedge \mathcal{M}\{\Lambda_2\}, \\
\mathcal{M}\{\Lambda_1 \cap \Lambda_2^c\} &= \mathcal{M}\{\Lambda_1\} \wedge \mathcal{M}\{\Lambda_2^c\}, \\
\mathcal{M}\{\Lambda_1^c \cap \Lambda_2^c\} &= \mathcal{M}\{\Lambda_1^c\} \wedge \mathcal{M}\{\Lambda_2^c\}.
\end{aligned} \tag{1.16}$$

Theorem 1.7. *The events* $\Lambda_1, \Lambda_2, \cdots, \Lambda_n$ *are independent if and only if*

$$\mathcal{M}\left\{\bigcup_{i=1}^{n} \Lambda_i^*\right\} = \max_{1 \le i \le n} \mathcal{M}\{\Lambda_i^*\} \tag{1.17}$$

where Λ_i^* *are arbitrarily chosen from* $\{\Lambda_i, \Lambda_i^c\}$, $i = 1, 2, \cdots, n$, *respectively.*

Proof: Assume $\Lambda_1, \Lambda_2, \cdots, \Lambda_n$ are independent events. It follows from the self-duality of uncertain measure that

$$\mathcal{M}\left\{\bigcup_{i=1}^{n} \Lambda_i^*\right\} = 1 - \mathcal{M}\left\{\bigcap_{i=1}^{n} (\Lambda_i^*)^c\right\} = 1 - \min_{1 \le i \le n} \mathcal{M}\{(\Lambda_i^*)^c\} = \max_{1 \le i \le n} \mathcal{M}\{\Lambda_i^*\}.$$

The equation (1.17) is proved. Conversely, assume (1.17). Then

$$\mathcal{M}\left\{\bigcap_{i=1}^{n} \Lambda_i^*\right\} = 1 - \mathcal{M}\left\{\bigcup_{i=1}^{n} (\Lambda_i^*)^c\right\} = 1 - \max_{1 \le i \le n} \mathcal{M}\{(\Lambda_i^*)^c\} = \min_{1 \le i \le n} \mathcal{M}\{\Lambda_i^*\}.$$

The equation (1.15) is true. The theorem is proved.

Uncertainty Space

Definition 1.3 *(Liu [120]).* *Let* Γ *be a nonempty set,* $\mathcal{L}$ *a* σ-algebra over Γ, *and* $\mathcal{M}$ *an uncertain measure. Then the triplet* $(\Gamma, \mathcal{L}, \mathcal{M})$ *is called an uncertainty space.*

Product Measure Axiom and Product Uncertain Measure

Product uncertain measure was defined by Liu [123] in 2009, thus producing the fifth axiom of uncertainty theory called *product measure axiom*. Let $(\Gamma_k, \mathcal{L}_k, \mathcal{M}_k)$ be uncertainty spaces for $k = 1, 2, \cdots, n$. Write

$$\Gamma = \Gamma_1 \times \Gamma_2 \times \cdots \times \Gamma_n, \quad \mathcal{L} = \mathcal{L}_1 \times \mathcal{L}_2 \times \cdots \times \mathcal{L}_n. \tag{1.18}$$

Then there is an uncertain measure $\mathcal{M}$ on the product σ-algebra $\mathcal{L}$ such that

$$\mathcal{M}\{\Lambda_1 \times \Lambda_2 \times \cdots \times \Lambda_n\} = \mathcal{M}_1\{\Lambda_1\} \wedge \mathcal{M}_2\{\Lambda_2\} \wedge \cdots \wedge \mathcal{M}_n\{\Lambda_n\} \tag{1.19}$$

for any measurable rectangle $\Lambda_1 \times \Lambda_2 \times \cdots \times \Lambda_n$. Such an uncertain measure is called the product uncertain measure denoted by

$$\mathcal{M} = \mathcal{M}_1 \wedge \mathcal{M}_2 \wedge \cdots \wedge \mathcal{M}_n. \tag{1.20}$$

In fact, the extension from the class of rectangles to the product σ-algebra $\mathcal{L}$ may be represented as follows.

Axiom 5. *(Liu [123], Product Measure Axiom) Let Γ_k be nonempty sets on which $\mathcal{M}_k$ are uncertain measures, $k = 1, 2, \cdots, n$, respectively. Then the product uncertain measure $\mathcal{M}$ is an uncertain measure on the product σ-algebra $\mathcal{L}_1 \times \mathcal{L}_2 \times \cdots \times \mathcal{L}_n$ satisfying*

$$\mathcal{M}\left\{ \prod_{k=1}^{n} \Lambda_k \right\} = \min_{1 \leq k \leq n} \mathcal{M}_k\{\Lambda_k\}. \tag{1.21}$$

That is, for each event $\Lambda \in \mathcal{L}$, we have

$$\mathcal{M}\{\Lambda\} = \begin{cases} \displaystyle\sup_{\Lambda_1 \times \Lambda_2 \times \cdots \times \Lambda_n \subset \Lambda} \min_{1 \leq k \leq n} \mathcal{M}_k\{\Lambda_k\}, \\ \qquad \text{if } \displaystyle\sup_{\Lambda_1 \times \Lambda_2 \times \cdots \times \Lambda_n \subset \Lambda} \min_{1 \leq k \leq n} \mathcal{M}_k\{\Lambda_k\} > 0.5 \\ 1 - \displaystyle\sup_{\Lambda_1 \times \Lambda_2 \times \cdots \times \Lambda_n \subset \Lambda^c} \min_{1 \leq k \leq n} \mathcal{M}_k\{\Lambda_k\}, \\ \qquad \text{if } \displaystyle\sup_{\Lambda_1 \times \Lambda_2 \times \cdots \times \Lambda_n \subset \Lambda^c} \min_{1 \leq k \leq n} \mathcal{M}_k\{\Lambda_k\} > 0.5 \\ 0.5, \qquad \text{otherwise.} \end{cases} \tag{1.22}$$

Theorem 1.8 *(Peng [176]). The product uncertain measure (1.22) is an uncertain measure.*

Proof: In order to prove that the product uncertain measure (1.22) is indeed an uncertain measure, we should verify that the product uncertain measure

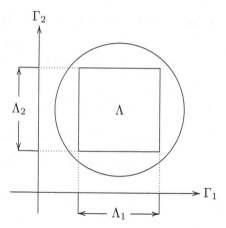

Figure 1.1: Graphical Illustration of Extension from the Class of Rectangles to the Product σ-Algebra. The uncertain measure of Λ (the disk) is essentially the acreage of its inscribed rectangle $\Lambda_1 \times \Lambda_2$ if it is greater than 0.5. Otherwise, we have to examine its complement Λ^c. If the inscribed rectangle of Λ^c is greater than 0.5, then $\mathcal{M}\{\Lambda^c\}$ is just its inscribed rectangle and $\mathcal{M}\{\Lambda\} = 1 - \mathcal{M}\{\Lambda^c\}$. If there does not exist an inscribed rectangle of Λ or Λ^c greater than 0.5, then we set $\mathcal{M}\{\Lambda\} = 0.5$.

satisfies the normality, monotonicity, self-duality and countable subadditivity axioms.

STEP 1: At first, for any event $\Lambda \in \mathcal{L}$, it is easy to verify that

$$\sup_{\Lambda_1 \times \Lambda_2 \times \cdots \times \Lambda_n \subset \Lambda} \min_{1 \leq k \leq n} \mathcal{M}_k\{\Lambda_k\} + \sup_{\Lambda_1 \times \Lambda_2 \times \cdots \times \Lambda_n \subset \Lambda^c} \min_{1 \leq k \leq n} \mathcal{M}_k\{\Lambda_k\} \leq 1.$$

This means that at most one of

$$\sup_{\Lambda_1 \times \Lambda_2 \times \cdots \times \Lambda_n \subset \Lambda} \min_{1 \leq k \leq n} \mathcal{M}_k\{\Lambda_k\} \quad \text{and} \quad \sup_{\Lambda_1 \times \Lambda_2 \times \cdots \times \Lambda_n \subset \Lambda^c} \min_{1 \leq k \leq n} \mathcal{M}_k\{\Lambda_k\}$$

is greater than 0.5. Thus the expression (1.22) is reasonable.

STEP 2: The product uncertain measure is clearly normal, i.e., $\mathcal{M}\{\Gamma\} = 1$.

STEP 3: We prove the self-duality, i.e., $\mathcal{M}\{\Lambda\} + \mathcal{M}\{\Lambda^c\} = 1$. The argument breaks down into three cases. Case 1: Assume

$$\sup_{\Lambda_1 \times \Lambda_2 \times \cdots \times \Lambda_n \subset \Lambda} \min_{1 \leq k \leq n} \mathcal{M}_k\{\Lambda_k\} > 0.5.$$

Then we immediately have

$$\sup_{\Lambda_1 \times \Lambda_2 \times \cdots \times \Lambda_n \subset \Lambda^c} \min_{1 \leq k \leq n} \mathcal{M}_k\{\Lambda_k\} < 0.5.$$

It follows from (1.22) that

$$\mathcal{M}\{\Lambda\} = \sup_{\Lambda_1 \times \Lambda_2 \times \cdots \times \Lambda_n \subset \Lambda} \min_{1 \leq k \leq n} \mathcal{M}_k\{\Lambda_k\},$$

$$\mathcal{M}\{\Lambda^c\} = 1 - \sup_{\Lambda_1 \times \Lambda_2 \times \cdots \times \Lambda_n \subset (\Lambda^c)^c} \min_{1 \leq k \leq n} \mathcal{M}_k\{\Lambda_k\} = 1 - \mathcal{M}\{\Lambda\}.$$

The self-duality is proved. Case 2: Assume

$$\sup_{\Lambda_1 \times \Lambda_2 \times \cdots \times \Lambda_n \subset \Lambda^c} \min_{1 \leq k \leq n} \mathcal{M}_k\{\Lambda_k\} > 0.5.$$

This case may be proved by a similar process. Case 3: Assume

$$\sup_{\Lambda_1 \times \Lambda_2 \times \cdots \times \Lambda_n \subset \Lambda} \min_{1 \leq k \leq n} \mathcal{M}_k\{\Lambda_k\} \leq 0.5$$

and

$$\sup_{\Lambda_1 \times \Lambda_2 \times \cdots \times \Lambda_n \subset \Lambda^c} \min_{1 \leq k \leq n} \mathcal{M}_k\{\Lambda_k\} \leq 0.5.$$

It follows from (1.22) that $\mathcal{M}\{\Lambda\} = \mathcal{M}\{\Lambda^c\} = 0.5$ which proves the self-duality.

Step 4: Let us prove that $\mathcal{M}$ is increasing. Suppose Λ and Δ are two events in $\mathcal{L}$ with $\Lambda \subset \Delta$. The argument breaks down into three cases. Case 1: Assume

$$\sup_{\Lambda_1 \times \Lambda_2 \times \cdots \times \Lambda_n \subset \Lambda} \min_{1 \leq k \leq n} \mathcal{M}_k\{\Lambda_k\} > 0.5.$$

Then

$$\sup_{\Delta_1 \times \Delta_2 \times \cdots \times \Delta_n \subset \Delta} \min_{1 \leq k \leq n} \mathcal{M}_k\{\Delta_k\} \geq \sup_{\Lambda_1 \times \Lambda_2 \times \cdots \times \Lambda_n \subset \Lambda} \min_{1 \leq k \leq n} \mathcal{M}_k\{\Lambda_k\} > 0.5.$$

It follows from (1.22) that $\mathcal{M}\{\Lambda\} \leq \mathcal{M}\{\Delta\}$. Case 2: Assume

$$\sup_{\Delta_1 \times \Delta_2 \times \cdots \times \Delta_n \subset \Delta^c} \min_{1 \leq k \leq n} \mathcal{M}_k\{\Delta_k\} > 0.5.$$

Then

$$\sup_{\Lambda_1 \times \Lambda_2 \times \cdots \times \Lambda_n \subset \Lambda^c} \min_{1 \leq k \leq n} \mathcal{M}_k\{\Lambda_k\} \geq \sup_{\Delta_1 \times \Delta_2 \times \cdots \times \Delta_n \subset \Delta^c} \min_{1 \leq k \leq n} \mathcal{M}_k\{\Delta_k\} > 0.5.$$

Thus

$$\mathcal{M}\{\Lambda\} = 1 - \sup_{\Lambda_1 \times \Lambda_2 \times \cdots \times \Lambda_n \subset \Lambda^c} \min_{1 \leq k \leq n} \mathcal{M}_k\{\Lambda_k\}$$

$$\leq 1 - \sup_{\Delta_1 \times \Delta_2 \times \cdots \times \Delta_n \subset \Delta^c} \min_{1 \leq k \leq n} \mathcal{M}_k\{\Delta_k\} = \mathcal{M}\{\Delta\}.$$

Case 3: Assume

$$\sup_{\Lambda_1 \times \Lambda_2 \times \cdots \times \Lambda_n \subset \Lambda} \min_{1 \leq k \leq n} \mathcal{M}_k\{\Lambda_k\} \leq 0.5$$

and

$$\sup_{\Lambda_1 \times \Lambda_2 \times \cdots \times \Lambda_n \subset \Delta^c} \min_{1 \le k \le n} \mathcal{M}_k\{\Delta_k\} \le 0.5.$$

Then

$$\mathcal{M}\{\Lambda\} \le 0.5 \le 1 - \mathcal{M}\{\Delta^c\} = \mathcal{M}\{\Delta\}.$$

STEP 5: Finally, we prove the countable subadditivity of $\mathcal{M}$. For simplicity, we only prove the case of two events Λ and Δ. The argument breaks down into three cases. Case 1: Assume $\mathcal{M}\{\Lambda\} < 0.5$ and $\mathcal{M}\{\Delta\} < 0.5$. For any given $\varepsilon > 0$, there are two rectangles

$$\Lambda_1 \times \Lambda_2 \times \cdots \times \Lambda_n \subset \Lambda^c, \quad \Delta_1 \times \Delta_2 \times \cdots \times \Delta_n \subset \Delta^c$$

such that

$$1 - \min_{1 \le k \le n} \mathcal{M}_k\{\Lambda_k\} \le \mathcal{M}\{\Lambda\} + \varepsilon/2,$$

$$1 - \min_{1 \le k \le n} \mathcal{M}_k\{\Delta_k\} \le \mathcal{M}\{\Delta\} + \varepsilon/2.$$

Note that

$$(\Lambda_1 \cap \Delta_1) \times (\Lambda_2 \cap \Delta_2) \times \cdots \times (\Lambda_n \cap \Delta_n) \subset (\Lambda \cup \Delta)^c.$$

It follows from Theorem 1.4 that

$$\mathcal{M}_k\{\Lambda_k \cap \Delta_k\} \ge \mathcal{M}_k\{\Lambda_k\} + \mathcal{M}_k\{\Delta_k\} - 1$$

for any k. Thus

$$\mathcal{M}\{\Lambda \cup \Delta\} \le 1 - \min_{1 \le k \le n} \mathcal{M}_k\{\Lambda_k \cap \Delta_k\}$$

$$\le 1 - \min_{1 \le k \le n} \mathcal{M}_k\{\Lambda_k\} + 1 - \min_{1 \le k \le n} \mathcal{M}_k\{\Delta_k\}$$

$$\le \mathcal{M}\{\Lambda\} + \mathcal{M}\{\Delta\} + \varepsilon.$$

Letting $\varepsilon \to 0$, we obtain

$$\mathcal{M}\{\Lambda \cup \Delta\} \le \mathcal{M}\{\Lambda\} + \mathcal{M}\{\Delta\}.$$

Case 2: Assume $\mathcal{M}\{\Lambda\} \ge 0.5$ and $\mathcal{M}\{\Delta\} < 0.5$. When $\mathcal{M}\{\Lambda \cup \Delta\} = 0.5$, the subadditivity is obvious. Now we consider the case $\mathcal{M}\{\Lambda \cup \Delta\} > 0.5$, i.e., $\mathcal{M}\{\Lambda^c \cap \Delta^c\} < 0.5$. By using $\Lambda^c \cup \Delta = (\Lambda^c \cap \Delta^c) \cup \Delta$ and Case 1, we get

$$\mathcal{M}\{\Lambda^c \cup \Delta\} \le \mathcal{M}\{\Lambda^c \cap \Delta^c\} + \mathcal{M}\{\Delta\}.$$

Thus

$$\mathcal{M}\{\Lambda \cup \Delta\} = 1 - \mathcal{M}\{\Lambda^c \cap \Delta^c\} \le 1 - \mathcal{M}\{\Lambda^c \cup \Delta\} + \mathcal{M}\{\Delta\}$$

$$\le 1 - \mathcal{M}\{\Lambda^c\} + \mathcal{M}\{\Delta\} = \mathcal{M}\{\Lambda\} + \mathcal{M}\{\Delta\}.$$

Case 3: If both $\mathcal{M}\{\Lambda\} \ge 0.5$ and $\mathcal{M}\{\Delta\} \ge 0.5$, then the subadditivity is obvious because $\mathcal{M}\{\Lambda\} + \mathcal{M}\{\Delta\} \ge 1$. The theorem is proved.

Definition 1.4. *Let $(\Gamma_k, \mathcal{L}_k, \mathcal{M}_k), k = 1, 2, \cdots, n$ be uncertainty spaces, $\Gamma = \Gamma_1 \times \Gamma_2 \times \cdots \times \Gamma_n$, $\mathcal{L} = \mathcal{L}_1 \times \mathcal{L}_2 \times \cdots \times \mathcal{L}_n$, and $\mathcal{M} = \mathcal{M}_1 \wedge \mathcal{M}_2 \wedge \cdots \wedge \mathcal{M}_n$. Then $(\Gamma, \mathcal{L}, \mathcal{M})$ is called the product uncertainty space of $(\Gamma_k, \mathcal{L}_k, \mathcal{M}_k), k = 1, 2, \cdots, n$.*

1.2 Uncertain Variable

This section introduces a concept of uncertain variable (neither random variable nor fuzzy variable) in order to describe imprecise quantities in human systems.

Definition 1.5 *(Liu [120]). An uncertain variable is a measurable function ξ from an uncertainty space $(\Gamma, \mathcal{L}, \mathcal{M})$ to the set of real numbers, i.e., for any Borel set B of real numbers, the set*

$$\{\xi \in B\} = \{\gamma \in \Gamma \mid \xi(\gamma) \in B\} \qquad (1.23)$$

is an event.

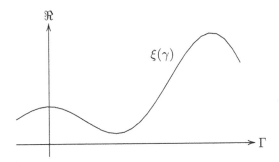

Figure 1.2: An Uncertain Variable

Example 1.6: Take $(\Gamma, \mathcal{L}, \mathcal{M})$ to be $\{\gamma_1, \gamma_2\}$ with $\mathcal{M}\{\gamma_1\} = \mathcal{M}\{\gamma_2\} = 0.5$. Then the function

$$\xi(\gamma) = \begin{cases} 0, \text{ if } \gamma = \gamma_1 \\ 1, \text{ if } \gamma = \gamma_2 \end{cases}$$

is an uncertain variable.

Example 1.7: A crisp number c may be regarded as a special uncertain variable. In fact, it is the constant function $\xi(\gamma) \equiv c$ on the uncertainty space $(\Gamma, \mathcal{L}, \mathcal{M})$.

Definition 1.6. *Let ξ and η be uncertain variables defined on the uncertainty space $(\Gamma, \mathcal{L}, \mathcal{M})$. We say $\xi = \eta$ if $\xi(\gamma) = \eta(\gamma)$ for almost all $\gamma \in \Gamma$.*

Definition 1.7. *The uncertain variables ξ and η are identically distributed if*

$$\mathcal{M}\{\xi \in B\} = \mathcal{M}\{\eta \in B\} \tag{1.24}$$

for any Borel set B of real numbers.

It is clear that uncertain variables ξ and η are identically distributed if $\xi = \eta$. However, identical distribution does not imply $\xi = \eta$. For example, let $(\Gamma, \mathcal{L}, \mathcal{M})$ be $\{\gamma_1, \gamma_2\}$ with $\mathcal{M}\{\gamma_1\} = \mathcal{M}\{\gamma_2\} = 0.5$. Define

$$\xi(\gamma) = \begin{cases} 1, & \text{if } \gamma = \gamma_1 \\ -1, & \text{if } \gamma = \gamma_2, \end{cases} \qquad \eta(\gamma) = \begin{cases} -1, & \text{if } \gamma = \gamma_1 \\ 1, & \text{if } \gamma = \gamma_2. \end{cases}$$

The two uncertain variables ξ and η are identically distributed but $\xi \neq \eta$.

Uncertain Vector

Definition 1.8. *An n-dimensional uncertain vector is a measurable function from an uncertainty space $(\Gamma, \mathcal{L}, \mathcal{M})$ to the set of n-dimensional real vectors, i.e., for any Borel set B of $\Re^n$, the set*

$$\{\boldsymbol{\xi} \in B\} = \{\gamma \in \Gamma \mid \boldsymbol{\xi}(\gamma) \in B\} \tag{1.25}$$

is an event.

Theorem 1.9. *The vector $(\xi_1, \xi_2, \cdots, \xi_n)$ is an uncertain vector if and only if $\xi_1, \xi_2, \cdots, \xi_n$ are uncertain variables.*

Proof: Write $\boldsymbol{\xi} = (\xi_1, \xi_2, \cdots, \xi_n)$. Suppose that $\boldsymbol{\xi}$ is an uncertain vector on the uncertainty space $(\Gamma, \mathcal{L}, \mathcal{M})$. For any Borel set B of $\Re$, the set $B \times \Re^{n-1}$ is a Borel set of $\Re^n$. Thus the set

$$\{\xi_1 \in B\} = \{\xi_1 \in B, \xi_2 \in \Re, \cdots, \xi_n \in \Re\} = \{\boldsymbol{\xi} \in B \times \Re^{n-1}\}$$

is an event. Hence ξ_1 is an uncertain variable. A similar process may prove that $\xi_2, \xi_3, \cdots, \xi_n$ are uncertain variables. Conversely, suppose that all $\xi_1, \xi_2, \cdots, \xi_n$ are uncertain variables on the uncertainty space $(\Gamma, \mathcal{L}, \mathcal{M})$. We define

$$\mathcal{B} = \{B \subset \Re^n \mid \{\boldsymbol{\xi} \in B\} \text{ is an event}\}.$$

The vector $\boldsymbol{\xi} = (\xi_1, \xi_2, \cdots, \xi_n)$ is proved to be an uncertain vector if we can prove that $\mathcal{B}$ contains all Borel sets of $\Re^n$. First, the class $\mathcal{B}$ contains all open intervals of $\Re^n$ because

$$\left\{\boldsymbol{\xi} \in \prod_{i=1}^{n}(a_i, b_i)\right\} = \bigcap_{i=1}^{n}\{\xi_i \in (a_i, b_i)\}$$

is an event. Next, the class $\mathcal{B}$ is a σ-algebra of $\Re^n$ because (i) we have $\Re^n \in \mathcal{B}$ since $\{\boldsymbol{\xi} \in \Re^n\} = \Gamma$; (ii) if $B \in \mathcal{B}$, then $\{\boldsymbol{\xi} \in B\}$ is an event, and

$$\{\boldsymbol{\xi} \in B^c\} = \{\boldsymbol{\xi} \in B\}^c$$

is an event. This means that $B^c \in \mathcal{B}$; (iii) if $B_i \in \mathcal{B}$ for $i = 1, 2, \cdots$, then $\{\boldsymbol{\xi} \in B_i\}$ are events and

$$\left\{ \boldsymbol{\xi} \in \bigcup_{i=1}^{\infty} B_i \right\} = \bigcup_{i=1}^{\infty} \{\boldsymbol{\xi} \in B_i\}$$

is an event. This means that $\cup_i B_i \in \mathcal{B}$. Since the smallest σ-algebra containing all open intervals of $\Re^n$ is just the Borel algebra of $\Re^n$, the class $\mathcal{B}$ contains all Borel sets of $\Re^n$. The theorem is proved.

Uncertain Arithmetic

Definition 1.9. *Suppose that $f : \Re^n \to \Re$ is a measurable function, and $\xi_1, \xi_2, \cdots, \xi_n$ uncertain variables on the uncertainty space $(\Gamma, \mathcal{L}, \mathcal{M})$. Then $\xi = f(\xi_1, \xi_2, \cdots, \xi_n)$ is an uncertain variable defined as*

$$\xi(\gamma) = f(\xi_1(\gamma), \xi_2(\gamma), \cdots, \xi_n(\gamma)), \quad \forall \gamma \in \Gamma. \tag{1.26}$$

Example 1.8: Let ξ_1 and ξ_2 be two uncertain variables. Then the sum $\xi = \xi_1 + \xi_2$ is an uncertain variable defined by

$$\xi(\gamma) = \xi_1(\gamma) + \xi_2(\gamma), \quad \forall \gamma \in \Gamma.$$

The product $\xi = \xi_1 \xi_2$ is also an uncertain variable defined by

$$\xi(\gamma) = \xi_1(\gamma) \cdot \xi_2(\gamma), \quad \forall \gamma \in \Gamma.$$

The reader may wonder whether $\xi(\gamma_1, \gamma_2, \cdots, \gamma_n)$ defined by (1.26) is an uncertain variable. The following theorem answers this question.

Theorem 1.10. *Let $\boldsymbol{\xi}$ be an n-dimensional uncertain vector, and $f : \Re^n \to \Re$ a measurable function. Then $f(\boldsymbol{\xi})$ is an uncertain variable such that*

$$\mathcal{M}\{f(\boldsymbol{\xi}) \in B\} = \mathcal{M}\{\boldsymbol{\xi} \in f^{-1}(B)\} \tag{1.27}$$

for any Borel set B of real numbers.

Proof: Assume that $\boldsymbol{\xi}$ is an uncertain vector on the uncertainty space $(\Gamma, \mathcal{L}, \mathcal{M})$. For any Borel set B of $\Re$, since f is a measurable function, the $f^{-1}(B)$ is a Borel set of $\Re^n$. Thus the set $\{f(\boldsymbol{\xi}) \in B\} = \{\boldsymbol{\xi} \in f^{-1}(B)\}$ is an event for any Borel set B. Hence $f(\boldsymbol{\xi})$ is an uncertain variable.

1.3   Uncertainty Distribution

This section introduces a concept of uncertainty distribution in order to describe uncertain variables. In many cases, it is sufficient to know the uncertainty distribution rather than the uncertain variable itself.

Definition 1.10 *(Liu [120]). The uncertainty distribution* $\Phi\colon \Re \to [0,1]$ *of an uncertain variable ξ is defined by*

$$\Phi(x) = \mathcal{M}\{\xi \le x\} \tag{1.28}$$

for any real number x.

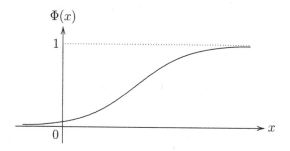

Figure 1.3: An Uncertainty Distribution

Theorem 1.11 *(Peng and Iwamura [177], Sufficient and Necessary Condition for Uncertainty Distribution). A function $\Phi : \Re \to [0,1]$ is an uncertainty distribution if and only if it is an increasing function except $\Phi(x) \equiv 0$ and $\Phi(x) \equiv 1$.*

Proof: It is obvious that an uncertainty distribution Φ is an increasing function. In addition, both $\Phi(x) \not\equiv 0$ and $\Phi(x) \not\equiv 1$ follow from the asymptotic theorem immediately. Conversely, suppose that Φ is an increasing function but $\Phi(x) \not\equiv 0$ and $\Phi(x) \not\equiv 1$. We will prove that there is an uncertain variable whose uncertainty distribution is just Φ. Let $\mathcal{C}$ be a collection of all intervals of the form $(-\infty, a]$, (b, ∞), $\emptyset$ and $\Re$. We define a set function on $\Re$ as follows,

$$\mathcal{M}\{(-\infty, a]\} = \Phi(a),$$

$$\mathcal{M}\{(b, +\infty)\} = 1 - \Phi(b),$$

$$\mathcal{M}\{\emptyset\} = 0, \quad \mathcal{M}\{\Re\} = 1.$$

For an arbitrary Borel set B of real numbers, there exists a sequence $\{A_i\}$ in $\mathcal{C}$ such that

$$B \subset \bigcup_{i=1}^{\infty} A_i.$$

Note that such a sequence is not unique. Thus the set function $\mathcal{M}\{B\}$ is defined by

$$\mathcal{M}\{B\} = \begin{cases} \inf\limits_{B \subset \cup A_i} \sum\limits_{i=1}^{\infty} \mathcal{M}\{A_i\}, & \text{if } \inf\limits_{B \subset \cup A_i} \sum\limits_{i=1}^{\infty} \mathcal{M}\{A_i\} < 0.5 \\ 1 - \inf\limits_{B^c \subset \cup A_i} \sum\limits_{i=1}^{\infty} \mathcal{M}\{A_i\}, & \text{if } \inf\limits_{B^c \subset \cup A_i} \sum\limits_{i=1}^{\infty} \mathcal{M}\{A_i\} < 0.5 \\ 0.5, & \text{otherwise.} \end{cases}$$

We may prove that the set function $\mathcal{M}$ is indeed an uncertain measure on $\Re$, and the uncertain variable defined by the identity function $\xi(\gamma) = \gamma$ from the uncertainty space $(\Re, \mathcal{L}, \mathcal{M})$ to $\Re$ has the uncertainty distribution Φ.

Example 1.9: Let c be a number with $0 < c < 1$. Then $\Phi(x) \equiv c$ is an uncertainty distribution. When $c \leq 0.5$, we define a set function over $\Re$ as follows,

$$\mathcal{M}\{\Lambda\} = \begin{cases} 0, & \text{if } \Lambda = \emptyset \\ c, & \text{if } \Lambda \text{ is upper bounded} \\ 0.5, & \text{if both } \Lambda \text{ and } \Lambda^c \text{ are upper unbounded} \\ 1 - c, & \text{if } \Lambda^c \text{ is upper bounded} \\ 1, & \text{if } \Lambda = \Gamma. \end{cases}$$

Then $(\Re, \mathcal{L}, \mathcal{M})$ is an uncertainty space. It is easy to verify that the identity function $\xi(\gamma) = \gamma$ is an uncertain variable whose uncertainty distribution is just $\Phi(x) \equiv c$. When $c > 0.5$, we define

$$\mathcal{M}\{\Lambda\} = \begin{cases} 0, & \text{if } \Lambda = \emptyset \\ 1 - c, & \text{if } \Lambda \text{ is upper bounded} \\ 0.5, & \text{if both } \Lambda \text{ and } \Lambda^c \text{ are upper unbounded} \\ c, & \text{if } \Lambda^c \text{ is upper bounded} \\ 1, & \text{if } \Lambda = \Gamma. \end{cases}$$

Then the function $\xi(\gamma) = -\gamma$ is an uncertain variable whose uncertainty distribution is just $\Phi(x) \equiv c$.

Example 1.10: Assume that two uncertain variables ξ and η have the same uncertainty distribution. One question is whether $\xi = \eta$ or not. Generally speaking, it is not true. Take $(\Gamma, \mathcal{L}, \mathcal{M})$ to be $\{\gamma_1, \gamma_2\}$ with

$$\mathcal{M}\{\gamma_1\} = \mathcal{M}\{\gamma_2\} = 0.5.$$

We now define two uncertain variables as follows,

$$\xi(\gamma) = \begin{cases} -1, & \text{if } \gamma = \gamma_1 \\ 1, & \text{if } \gamma = \gamma_2, \end{cases} \qquad \eta(\gamma) = \begin{cases} 1, & \text{if } \gamma = \gamma_1 \\ -1, & \text{if } \gamma = \gamma_2. \end{cases}$$

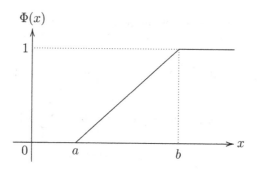

Figure 1.4: Linear Uncertainty Distribution

Then ξ and η have the same uncertainty distribution,

$$\Phi(x) = \begin{cases} 0, & \text{if } x < -1 \\ 0.5, & \text{if } -1 \leq x < 1 \\ 1, & \text{if } x \geq 1. \end{cases}$$

However, it is clear that $\xi \neq \eta$ in the sense of Definition 1.6.

Definition 1.11. *An uncertain variable ξ is called linear if it has a linear uncertainty distribution*

$$\Phi(x) = \begin{cases} 0, & \text{if } x \leq a \\ (x - a)/(b - a), & \text{if } a \leq x \leq b \\ 1, & \text{if } x \geq b \end{cases} \tag{1.29}$$

denoted by $\mathcal{L}(a, b)$ where a and b are real numbers with $a < b$.

Definition 1.12. *An uncertain variable ξ is called zigzag if it has a zigzag uncertainty distribution*

$$\Phi(x) = \begin{cases} 0, & \text{if } x \leq a \\ (x - a)/2(b - a), & \text{if } a \leq x \leq b \\ (x + c - 2b)/2(c - b), & \text{if } b \leq x \leq c \\ 1, & \text{if } x \geq c \end{cases} \tag{1.30}$$

denoted by $\mathcal{Z}(a, b, c)$ where a, b, c are real numbers with $a < b < c$.

Definition 1.13. *An uncertain variable ξ is called normal if it has a normal uncertainty distribution*

$$\Phi(x) = \left(1 + \exp\left(\frac{\pi(e - x)}{\sqrt{3}\sigma}\right)\right)^{-1}, \quad x \in \Re \tag{1.31}$$

denoted by $\mathcal{N}(e, \sigma)$ where e and σ are real numbers with $\sigma > 0$.

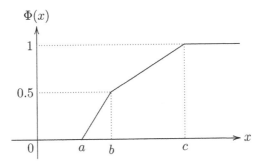

Figure 1.5: Zigzag Uncertainty Distribution

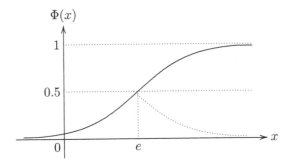

Figure 1.6: Normal Uncertainty Distribution

Definition 1.14. *An uncertain variable ξ is called lognormal if $\ln \xi$ is a normal uncertain variable $\mathcal{N}(e, \sigma)$. In other words, a lognormal uncertain variable has an uncertainty distribution*

$$\Phi(x) = \left(1 + \exp \left(\frac{\pi(e - \ln x)}{\sqrt{3}\sigma} \right) \right)^{-1}, \quad x \geq 0 \qquad (1.32)$$

denoted by $\mathcal{LOGN}(e, \sigma)$, where e and σ are real numbers with $\sigma > 0$.

Definition 1.15. *An uncertain variable ξ is called discrete if it takes values in $\{x_1, x_2, \cdots, x_m\}$ and*

$$\Phi(x_i) = \alpha_i, \quad i = 1, 2, \cdots, m \qquad (1.33)$$

where $x_1 < x_2 < \cdots < x_m$ and $0 \leq \alpha_1 \leq \alpha_2 \leq \cdots \leq \alpha_m = 1$. For simplicity, the discrete uncertain variable will be denoted by

$$\xi = \begin{array}{|c|c|c|c|} \hline \alpha_1 & \alpha_2 & \cdots & \alpha_m \\ \hline x_1 & x_2 & \cdots & x_m \\ \hline \end{array} \qquad (1.34)$$

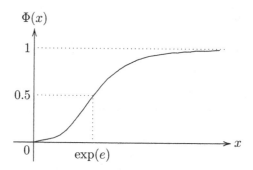

Figure 1.7: Lognormal Uncertainty Distribution

The uncertainty distribution Φ of the discrete uncertain variable (1.34) is a step function jumping only at $x_1, x_2, \cdots, x_m$, i.e.,

$$\Phi(x) = \begin{cases} \alpha_0, & \text{if } x < x_1 \\ \alpha_i, & \text{if } x_i \leq x < x_{i+1}, \ i = 1, 2, \cdots, m \\ \alpha_m, & \text{if } x \geq x_m \end{cases} \tag{1.35}$$

where $\alpha_0 \equiv 0$ and $\alpha_m \equiv 1$.

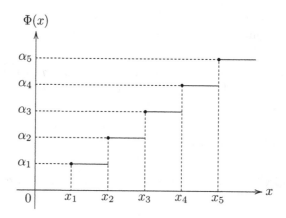

Figure 1.8: Discrete Uncertainty Distribution

Measure Inversion Theorem

Theorem 1.12 *(Measure Inversion Theorem). Let ξ be an uncertain variable with continuous uncertainty distribution Φ. Then for any real number x, we have*

$$\mathcal{M}\{\xi \leq x\} = \Phi(x), \quad \mathcal{M}\{\xi \geq x\} = 1 - \Phi(x). \tag{1.36}$$

Proof: The equation $\mathcal{M}\{\xi \le x\} = \Phi(x)$ follows from the definition of uncertainty distribution immediately. By using the self-duality of uncertain measure and continuity of uncertainty distribution, we get $\mathcal{M}\{\xi \ge x\} = 1 - \mathcal{M}\{\xi < x\} = 1 - \Phi(x)$.

Theorem 1.13. *Let ξ be an uncertain variable with continuous uncertainty distribution Φ. Then for any interval $[a, b]$, we have*

$$\Phi(b) - \Phi(a) \le \mathcal{M}\{a \le \xi \le b\} \le \Phi(b) \wedge (1 - \Phi(a)). \qquad (1.37)$$

Proof: It follows from the subadditivity of uncertain measure and the measure inversion theorem that

$$\mathcal{M}\{a \le \xi \le b\} + \mathcal{M}\{\xi \le a\} \ge \mathcal{M}\{\xi \le b\}.$$

That is,

$$\mathcal{M}\{a \le \xi \le b\} + \Phi(a) \ge \Phi(b).$$

Thus the inequality on the left hand side is verified. It follows from the monotonicity of uncertain measure and the measure inversion theorem that

$$\mathcal{M}\{a \le \xi \le b\} \le \mathcal{M}\{\xi \in (-\infty, b]\} = \Phi(b).$$

On the other hand,

$$\mathcal{M}\{a \le x \le b\} \le \mathcal{M}\{\xi \in [a, +\infty)\} = 1 - \Phi(a).$$

Hence the inequality on the right hand side is proved.

Perhaps some readers would like to get an exactly scalar value of the uncertain measure $\mathcal{M}\{a \le x \le b\}$. Generally speaking, it is an impossible job (except $a = -\infty$ or $b = +\infty$) if only an uncertainty distribution is available. I would like to ask if there is a need to know it. In fact, it is not a must for practical purpose. Would you believe?

Regular Uncertainty Distribution

Definition 1.16. *An uncertainty distribution Φ is said to be regular if its inverse function $\Phi^{-1}(\alpha)$ exists and is unique for each $\alpha \in (0, 1)$.*

It is easy to verify that a regular uncertainty distribution Φ is a continuous function. In addition, Φ is strictly increasing at each point x with $0 < \Phi(x) < 1$. Furthermore,

$$\lim_{x \to -\infty} \Phi(x) = 0, \quad \lim_{x \to +\infty} \Phi(x) = 1. \qquad (1.38)$$

For example, linear uncertainty distribution, zigzag uncertainty distribution, normal uncertainty distribution, and lognormal uncertainty distribution are all regular.

In this book we will assume all uncertainty distributions are regular. Otherwise, we may give the uncertainty distribution a small perturbation such that it becomes regular.

Inverse Uncertainty Distribution

Definition 1.17. *Let ξ be an uncertain variable with uncertainty distribution Φ. Then the inverse function Φ^{-1} is called the inverse uncertainty distribution of ξ.*

Note that the inverse uncertainty distribution $\Phi^{-1}(\alpha)$ is well defined on the open interval $(0, 1)$. If needed, we may extend the domain via

$$\Phi^{-1}(0) = \lim_{\alpha \to 0} \Phi^{-1}(\alpha), \quad \Phi^{-1}(1) = \lim_{\alpha \to 1} \Phi^{-1}(\alpha). \tag{1.39}$$

It is easy to verify that inverse uncertainty distribution is a monotone increasing function on $[0, 1]$.

Example 1.11: The inverse uncertainty distribution of linear uncertain variable $\mathcal{L}(a, b)$ is

$$\Phi^{-1}(\alpha) = (1 - \alpha)a + \alpha b. \tag{1.40}$$

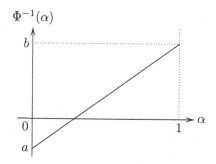

Figure 1.9: Inverse Linear Uncertainty Distribution

Example 1.12: The inverse uncertainty distribution of zigzag uncertain variable $\mathcal{Z}(a, b, c)$ is

$$\Phi^{-1}(\alpha) = \begin{cases} (1 - 2\alpha)a + 2\alpha b, & \text{if } \alpha < 0.5 \\ (2 - 2\alpha)b + (2\alpha - 1)c, & \text{if } \alpha \geq 0.5. \end{cases} \tag{1.41}$$

Example 1.13: The inverse uncertainty distribution of normal uncertain variable $\mathcal{N}(e, \sigma)$ is

$$\Phi^{-1}(\alpha) = e + \frac{\sigma \sqrt{3}}{\pi} \ln \frac{\alpha}{1 - \alpha}. \tag{1.42}$$

Example 1.14: The inverse uncertainty distribution of lognormal uncertain variable $\mathcal{LOGN}(e, \sigma)$ is

$$\Phi^{-1}(\alpha) = \exp(e) \left(\frac{\alpha}{1 - \alpha} \right)^{\sqrt{3}\sigma/\pi}. \tag{1.43}$$

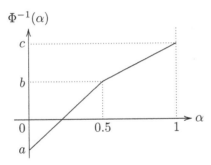

Figure 1.10: Inverse Zigzag Uncertainty Distribution

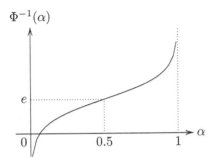

Figure 1.11: Inverse Normal Uncertainty Distribution

Joint Uncertainty Distribution

Definition 1.18 *Let $(\xi_1, \xi_2, \cdots, \xi_n)$ be an uncertain vector. Then the joint uncertainty distribution $\Phi : \Re^n \to [0, 1]$ is defined by*

$$\Phi(x_1, x_2, \cdots, x_n) = \mathcal{M}\{\xi_1 \leq x_1, \xi_2 \leq x_2, \cdots, \xi_n \leq x_n\} \qquad (1.44)$$

for any real numbers $x_1, x_2, \cdots, x_n$.

1.4 Independence

Independence has been explained in many ways. However, the essential feature is that those uncertain variables may be separately defined on different uncertainty spaces. In order to ensure that we are able to do so, we may define independence in the following mathematical form.

Definition 1.19 *(Liu [123]). The uncertain variables $\xi_1, \xi_2, \cdots, \xi_m$ are said to be independent if*

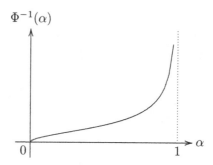

Figure 1.12: Inverse Lognormal Uncertainty Distribution

$$\mathcal{M}\left\{\bigcap_{i=1}^{m}(\xi_i \in B_i)\right\} = \min_{1\le i \le m} \mathcal{M}\{\xi_i \in B_i\} \tag{1.45}$$

for any Borel sets $B_1, B_2, \cdots, B_m$ of real numbers.

Example 1.15: Let ξ_1 be an uncertain variable and let ξ_2 be a constant c. For any Borel sets B_1 and B_2, if $c \in B_2$, then $\mathcal{M}\{\xi_2 \in B_2\} = 1$ and

$$\mathcal{M}\{(\xi_1 \in B_1) \cap (\xi_2 \in B_2)\} = \mathcal{M}\{\xi_1 \in B_1\} = \mathcal{M}\{\xi_1 \in B_1\} \wedge \mathcal{M}\{\xi_2 \in B_2\}.$$

If $c \notin B_2$, then $\mathcal{M}\{\xi_2 \in B_2\} = 0$ and

$$\mathcal{M}\{(\xi_1 \in B_1) \cap (\xi_2 \in B_2)\} = \mathcal{M}\{\emptyset\} = 0 = \mathcal{M}\{\xi_1 \in B_1\} \wedge \mathcal{M}\{\xi_2 \in B_2\}.$$

It follows from the definition of independence that an uncertain variable is always independent of a constant.

Theorem 1.14. *The uncertain variables $\xi_1, \xi_2, \cdots, \xi_m$ are independent if and only if*

$$\mathcal{M}\left\{\bigcup_{i=1}^{m}(\xi_i \in B_i)\right\} = \max_{1\le i \le m} \mathcal{M}\{\xi_i \in B_i\} \tag{1.46}$$

for any Borel sets $B_1, B_2, \cdots, B_m$ of real numbers.

Proof: It follows from the self-duality of uncertain measure that $\xi_1, \xi_2, \cdots,$ ξ_m are independent if and only if

$$\mathcal{M}\left\{\bigcup_{i=1}^{m}(\xi_i \in B_i)\right\} = 1 - \mathcal{M}\left\{\bigcap_{i=1}^{m}(\xi_i \in B_i^c)\right\}$$

$$= 1 - \min_{1\le i \le m} \mathcal{M}\{\xi_i \in B_i^c\} = \max_{1\le i \le m} \mathcal{M}\{\xi_i \in B_i\}.$$

Thus the proof is complete.

Theorem 1.15. *Let Φ_i be uncertainty distributions of uncertain variables ξ_i, $i = 1, 2, \cdots, m$, respectively, and Φ the joint uncertainty distribution of uncertain vector $(\xi_1, \xi_2, \cdots, \xi_m)$. If $\xi_1, \xi_2, \cdots, \xi_m$ are independent, then we have*

$$\Phi(x_1, x_2, \cdots, x_m) = \min_{1 \leq i \leq m} \Phi_i(x_i) \tag{1.47}$$

for any real numbers $x_1, x_2, \cdots, x_m$.

Proof: Since $\xi_1, \xi_2, \cdots, \xi_m$ are independent uncertain variables, we have

$$\Phi(x_1, x_2, \cdots, x_m) = \mathcal{M}\left\{ \bigcap_{i=1}^{m} (\xi_i \leq x_i) \right\} = \min_{1 \leq i \leq m} \mathcal{M}\{\xi_i \leq x_i\} = \min_{1 \leq i \leq m} \Phi_i(x_i)$$

for any real numbers $x_1, x_2, \cdots, x_m$. The theorem is proved.

Example 1.16: However, the equation (1.47) does not imply that the uncertain variables are independent. For example, let ξ be an uncertain variable with uncertainty distribution Φ. Then the joint uncertainty distribution Ψ of uncertain vector (ξ, ξ) is

$$\Psi(x_1, x_2) = \mathcal{M}\{\xi \leq x_1, \xi \leq x_2\} = \mathcal{M}\{\xi \leq x_1\} \wedge \mathcal{M}\{\xi \leq x_2\} = \Phi(x_1) \wedge \Phi(x_2)$$

for any real numbers x_1 and x_2. But, generally speaking, an uncertain variable is not independent with itself.

Theorem 1.16. *Let $\xi_1, \xi_2, \cdots, \xi_m$ be independent uncertain variables, and $f_1, f_2, \cdots, f_n$ measurable functions. Then $f_1(\xi_1), f_2(\xi_2), \cdots, f_m(\xi_m)$ are independent uncertain variables.*

Proof: For any Borel sets $B_1, B_2, \cdots, B_m$ of $\Re$, it follows from the definition of independence that

$$\mathcal{M}\left\{ \bigcap_{i=1}^{m} (f_i(\xi_i) \in B_i) \right\} = \mathcal{M}\left\{ \bigcap_{i=1}^{m} (\xi_i \in f_i^{-1}(B_i)) \right\}$$

$$= \min_{1 \leq i \leq m} \mathcal{M}\{\xi_i \in f_i^{-1}(B_i)\} = \min_{1 \leq i \leq m} \mathcal{M}\{f_i(\xi_i) \in B_i\}.$$

Thus $f_1(\xi_1), f_2(\xi_2), \cdots, f_m(\xi_m)$ are independent uncertain variables.

1.5   Operational Law

This section will introduce an operational law of independent uncertain variables and present a 99-method for calculating the uncertainty distribution of monotone function of uncertain variables.

Theorem 1.17 *(Liu [123], Operational Law). Let $\xi_1, \xi_2, \cdots, \xi_n$ be independent uncertain variables, and $f : \Re^n \to \Re$ a measurable function. Then $\xi = f(\xi_1, \xi_2, \cdots, \xi_n)$ is an uncertain variable such that*

$$\mathcal{M}\{\xi \in B\} = \begin{cases} \sup_{f(B_1, B_2, \cdots, B_n) \subset B} \min_{1 \leq k \leq n} \mathcal{M}_k\{\xi_k \in B_k\}, \\ \quad if \quad \sup_{f(B_1, B_2, \cdots, B_n) \subset B} \min_{1 \leq k \leq n} \mathcal{M}_k\{\xi_k \in B_k\} > 0.5 \\ 1 - \sup_{f(B_1, B_2, \cdots, B_n) \subset B^c} \min_{1 \leq k \leq n} \mathcal{M}_k\{\xi_k \in B_k\}, \\ \quad if \quad \sup_{f(B_1, B_2, \cdots, B_n) \subset B^c} \min_{1 \leq k \leq n} \mathcal{M}_k\{\xi_k \in B_k\} > 0.5 \\ 0.5, \quad otherwise \end{cases}$$

where $B, B_1, B_2, \cdots, B_n$ are Borel sets, and $f(B_1, B_2, \cdots, B_n) \subset B$ means $f(x_1, x_2, \cdots, x_n) \in B$ for any $x_1 \in B_1, x_2 \in B_2, \cdots, x_n \in B_n$.

Proof: Write $\Lambda = \{\xi \in B\}$ and $\Lambda_k = \{\xi_k \in B_k\}$ for $k = 1, 2, \cdots, n$. It is easy to verify that

$$\Lambda_1 \times \Lambda_2 \times \cdots \times \Lambda_n \subset \Lambda \text{ if and only if } f(B_1, B_2, \cdots, B_n) \subset B,$$

$$\Lambda_1 \times \Lambda_2 \times \cdots \times \Lambda_n \subset \Lambda^c \text{ if and only if } f(B_1, B_2, \cdots, B_n) \subset B^c.$$

Thus the operational law follows from the product measure axiom immediately.

Increasing Function of Single Uncertain Variable

Theorem 1.18. *Let ξ be an uncertain variable with uncertainty distribution Φ, and let f be a strictly increasing function. Then $f(\xi)$ is an uncertain variable with inverse uncertainty distribution*

$$\Psi^{-1}(\alpha) = f(\Phi^{-1}(\alpha)). \tag{1.48}$$

Proof: Since f is a strictly increasing function, we have, for each $\alpha \in (0, 1)$,

$$\mathcal{M}\{f(\xi) \leq f(\Phi^{-1}(\alpha))\} = \mathcal{M}\{\xi \leq \Phi^{-1}(\alpha)\} = \alpha.$$

Thus we have $\Psi^{-1}(\alpha) = f(\Phi^{-1}(\alpha))$. In fact, the uncertainty distribution of $f(\xi)$ is

$$\Psi(x) = \Phi(f^{-1}(x)).$$

The theorem is proved.

99-Method 1.1. *It is suggested that an uncertain variable ξ with uncertainty distribution Φ is represented by a 99-table,*

0.01	0.02	0.03	$\cdots$	0.99
x_1	x_2	x_3	$\cdots$	x_{99}

$$\tag{1.49}$$

where $0.01, 0.02, 0.03, \cdots, 0.99$ *in the first row are the values of uncertainty distribution* Φ, *and* $x_1, x_2, x_3, \cdots, x_{99}$ *in the second row are the corresponding values of* $\Phi^{-1}(0.01), \Phi^{-1}(0.02), \Phi^{-1}(0.03), \cdots, \Phi^{-1}(0.99)$. *Essentially, the 99-table is a discrete representation of uncertainty distribution* Φ. *Then for any strictly increasing function* $f(x)$, *the uncertain variable* $f(\xi)$ *has a 99-table,*

0.01	0.02	0.03	$\cdots$	0.99
$f(x_1)$	$f(x_2)$	$f(x_3)$	$\cdots$	$f(x_{99})$

$$(1.50)$$

The 99-method may be extended to the 999-method if a more precise result is needed.

Example 1.17: Let ξ be an uncertain variable with uncertainty distribution Φ. Then for any number $k > 0$, the inverse uncertainty distribution of $k\xi$ is

$$\Psi^{-1}(\alpha) = k\Phi^{-1}(\alpha). \tag{1.51}$$

If ξ is represented by a 99-table,

0.01	0.02	0.03	$\cdots$	0.99
x_1	x_2	x_3	$\cdots$	x_{99}

$$(1.52)$$

then the 99-method yields that $k\xi$ has a 99-table,

0.01	0.02	0.03	$\cdots$	0.99
kx_1	kx_2	kx_3	$\cdots$	kx_{99}

$$(1.53)$$

Example 1.18: If ξ is an uncertain variable with uncertainty distribution Φ and k is a constant, then $\xi + k$ is an uncertain variable with inverse uncertainty distribution

$$\Psi^{-1}(\alpha) = \Phi^{-1}(\alpha) + k. \tag{1.54}$$

If ξ is represented by a 99-table,

0.01	0.02	0.03	$\cdots$	0.99
x_1	x_2	x_3	$\cdots$	x_{99}

$$(1.55)$$

then the 99-method yields that $\xi + k$ has a 99-table,

0.01	0.02	0.03	$\cdots$	0.99
$x_1 + k$	$x_2 + k$	$x_3 + k$	$\cdots$	$x_{99} + k$

$$(1.56)$$

Example 1.19: Let ξ be a nonnegative uncertain variable with uncertainty distribution Φ. Since x^2 is a strictly increasing function on $[0, +\infty)$, the square ξ^2 is an uncertain variable with inverse uncertainty distribution

$$\Psi^{-1}(\alpha) = \left(\Phi^{-1}(\alpha)\right)^2. \tag{1.57}$$

If ξ is represented by a 99-table,

0.01	0.02	0.03	$\cdots$	0.99
x_1	x_2	x_3	$\cdots$	x_{99}

$$(1.58)$$

then the 99-method yields that the uncertain variable ξ^2 has a 99-table,

0.01	0.02	0.03	$\cdots$	0.99
x_1^2	x_2^2	x_3^2	$\cdots$	x_{99}^2

$$(1.59)$$

Example 1.20: Let ξ be an uncertain variable with uncertainty distribution Φ. Since $\exp(x)$ is a strictly increasing function, $\exp(\xi)$ is an uncertain variable with inverse uncertainty distribution

$$\Psi^{-1}(\alpha) = \exp\left(\Phi^{-1}(\alpha)\right). \qquad (1.60)$$

If ξ is represented by a 99-table,

0.01	0.02	0.03	$\cdots$	0.99
x_1	x_2	x_3	$\cdots$	x_{99}

$$(1.61)$$

then the 99-method yields that the uncertain variable $\exp(\xi)$ has a 99-table,

0.01	0.02	0.03	$\cdots$	0.99
$\exp(x_1)$	$\exp(x_2)$	$\exp(x_3)$	$\cdots$	$\exp(x_{99})$

$$(1.62)$$

Decreasing Function of Single Uncertain Variable

Theorem 1.19. *Let ξ be an uncertain variable with uncertainty distribution Φ, and let f be a strictly decreasing function. Then $f(\xi)$ is an uncertainty distribution with inverse uncertainty distribution*

$$\Psi^{-1}(\alpha) = f(\Phi^{-1}(1 - \alpha)). \qquad (1.63)$$

Proof: Since f is a strictly decreasing function, we have, for each $\alpha \in (0, 1)$,

$$\mathcal{M}\{f(\xi) \le f(\Phi^{-1}(1 - \alpha))\} = \mathcal{M}\{\xi \ge \Phi^{-1}(1 - \alpha)\} = \alpha.$$

Thus we have $\Psi^{-1}(\alpha) = f(\Phi^{-1}(1 - \alpha))$. In fact, the uncertainty distribution of $f(\xi)$ is

$$\Psi(x) = 1 - \Phi(f^{-1}(x)).$$

The theorem is proved.

99-Method 1.2. *Let ξ be an uncertain variable represented by a 99-table,*

0.01	0.02	0.03	$\cdots$	0.99
x_1	x_2	x_3	$\cdots$	x_{99}

$$(1.64)$$

Then for any strictly decreasing function $f(x)$, the uncertain variable $f(\xi)$ has a 99-table,

0.01	0.02	0.03	$\cdots$	0.99
$f(x_{99})$	$f(x_{98})$	$f(x_{97})$	$\cdots$	$f(x_1)$

(1.65)

Example 1.21: Let ξ be an uncertain variable with uncertainty distribution Φ. Then $-\xi$ has an inverse uncertainty distribution

$$\Psi^{-1}(\alpha) = -\Phi^{-1}(1-\alpha).$$

(1.66)

If ξ is represented by a 99-table,

0.01	0.02	0.03	$\cdots$	0.99
x_1	x_2	x_3	$\cdots$	x_{99}

(1.67)

then the 99-method yields that the uncertain variable $-\xi$ has a 99-table,

0.01	0.02	0.03	$\cdots$	0.99
$-x_{99}$	$-x_{98}$	$-x_{97}$	$\cdots$	$-x_1$

(1.68)

Example 1.22: Let ξ be a positive uncertain variable with uncertainty distribution Φ. Since $1/x$ is a strictly decreasing function on $(0, +\infty)$, the reciprocal $1/\xi$ is an uncertain variable with inverse uncertainty distribution

$$\Psi^{-1}(\alpha) = \frac{1}{\Phi^{-1}(1-\alpha)}.$$

(1.69)

If ξ is represented by a 99-table,

0.01	0.02	0.03	$\cdots$	0.99
x_1	x_2	x_3	$\cdots$	x_{99}

(1.70)

then the 99-method yields that the uncertain variable $1/\xi$ has a 99-table,

0.01	0.02	0.03	$\cdots$	0.99
$1/x_{99}$	$1/x_{98}$	$1/x_{97}$	$\cdots$	$1/x_1$

(1.71)

Example 1.23: Let ξ be an uncertain variable with uncertainty distribution Φ. Since $\exp(-x)$ is a strictly decreasing function, $\exp(-\xi)$ is an uncertain variable with inverse uncertainty distribution

$$\Psi^{-1}(\alpha) = \exp\left(-\Phi^{-1}(1-\alpha)\right).$$

(1.72)

If ξ is represented by a 99-table,

0.01	0.02	0.03	$\cdots$	0.99
x_1	x_2	x_3	$\cdots$	x_{99}

(1.73)

then the 99-method yields that the uncertain variable $\exp(-\xi)$ has a 99-table,

0.01	0.02	0.03	$\cdots$	0.99
$\exp(-x_{99})$	$\exp(-x_{98})$	$\exp(-x_{97})$	$\cdots$	$\exp(-x_1)$

(1.74)

Increasing Function of Multiple Uncertain Variables

A real-valued function $f(x_1, x_2, \cdots, x_n)$ is said to be strictly increasing if

$$f(x_1, x_2, \cdots, x_n) < f(y_1, y_2, \cdots, y_n) \tag{1.75}$$

whenever $x_i \leq y_i$ for $i = 1, 2, \cdots, n$ and $x_j < y_j$ for at least one index j.

Theorem 1.20. *Let $\xi_1, \xi_2, \cdots, \xi_n$ be independent uncertain variables with uncertainty distributions $\Phi_1, \Phi_2, \cdots, \Phi_n$, respectively. If $f : \Re^n \to \Re$ is a strictly increasing function, then*

$$\xi = f(\xi_1, \xi_2, \cdots, \xi_n) \tag{1.76}$$

is an uncertain variable with inverse uncertainty distribution

$$\Psi^{-1}(\alpha) = f(\Phi_1^{-1}(\alpha), \Phi_2^{-1}(\alpha), \cdots, \Phi_n^{-1}(\alpha)). \tag{1.77}$$

Proof: Since $\xi_1, \xi_2, \cdots, \xi_n$ are independent uncertain variables and f is a strictly increasing function, we have

$$\mathcal{M}\{\xi \leq \Psi^{-1}(\alpha)\}$$

$$= \mathcal{M}\{f(\xi_1, \xi_2, \cdots, \xi_n) \leq f(\Phi_1^{-1}(\alpha), \Phi_2^{-1}(\alpha), \cdots, \Phi_n^{-1}(\alpha))\}$$

$$\geq \mathcal{M}\{(\xi_1 \leq \Phi_1^{-1}(\alpha)) \cap (\xi_2 \leq \Phi_2^{-1}(\alpha)) \cap \cdots \cap (\xi_n \leq \Phi_n^{-1}(\alpha))\}$$

$$= \mathcal{M}\{\xi_1 \leq \Phi_1^{-1}(\alpha)\} \wedge \mathcal{M}\{\xi_2 \leq \Phi_2^{-1}(\alpha)\} \wedge \cdots \wedge \mathcal{M}\{\xi_n \leq \Phi_n^{-1}(\alpha)\}$$

$$= \alpha \wedge \alpha \wedge \cdots \wedge \alpha = \alpha.$$

On the other hand, there exists some index i such that

$$\{f(\xi_1, \xi_2, \cdots, \xi_n) \leq f(\Phi_1^{-1}(\alpha), \Phi_2^{-1}(\alpha), \cdots, \Phi_n^{-1}(\alpha))\} \subset \{\xi_i \leq \Phi_i^{-1}(\alpha)\}.$$

Thus

$$\mathcal{M}\{\xi \leq \Psi^{-1}(\alpha)\} \leq \mathcal{M}\{\xi_i \leq \Phi_i^{-1}(\alpha)\} = \alpha.$$

It follows that $\mathcal{M}\{\xi \leq \Psi^{-1}(\alpha)\} = \alpha$. In other words, Ψ is just the uncertainty distribution of ξ. In fact, we also have

$$\Psi(x) = \sup_{f(x_1, x_2, \cdots, x_n) = x} \min_{1 \leq i \leq n} \Phi_i(x_i). \tag{1.78}$$

The theorem is proved.

99-Method 1.3. *Assume $\xi_1, \xi_2, \cdots, \xi_n$ are uncertain variables, and each ξ_i is represented by a 99-table,*

0.01	0.02	0.03	$\cdots$	0.99
x_1^i	x_2^i	x_3^i	$\cdots$	x_{99}^i

(1.79)

Then for any strictly increasing function $f(x_1, x_2, \cdots, x_n)$, the uncertain variable $f(\xi_1, \xi_2, \cdots, \xi_n)$ has a 99-table,

0.01	0.02	$\cdots$	0.99
$f(x_1^1, x_1^2, \cdots, x_1^n)$	$f(x_2^1, x_2^2, \cdots, x_2^n)$	$\cdots$	$f(x_{99}^1, x_{99}^2, \cdots, x_{99}^n)$

(1.80)

Example 1.24: Let $\xi_1, \xi_2, \cdots, \xi_n$ be independent uncertain variables with uncertainty distributions $\Phi_1, \Phi_2, \cdots, \Phi_n$, respectively. Then the sum

$$\xi = \xi_1 + \xi_2 + \cdots + \xi_n \tag{1.81}$$

is an uncertain variable with inverse uncertainty distribution

$$\Psi^{-1}(\alpha) = \Phi_1^{-1}(\alpha) + \Phi_2^{-1}(\alpha) + \cdots + \Phi_n^{-1}(\alpha). \tag{1.82}$$

If each ξ_i $(1 \le i \le n)$ is represented by a 99-table,

0.01	0.02	0.03	$\cdots$	0.99
x_1^i	x_2^i	x_3^i	$\cdots$	x_{99}^i

$$\tag{1.83}$$

then the 99-method yields that the sum $\xi_1 + \xi_2 + \cdots + \xi_n$ has a 99-table,

0.01	0.02	0.03	$\cdots$	0.99
$\sum\limits_{i=1}^{n} x_1^i$	$\sum\limits_{i=1}^{n} x_2^i$	$\sum\limits_{i=1}^{n} x_3^i$	$\cdots$	$\sum\limits_{i=1}^{n} x_{99}^i$

$$\tag{1.84}$$

Example 1.25: Let $\xi_1, \xi_2, \cdots, \xi_n$ be independent and nonnegative uncertain variables with uncertainty distributions $\Phi_1, \Phi_2, \cdots, \Phi_n$, respectively. Then the product

$$\xi = \xi_1 \times \xi_2 \times \cdots \times \xi_n \tag{1.85}$$

is an uncertain variable with inverse uncertainty distribution

$$\Psi^{-1}(\alpha) = \Phi_1^{-1}(\alpha) \times \Phi_2^{-1}(\alpha) \times \cdots \times \Phi_n^{-1}(\alpha). \tag{1.86}$$

If each ξ_i $(1 \le i \le n)$ is represented by a 99-table,

0.01	0.02	0.03	$\cdots$	0.99
x_1^i	x_2^i	x_3^i	$\cdots$	x_{99}^i

$$\tag{1.87}$$

then the 99-method yields that the product $\xi_1 \times \xi_2 \times \cdots \times \xi_n$ has a 99-table,

0.01	0.02	0.03	$\cdots$	0.99
$\prod\limits_{i=1}^{n} x_1^i$	$\prod\limits_{i=1}^{n} x_2^i$	$\prod\limits_{i=1}^{n} x_3^i$	$\cdots$	$\prod\limits_{i=1}^{n} x_{99}^i$

$$\tag{1.88}$$

Example 1.26: Assume ξ_1, ξ_2, ξ_3 are independent and nonnegative uncertain variables with uncertainty distributions Φ_1, Φ_2, Φ_3, respectively. Then the inverse uncertainty distribution of $(\xi_1 + \xi_2)\xi_3$ is

$$\Psi^{-1}(\alpha) = \left(\Phi_1^{-1}(\alpha) + \Phi_2^{-1}(\alpha)\right) \Phi_3^{-1}(\alpha). \tag{1.89}$$

If ξ_1, ξ_2, ξ_3 are respectively represented by 99-tables,

0.01	0.02	0.03	$\cdots$	0.99
x_1^1	x_2^1	x_3^1	$\cdots$	x_{99}^1

0.01	0.02	0.03	$\cdots$	0.99
x_1^2	x_2^2	x_3^2	$\cdots$	x_{99}^2

0.01	0.02	0.03	$\cdots$	0.99
x_1^3	x_2^3	x_3^3	$\cdots$	x_{99}^3

$$(1.90)$$

then the 99-method yields that the uncertain variable $(\xi_1 + \xi_2)\xi_3$ has a 99-table,

0.01	0.02	$\cdots$	0.99
$(x_1^1 + x_1^2)x_1^3$	$(x_2^1 + x_2^2)x_2^3$	$\cdots$	$(x_{99}^1 + x_{99}^2)x_{99}^3$

$$(1.91)$$

Theorem 1.21. *Assume that ξ_1 and ξ_2 are independent linear uncertain variables $\mathcal{L}(a_1, b_1)$ and $\mathcal{L}(a_2, b_2)$, respectively. Then the sum $\xi_1 + \xi_2$ is also a linear uncertain variable $\mathcal{L}(a_1 + a_2, b_1 + b_2)$, i.e.,*

$$\mathcal{L}(a_1, b_1) + \mathcal{L}(a_2, b_2) = \mathcal{L}(a_1 + a_2, b_1 + b_2). \tag{1.92}$$

The product of a linear uncertain variable $\mathcal{L}(a, b)$ and a scalar number $k > 0$ is also a linear uncertain variable $\mathcal{L}(ka, kb)$, i.e.,

$$k \cdot \mathcal{L}(a, b) = \mathcal{L}(ka, kb). \tag{1.93}$$

Proof: Assume that the uncertain variables ξ_1 and ξ_2 have uncertainty distributions Φ_1 and Φ_2, respectively. Then

$$\Phi_1^{-1}(\alpha) = (1 - \alpha)a_1 + \alpha b_1,$$

$$\Phi_2^{-1}(\alpha) = (1 - \alpha)a_2 + \alpha b_2.$$

It follows from the operational law that the inverse uncertainty distribution of $\xi_1 + \xi_2$ is

$$\Psi^{-1}(\alpha) = \Phi_1^{-1}(\alpha) + \Phi_2^{-1}(\alpha) = (1 - \alpha)(a_1 + a_2) + \alpha(b_1 + b_2).$$

Hence the sum is also a linear uncertain variable $\mathcal{L}(a_1 + a_2, b_1 + b_2)$. The first part is verified. Next, suppose that the uncertainty distribution of the uncertain variable $\xi \sim \mathcal{L}(a, b)$ is Φ. It follows from the operational law that when $k > 0$, the inverse uncertainty distribution of $k\xi$ is

$$\Psi^{-1}(\alpha) = k\Phi^{-1}(\alpha) = (1 - \alpha)(ka) + \alpha(kb).$$

Hence $k\xi$ is just a linear uncertain variable $\mathcal{L}(ka, kb)$.

Theorem 1.22. *Assume that ξ_1 and ξ_2 are independent zigzag uncertain variables $\mathcal{Z}(a_1, b_1, c_1)$ and $\mathcal{Z}(a_2, b_2, c_3)$, respectively. Then the sum $\xi_1 + \xi_2$ is also a zigzag uncertain variable $\mathcal{Z}(a_1 + a_2, b_1 + b_2, c_1 + c_2)$, i.e.,*

$$\mathcal{Z}(a_1, b_1, c_1) + \mathcal{Z}(a_2, b_2, c_2) = \mathcal{Z}(a_1 + a_2, b_1 + b_2, c_1 + c_2). \qquad (1.94)$$

The product of a zigzag uncertain variable $\mathcal{Z}(a, b, c)$ and a scalar number $k > 0$ is also a zigzag uncertain variable $\mathcal{Z}(ka, kb, kc)$, i.e.,

$$k \cdot \mathcal{Z}(a, b, c) = \mathcal{Z}(ka, kb, kc). \qquad (1.95)$$

Proof: Assume that the uncertain variables ξ_1 and ξ_2 have uncertainty distributions Φ_1 and Φ_2, respectively. Then

$$\Phi_1^{-1}(\alpha) = \begin{cases} (1 - 2\alpha)a_1 + 2\alpha b_1, & \text{if } \alpha < 0.5 \\ (2 - 2\alpha)b_1 + (2\alpha - 1)c_1, & \text{if } \alpha \geq 0.5, \end{cases}$$

$$\Phi_2^{-1}(\alpha) = \begin{cases} (1 - 2\alpha)a_2 + 2\alpha b_2, & \text{if } \alpha < 0.5 \\ (2 - 2\alpha)b_2 + (2\alpha - 1)c_2, & \text{if } \alpha \geq 0.5. \end{cases}$$

It follows from the operational law that the inverse uncertainty distribution of $\xi_1 + \xi_2$ is

$$\Psi^{-1}(\alpha) = \begin{cases} (1 - 2\alpha)(a_1 + a_2) + 2\alpha(b_1 + b_2), & \text{if } \alpha < 0.5 \\ (2 - 2\alpha)(b_1 + b_2) + (2\alpha - 1)(c_1 + c_2), & \text{if } \alpha \geq 0.5. \end{cases}$$

Hence the sum is also a zigzag uncertain variable $\mathcal{Z}(a_1 + a_2, b_1 + b_2, c_1 + c_2)$. The first part is verified. Next, suppose that the uncertainty distribution of the uncertain variable $\xi \sim \mathcal{Z}(a, b, c)$ is Φ. It follows from the operational law that when $k > 0$, the inverse uncertainty distribution of $k\xi$ is

$$\Psi^{-1}(\alpha) = k\Phi^{-1}(\alpha) = \begin{cases} (1 - 2\alpha)(ka) + 2\alpha(kb), & \text{if } \alpha < 0.5 \\ (2 - 2\alpha)(kb) + (2\alpha - 1)(kc), & \text{if } \alpha \geq 0.5. \end{cases}$$

Hence $k\xi$ is just a zigzag uncertain variable $\mathcal{Z}(ka, kb, kc)$.

Theorem 1.23. *Let ξ_1 and ξ_2 be independent normal uncertain variables $\mathcal{N}(e_1, \sigma_1)$ and $\mathcal{N}(e_2, \sigma_2)$, respectively. Then the sum $\xi_1 + \xi_2$ is also a normal uncertain variable $\mathcal{N}(e_1 + e_2, \sigma_1 + \sigma_2)$, i.e.,*

$$\mathcal{N}(e_1, \sigma_1) + \mathcal{N}(e_2, \sigma_2) = \mathcal{N}(e_1 + e_2, \sigma_1 + \sigma_2). \qquad (1.96)$$

The product of a normal uncertain variable $\mathcal{N}(e, \sigma)$ and a scalar number $k > 0$ is also a normal uncertain variable $\mathcal{N}(ke, k\sigma)$, i.e.,

$$k \cdot \mathcal{N}(e, \sigma) = \mathcal{N}(ke, k\sigma). \qquad (1.97)$$

Proof: Assume that the uncertain variables ξ_1 and ξ_2 have uncertainty distributions Φ_1 and Φ_2, respectively. Then

$$\Phi_1^{-1}(\alpha) = e_1 + \frac{\sigma_1\sqrt{3}}{\pi}\ln\frac{\alpha}{1-\alpha},$$

$$\Phi_2^{-1}(\alpha) = e_2 + \frac{\sigma_2\sqrt{3}}{\pi}\ln\frac{\alpha}{1-\alpha}.$$

It follows from the operational law that the inverse uncertainty distribution of $\xi_1 + \xi_2$ is

$$\Psi^{-1}(\alpha) = \Phi_1^{-1}(\alpha) + \Phi_2^{-1}(\alpha) = (e_1 + e_2) + \frac{(\sigma_1 + \sigma_2)\sqrt{3}}{\pi}\ln\frac{\alpha}{1-\alpha}.$$

Hence the sum is also a normal uncertain variable $\mathcal{N}(e_1 + e_2, \sigma_1 + \sigma_2)$. The first part is verified. Next, suppose that the uncertainty distribution of the uncertain variable $\xi \sim \mathcal{N}(e, \sigma)$ is Φ. It follows from the operational law that, when $k > 0$, the inverse uncertainty distribution of $k\xi$ is

$$\Psi^{-1}(\alpha) = k\Phi^{-1}(\alpha) = (ke) + \frac{(k\sigma)\sqrt{3}}{\pi}\ln\frac{\alpha}{1-\alpha}.$$

Hence $k\xi$ is just a normal uncertain variable $\mathcal{N}(ke, k\sigma)$.

Theorem 1.24. *Assume that ξ_1 and ξ_2 are independent lognormal uncertain variables $\mathcal{LOGN}(e_1, \sigma_1)$ and $\mathcal{LOGN}(e_2, \sigma_2)$, respectively. Then the product $\xi_1 \cdot \xi_2$ is also a lognormal uncertain variable $\mathcal{LOGN}(e_1 + e_2, \sigma_1 + \sigma_2)$, i.e.,*

$$\mathcal{LOGN}(e_1, \sigma_1) \cdot \mathcal{LOGN}(e_2, \sigma_2) = \mathcal{LOGN}(e_1 + e_2, \sigma_1 + \sigma_2). \tag{1.98}$$

The product of a lognormal uncertain variable $\mathcal{LOGN}(e, \sigma)$ and a scalar number $k > 0$ is also a lognormal uncertain variable $\mathcal{LOGN}(e + \ln k, \sigma)$, i.e.,

$$k \cdot \mathcal{LOGN}(e, \sigma) = \mathcal{LOGN}(e + \ln k, \sigma). \tag{1.99}$$

Proof: Assume that the uncertain variables ξ_1 and ξ_2 have uncertainty distributions Φ_1 and Φ_2, respectively. Then

$$\Phi_1^{-1}(\alpha) = \exp(e_1)\left(\frac{\alpha}{1-\alpha}\right)^{\sqrt{3}\sigma_1/\pi},$$

$$\Phi_2^{-1}(\alpha) = \exp(e_2)\left(\frac{\alpha}{1-\alpha}\right)^{\sqrt{3}\sigma_2/\pi}.$$

It follows from the operational law that the inverse uncertainty distribution of $\xi_1 \cdot \xi_2$ is

$$\Psi^{-1}(\alpha) = \Phi_1^{-1}(\alpha) \cdot \Phi_2^{-1}(\alpha) = \exp(e_1 + e_2)\left(\frac{\alpha}{1-\alpha}\right)^{\sqrt{3}(\sigma_1 + \sigma_2)/\pi}.$$

Hence the product is a lognormal uncertain variable $\mathcal{LOGN}(e_1 + e_2, \sigma_1 + \sigma_2)$. The first part is verified. Next, suppose that the uncertainty distribution of the uncertain variable $\xi \sim \mathcal{LOGN}(e, \sigma)$ is Φ. It follows from the operational law that, when $k > 0$, the inverse uncertainty distribution of $k\xi$ is

$$\Psi^{-1}(\alpha) = k\Phi^{-1}(\alpha) = \exp(e + \ln k) \left(\frac{\alpha}{1-\alpha}\right)^{\sqrt{3}\sigma/\pi}.$$

Hence $k\xi$ is just a lognormal uncertain variable $\mathcal{LOGN}(e + \ln k, \sigma)$.

Example 1.27: Let $\xi_1, \xi_2, \cdots, \xi_n$ be independent uncertain variables with uncertainty distributions $\Phi_1, \Phi_2, \cdots, \Phi_n$, respectively. Then the maximum

$$\xi = \xi_1 \vee \xi_2 \vee \cdots \vee \xi_n \tag{1.100}$$

is an uncertain variable with uncertainty distribution

$$\Psi(x) = \Phi_1(x) \wedge \Phi_2(x) \wedge \cdots \wedge \Phi_n(x) \tag{1.101}$$

whose inverse function is

$$\Psi^{-1}(\alpha) = \Phi_1^{-1}(\alpha) \vee \Phi_2^{-1}(\alpha) \vee \cdots \vee \Phi_n^{-1}(\alpha). \tag{1.102}$$

If each ξ_i $(1 \le i \le n)$ is represented by a 99-table,

0.01	0.02	0.03	$\cdots$	0.99
x_1^i	x_2^i	x_3^i	$\cdots$	x_{99}^i

$$\tag{1.103}$$

then the 99-method yields that the maximum $\xi_1 \vee \xi_2 \vee \cdots \vee \xi_n$ has a 99-table,

0.01	0.02	0.03	$\cdots$	0.99
$\bigvee\limits_{i=1}^{n} x_1^i$	$\bigvee\limits_{i=1}^{n} x_2^i$	$\bigvee\limits_{i=1}^{n} x_3^i$	$\cdots$	$\bigvee\limits_{i=1}^{n} x_{99}^i$

$$\tag{1.104}$$

Example 1.28: Let $\xi_1, \xi_2, \cdots, \xi_n$ be independent uncertain variables with uncertainty distributions $\Phi_1, \Phi_2, \cdots, \Phi_n$, respectively. Then the minimum

$$\xi = \xi_1 \wedge \xi_2 \wedge \cdots \wedge \xi_n \tag{1.105}$$

is an uncertain variable with uncertainty distribution

$$\Psi(x) = \Phi_1(x) \vee \Phi_2(x) \vee \cdots \vee \Phi_n(x) \tag{1.106}$$

whose inverse function is

$$\Psi^{-1}(\alpha) = \Phi_1^{-1}(\alpha) \wedge \Phi_2^{-1}(\alpha) \wedge \cdots \wedge \Phi_n^{-1}(\alpha). \tag{1.107}$$

If each ξ_i $(1 \leq i \leq n)$ is represented by a 99-table,

0.01	0.02	0.03	$\cdots$	0.99
x_1^i	x_2^i	x_3^i	$\cdots$	x_{99}^i

(1.108)

then the 99-method yields that the minimum $\xi_1 \wedge \xi_2 \wedge \cdots \wedge \xi_n$ has a 99-table,

0.01	0.02	0.03	$\cdots$	0.99
$\bigwedge\limits_{i=1}^{n} x_1^i$	$\bigwedge\limits_{i=1}^{n} x_2^i$	$\bigwedge\limits_{i=1}^{n} x_3^i$	$\cdots$	$\bigwedge\limits_{i=1}^{n} x_{99}^i$

(1.109)

Example 1.29: If ξ is an uncertain variable with uncertainty distribution Φ and k is a constant, then $\xi \vee k$ is an uncertain variable with inverse uncertainty distribution

$$\Psi^{-1}(\alpha) = \Phi^{-1}(\alpha) \vee k \qquad (1.110)$$

and has a 99-table,

0.01	0.02	0.03	$\cdots$	0.99
$x_1 \vee k$	$x_2 \vee k$	$x_3 \vee k$	$\cdots$	$x_{99} \vee k$

(1.111)

In addition, $\xi \wedge k$ is an uncertain variable with inverse uncertainty distribution

$$\Psi^{-1}(\alpha) = \Phi^{-1}(\alpha) \wedge k \qquad (1.112)$$

and has a 99-table,

0.01	0.02	0.03	$\cdots$	0.99
$x_1 \wedge k$	$x_2 \wedge k$	$x_3 \wedge k$	$\cdots$	$x_{99} \wedge k$

(1.113)

Decreasing Function of Multiple Uncertain Variables

A real-valued function $f(x_1, x_2, \cdots, x_n)$ is said to be strictly decreasing if

$$f(x_1, x_2, \cdots, x_n) > f(y_1, y_2, \cdots, y_n) \qquad (1.114)$$

whenever $x_i \leq y_i$ for $i = 1, 2, \cdots, n$ and $x_j < y_j$ for at least one index j.

Theorem 1.25. *Let $\xi_1, \xi_2, \cdots, \xi_n$ be independent uncertain variables with uncertainty distributions $\Phi_1, \Phi_2, \cdots, \Phi_n$, respectively. If $f : \Re^n \to \Re$ is a strictly decreasing function, then*

$$\xi = f(\xi_1, \xi_2, \cdots, \xi_n) \qquad (1.115)$$

is an uncertain variable with inverse uncertainty distribution

$$\Psi^{-1}(\alpha) = f(\Phi_1^{-1}(1 - \alpha), \Phi_2^{-1}(1 - \alpha), \cdots, \Phi_n^{-1}(1 - \alpha)). \qquad (1.116)$$

Proof: Since $\xi_1, \xi_2, \cdots, \xi_n$ are independent uncertain variables and f is a strictly decreasing function, we have

$$\mathcal{M}\{\xi \le \Psi^{-1}(\alpha)\}$$
$$= \mathcal{M}\{f(\xi_1, \xi_2, \cdots, \xi_n) \le f(\Phi_1^{-1}(1-\alpha), \Phi_2^{-1}(1-\alpha), \cdots, \Phi_n^{-1}(1-\alpha))\}$$
$$\ge \mathcal{M}\{(\xi_1 \ge \Phi_1^{-1}(1-\alpha)) \cap (\xi_2 \ge \Phi_2^{-1}(1-\alpha)) \cap \cdots \cap (\xi_n \ge \Phi_n^{-1}(1-\alpha))\}$$
$$= \mathcal{M}\{\xi_1 \ge \Phi_1^{-1}(1-\alpha)\} \wedge \mathcal{M}\{\xi_2 \ge \Phi_2^{-1}(1-\alpha)\} \wedge \cdots \wedge \mathcal{M}\{\xi_n \ge \Phi_n^{-1}(1-\alpha)\}$$
$$= \alpha \wedge \alpha \wedge \cdots \wedge \alpha = \alpha. \quad \text{(By the continuity of } \Phi_i\text{'s)}$$

On the other hand, there exists some index i such that

$$\{f(\xi_1, \cdots, \xi_n) \le f(\Phi_1^{-1}(1-\alpha), \cdots, \Phi_n^{-1}(1-\alpha))\} \subset \{\xi_i \ge \Phi_i^{-1}(1-\alpha)\}.$$

Thus

$$\mathcal{M}\{\xi \le \Psi^{-1}(\alpha)\} \le \mathcal{M}\{\xi_i \ge \Phi_i^{-1}(1-\alpha)\} = \alpha.$$

It follows that $\mathcal{M}\{\xi \le \Psi^{-1}(\alpha)\} = \alpha$. In other words, Ψ is just the uncertainty distribution of ξ. In fact, we also have

$$\Psi(x) = \sup_{f(x_1, x_2, \cdots, x_n)=x} \min_{1 \le i \le n} (1 - \Phi_i(x_i)). \qquad (1.117)$$

The theorem is proved.

99-Method 1.4. *Assume $\xi_1, \xi_2, \cdots, \xi_n$ are uncertain variables, and each ξ_i is represented by a 99-table,*

0.01	0.02	0.03	$\cdots$	0.99
x_1^i	x_2^i	x_3^i	$\cdots$	x_{99}^i

$\qquad (1.118)$

Then for any strictly decreasing function $f(x_1, x_2, \cdots, x_n)$, the uncertain variable $f(\xi_1, \xi_2, \cdots, \xi_n)$ has a 99-table,

0.01	0.02	$\cdots$	0.99
$f(x_{99}^1, x_{99}^2, \cdots, x_{99}^n)$	$f(x_{98}^1, x_{98}^2, \cdots, x_{98}^n)$	$\cdots$	$f(x_1^1, x_1^2, \cdots, x_1^n)$

Alternating Monotone Function of Multiple Uncertain Variables

A real-valued function $f(x_1, x_2, \cdots, x_n)$ is said to be alternating monotone if it is increasing with respect to some variables and decreasing with respect to other variables.

Theorem 1.26. *Let $\xi_1, \xi_2, \cdots, \xi_n$ be independent uncertain variables with uncertainty distributions $\Phi_1, \Phi_2, \cdots, \Phi_n$, respectively. If $f(x_1, x_2, \cdots, x_n)$ is strictly increasing with respect to $x_1, x_2, \cdots, x_m$ and strictly decreasing with respect to $x_{m+1}, x_{m+2}, \cdots, x_n$, then*

$$\xi = f(\xi_1, \xi_2, \cdots, \xi_n)$$

is an uncertain variable with inverse uncertainty distribution

$$\Psi^{-1}(\alpha) = f(\Phi_1^{-1}(\alpha), \cdots, \Phi_m^{-1}(\alpha), \Phi_{m+1}^{-1}(1-\alpha), \cdots, \Phi_n^{-1}(1-\alpha)). \quad (1.119)$$

Proof: We only prove the case of $m = 1$ and $n = 2$. Since ξ_1 and ξ_2 are independent uncertain variables and the function $f(x_1, x_2)$ is strictly increasing with respect to x_1 and strictly decreasing with x_2, we have

$$\mathcal{M}\{\xi \leq \Psi^{-1}(\alpha)\} = \mathcal{M}\{f(\xi_1, \xi_2) \leq f(\Phi_1^{-1}(\alpha), \Phi_2^{-1}(1-\alpha))\}$$
$$\geq \mathcal{M}\{(\xi_1 \leq \Phi_1^{-1}(\alpha)) \cap (\xi_2 \geq \Phi_2^{-1}(1-\alpha))\}$$
$$= \mathcal{M}\{\xi_1 \leq \Phi_1^{-1}(\alpha)\} \wedge \mathcal{M}\{\xi_2 \geq \Phi_2^{-1}(1-\alpha)\}$$
$$= \alpha \wedge \alpha = \alpha.$$

On the other hand, the event $\{\xi \leq \Psi^{-1}(\alpha)\}$ is a subset of either $\{\xi_1 \leq \Phi_1^{-1}(\alpha)\}$ or $\{\xi_2 \geq \Phi_2^{-1}(1-\alpha)\}$. Thus $\mathcal{M}\{\xi \leq \Psi^{-1}(\alpha)\} \leq \alpha$. It follows that

$$\mathcal{M}\{\xi \leq \Psi^{-1}(\alpha)\} = \alpha.$$

In other words, Ψ is just the uncertainty distribution of ξ. In fact, we also have

$$\Psi(x) = \sup_{f(x_1, x_2, \cdots, x_n) = x} \left(\min_{1 \leq i \leq m} \Phi_i(x_i) \wedge \min_{m+1 \leq i \leq n} (1 - \Phi_i(x_i)) \right). \quad (1.120)$$

The theorem is proved.

99-Method 1.5. *Assume $\xi_1, \xi_2, \cdots, \xi_n$ are independent uncertain variables, and each ξ_i is represented by a 99-table,*

0.01	0.02	0.03	$\cdots$	0.99
x_1^i	x_2^i	x_3^i	$\cdots$	x_{99}^i

(1.121)

If the function $f(x_1, x_2, \cdots, x_n)$ is strictly increasing with respect to $x_1, x_2, \cdots, x_m$ and strictly decreasing with $x_{m+1}, x_{m+2}, \cdots, x_n$, then the uncertain variable $f(\xi_1, \xi_2, \cdots, \xi_n)$ has a 99-table,

0.01	$\cdots$	0.99
$f(x_1^1, \cdots, x_1^m, x_{99}^{m+1}, \cdots, x_{99}^n)$	$\cdots$	$f(x_{99}^1, \cdots, x_{99}^m, x_1^{m+1}, \cdots, x_1^n)$

Example 1.30: Let ξ_1 and ξ_2 be independent uncertain variables with uncertainty distributions Φ_1 and Φ_2, respectively. Then the inverse uncertainty distribution of the difference $\xi_1 - \xi_2$ is

$$\Psi^{-1}(\alpha) = \Phi_1^{-1}(\alpha) - \Phi_2^{-1}(1-\alpha). \quad (1.122)$$

If ξ_1 and x_2 are respectively represented by 99-tables,

0.01	0.02	0.03	$\cdots$	0.99
x_1^1	x_2^1	x_3^1	$\cdots$	x_{99}^1

$$(1.123)$$

0.01	0.02	0.03	$\cdots$	0.99
x_1^2	x_2^2	x_3^2	$\cdots$	x_{99}^2

$$(1.124)$$

then the 99-method yields that $\xi_1 - \xi_2$ has a 99-table,

0.01	0.02	0.03	$\cdots$	0.99
$x_1^1 - x_{99}^2$	$x_2^1 - x_{98}^2$	$x_3^1 - x_{97}^2$	$\cdots$	$x_{99}^1 - x_1^2$

$$(1.125)$$

Example 1.31: Let ξ_1 and ξ_2 be independent and positive uncertain variables with uncertainty distributions Φ_1 and Φ_2, respectively. Then the inverse uncertainty distribution of the quotient ξ_1/ξ_2 is

$$\Psi^{-1}(\alpha) = \Phi_1^{-1}(\alpha)/\Phi_2^{-1}(1-\alpha). \qquad (1.126)$$

If ξ_1 and ξ_2 are respectively represented by 99-tables,

0.01	0.02	0.03	$\cdots$	0.99
x_1^1	x_2^1	x_3^1	$\cdots$	x_{99}^1

$$(1.127)$$

0.01	0.02	0.03	$\cdots$	0.99
x_1^2	x_2^2	x_3^2	$\cdots$	x_{99}^2

$$(1.128)$$

then the 99-method yields that ξ_1/ξ_2 has a 99-table,

0.01	0.02	0.03	$\cdots$	0.99
x_1^1/x_{99}^2	x_2^1/x_{98}^2	x_3^1/x_{97}^2	$\cdots$	x_{99}^1/x_1^2

$$(1.129)$$

Example 1.32: Assume ξ_1, ξ_2, ξ_3 are independent and positive uncertain variables with uncertainty distributions Φ_1, Φ_2, Φ_3, respectively. Then the inverse uncertainty distribution of $\xi_1/(\xi_2 + \xi_3)$ is

$$\Psi^{-1}(\alpha) = \Phi_1^{-1}(\alpha)/(\Phi_2^{-1}(1-\alpha) + \Phi_3^{-1}(1-\alpha)). \qquad (1.130)$$

If ξ_1, ξ_2, ξ_3 are respectively represented by 99-tables,

0.01	0.02	0.03	$\cdots$	0.99
x_1^1	x_2^1	x_3^1	$\cdots$	x_{99}^1

$$(1.131)$$

0.01	0.02	0.03	$\cdots$	0.99
x_1^2	x_2^2	x_3^2	$\cdots$	x_{99}^2

$$(1.132)$$

0.01	0.02	0.03	$\cdots$	0.99
x_1^3	x_2^3	x_3^3	$\cdots$	x_{99}^3

$$(1.133)$$

then the 99-method yields that $\xi_1/(\xi_2 + \xi_3)$ has a 99-table,

0.01	0.02	$\cdots$	0.99
$x_1^1/(x_{99}^2 + x_{99}^3)$	$x_2^1/(x_{98}^2 + x_{98}^3)$	$\cdots$	$x_{99}^1/(x_1^2 + x_1^3)$

$$(1.134)$$

Operational Law for Boolean Uncertain Variables

A function is said to be Boolean if it is a mapping from $\{0,1\}^n$ to $\{0,1\}$. For example, the following are Boolean functions,

$$f(x_1, x_2, \cdots, x_n) = x_1 \vee x_2 \vee \cdots \vee x_n, \qquad (1.135)$$

$$f(x_1, x_2, \cdots, x_n) = x_1 \wedge x_2 \wedge \cdots \wedge x_n. \qquad (1.136)$$

An uncertain variable is said to be Boolean if it takes values either 0 or 1. For example, the following is a Boolean uncertain variable,

$$\xi = \begin{cases} 1 & \text{with uncertain measure } a \\ 0 & \text{with uncertain measure } 1 - a \end{cases} \qquad (1.137)$$

where a is a number between 0 and 1. This subsection introduces an operational law for this type of uncertain variables.

Theorem 1.27. *Assume that $\xi_1, \xi_2, \cdots, \xi_n$ are independent Boolean uncertain variables, i.e.,*

$$\xi_i = \begin{cases} 1 & \text{with uncertain measure } a_i \\ 0 & \text{with uncertain measure } 1 - a_i \end{cases} \qquad (1.138)$$

for $i = 1, 2, \cdots, n$. If f is a Boolean function (not necessarily monotone), then $\xi = f(\xi_1, \xi_2, \cdots, \xi_n)$ is a Boolean uncertain variable such that

$$\mathcal{M}\{\xi = 1\} = \begin{cases} \sup\limits_{f(x_1, x_2, \cdots, x_n)=1} \min\limits_{1 \le i \le n} \nu_i(x_i), \\ \quad \text{if} \sup\limits_{f(x_1, x_2, \cdots, x_n)=1} \min\limits_{1 \le i \le n} \nu_i(x_i) < 0.5 \\ 1 - \sup\limits_{f(x_1, x_2, \cdots, x_n)=0} \min\limits_{1 \le i \le n} \nu_i(x_i), \\ \quad \text{if} \sup\limits_{f(x_1, x_2, \cdots, x_n)=1} \min\limits_{1 \le i \le n} \nu_i(x_i) \ge 0.5 \end{cases} \qquad (1.139)$$

and

$$\mathcal{M}\{\xi = 0\} = \begin{cases} \sup\limits_{f(x_1, x_2, \cdots, x_n)=0} \min\limits_{1 \le i \le n} \nu_i(x_i), \\ \quad \text{if} \sup\limits_{f(x_1, x_2, \cdots, x_n)=0} \min\limits_{1 \le i \le n} \nu_i(x_i) < 0.5 \\ 1 - \sup\limits_{f(x_1, x_2, \cdots, x_n)=1} \min\limits_{1 \le i \le n} \nu_i(x_i), \\ \quad \text{if} \sup\limits_{f(x_1, x_2, \cdots, x_n)=0} \min\limits_{1 \le i \le n} \nu_i(x_i) \ge 0.5 \end{cases} \qquad (1.140)$$

where x_i take values either 0 or 1, and ν_i are defined by

$$\nu_i(x_i) = \begin{cases} a_i, & \text{if } x_i = 1 \\ 1 - a_i, & \text{if } x_i = 0 \end{cases} \qquad (1.141)$$

for $i = 1, 2, \cdots, n$, respectively.

Proof: It follows from the operational law and independence of uncertain variables that

$$
\mathcal{M}\{\xi = 1\} = \begin{cases} \sup\limits_{f(B_1,B_2,\cdots,B_n)=1} \min\limits_{1\le i\le n} \mathcal{M}\{\xi_i \in B_i\}, \\ \quad \text{if} \quad \sup\limits_{f(B_1,B_2,\cdots,B_n)=1} \min\limits_{1\le i\le n} \mathcal{M}\{\xi_i \in B_i\} > 0.5 \\ 1 - \sup\limits_{f(B_1,B_2,\cdots,B_n)=0} \min\limits_{1\le i\le n} \mathcal{M}\{\xi_i \in B_i\}, \qquad (1.142) \\ \quad \text{if} \quad \sup\limits_{f(B_1,B_2,\cdots,B_n)=0} \min\limits_{1\le i\le n} \mathcal{M}\{\xi_i \in B_i\} > 0.5 \\ 0.5, \ \text{otherwise} \end{cases}
$$

where $B_1, B_2, \cdots, B_n$ are subsets of $\{0,1\}$, and $f(B_1, B_2, \cdots, B_n) = 1$ means $f(x_1, x_2, \cdots, x_n) = 1$ for any $x_1 \in B_1, x_2 \in B_2, \cdots, x_n \in B_n$. Please also note that

$$
\nu_i(1) = \mathcal{M}\{\xi_i = 1\}, \quad \nu_i(0) = \mathcal{M}\{\xi_i = 0\}
$$

for $i = 1, 2, \cdots, n$. The argument breaks down into four cases. Case 1: Assume

$$
\sup_{f(x_1,x_2,\cdots,x_n)=1} \min_{1\le i\le n} \nu_i(x_i) < 0.5.
$$

Then we have

$$
\sup_{f(B_1,B_2,\cdots,B_n)=0} \min_{1\le i\le n} \mathcal{M}\{\xi_i \in B_i\} = 1 - \sup_{f(x_1,x_2,\cdots,x_n)=1} \min_{1\le i\le n} \nu_i(x_i) > 0.5.
$$

It follows from (1.142) that

$$
\mathcal{M}\{\xi = 1\} = \sup_{f(x_1,x_2,\cdots,x_n)=1} \min_{1\le i\le n} \nu_i(x_i).
$$

Case 2: Assume

$$
\sup_{f(x_1,x_2,\cdots,x_n)=1} \min_{1\le i\le n} \nu_i(x_i) > 0.5.
$$

Then we have

$$
\sup_{f(B_1,B_2,\cdots,B_n)=1} \min_{1\le i\le n} \mathcal{M}\{\xi_i \in B_i\} = 1 - \sup_{f(x_1,x_2,\cdots,x_n)=0} \min_{1\le i\le n} \nu_i(x_i) > 0.5.
$$

It follows from (1.142) that

$$
\mathcal{M}\{\xi = 1\} = 1 - \sup_{f(x_1,x_2,\cdots,x_n)=0} \min_{1\le i\le n} \nu_i(x_i).
$$

Case 3: Assume

$$
\sup_{f(x_1,x_2,\cdots,x_n)=1} \min_{1\le i\le n} \nu_i(x_i) = 0.5,
$$

$$
\sup_{f(x_1,x_2,\cdots,x_n)=0} \min_{1\le i\le n} \nu_i(x_i) = 0.5.
$$

Then we have

$$\sup_{f(B_1,B_2,\cdots,B_n)=1} \min_{1\leq i\leq n} \mathcal{M}\{\xi_i \in B_i\} = 0.5,$$

$$\sup_{f(B_1,B_2,\cdots,B_n)=0} \min_{1\leq i\leq n} \mathcal{M}\{\xi_i \in B_i\} = 0.5.$$

It follows from (1.142) that

$$\mathcal{M}\{\xi = 1\} = 0.5 = 1 - \sup_{f(x_1,x_2,\cdots,x_n)=0} \min_{1\leq i\leq n} \nu_i(x_i).$$

Case 4: Assume

$$\sup_{f(x_1,x_2,\cdots,x_n)=1} \min_{1\leq i\leq n} \nu_i(x_i) = 0.5,$$

$$\sup_{f(x_1,x_2,\cdots,x_n)=0} \min_{1\leq i\leq n} \nu_i(x_i) < 0.5.$$

Then we have

$$\sup_{f(B_1,B_2,\cdots,B_n)=1} \min_{1\leq i\leq n} \mathcal{M}\{\xi_i \in B_i\} = 1 - \sup_{f(x_1,x_2,\cdots,x_n)=0} \min_{1\leq i\leq n} \nu_i(x_i) > 0.5.$$

It follows from (1.142) that

$$\mathcal{M}\{\xi = 1\} = 1 - \sup_{f(x_1,x_2,\cdots,x_n)=0} \min_{1\leq i\leq n} \nu_i(x_i).$$

Hence the equation (1.139) is proved for the four cases. Similarly, we may verify the equation (1.140).

Theorem 1.28. *Assume that $\xi_1, \xi_2, \cdots, \xi_n$ are independent Boolean uncertain variables, i.e.,*

$$\xi_i = \begin{cases} 1 & \text{with uncertain measure } a_i \\ 0 & \text{with uncertain measure } 1 - a_i \end{cases} \tag{1.143}$$

for $i = 1, 2, \cdots, n$. Then the minimum

$$\xi = \xi_1 \wedge \xi_2 \wedge \cdots \wedge \xi_n \tag{1.144}$$

is a Boolean uncertain variable such that

$$\mathcal{M}\{\xi = 1\} = a_1 \wedge a_2 \wedge \cdots \wedge a_n, \tag{1.145}$$

$$\mathcal{M}\{\xi = 0\} = (1 - a_1) \vee (1 - a_2) \vee \cdots \vee (1 - a_n). \tag{1.146}$$

Proof: Since ξ is the minimum of Boolean uncertain variables, the corresponding Boolean function is

$$f(x_1, x_2, \cdots, x_n) = x_1 \wedge x_2 \wedge \cdots \wedge x_n. \tag{1.147}$$

Without loss of generality, we assume $a_1 \geq a_2 \geq \cdots \geq a_n$. Then we have

$$\sup_{f(x_1,x_2,\cdots,x_n)=1} \min_{1\leq i\leq n} \nu_i(x_i) = \min_{1\leq i\leq n} \nu_i(1) = a_n,$$

$$\sup_{f(x_1,x_2,\cdots,x_n)=0} \min_{1\leq i\leq n} \nu_i(x_i) = (1-a_n) \wedge \min_{1\leq i<n} (a_i \vee (1-a_i))$$

where $\nu_i(x_i)$ are defined by (1.141) for $i = 1, 2, \cdots, n$, respectively. When $a_n < 0.5$, we have

$$\sup_{f(x_1,x_2,\cdots,x_n)=1} \min_{1\leq i\leq n} \nu_i(x_i) = a_n < 0.5.$$

It follows from Theorem 1.27 that

$$\mathcal{M}\{\xi = 1\} = \sup_{f(x_1,x_2,\cdots,x_n)=1} \min_{1\leq i\leq n} \nu_i(x_i) = a_n.$$

When $a_n \geq 0.5$, we have

$$\sup_{f(x_1,x_2,\cdots,x_n)=1} \min_{1\leq i\leq n} \nu_i(x_i) = a_n \geq 0.5.$$

It follows from Theorem 1.27 that

$$\mathcal{M}\{\xi = 1\} = 1 - \sup_{f(x_1,x_2,\cdots,x_n)=0} \min_{1\leq i\leq n} \nu_i(x_i) = 1 - (1-a_n) = a_n.$$

Thus $\mathcal{M}\{\xi = 1\}$ is always a_n, i.e., the minimum value of $a_1, a_2, \cdots, a_n$. Thus the equation (1.145) is proved. The equation (1.146) may be verified by the self-duality of uncertain measure.

Theorem 1.29. *Assume that $\xi_1, \xi_2, \cdots, \xi_n$ are independent Boolean uncertain variables, i.e.,*

$$\xi_i = \begin{cases} 1 \text{ with uncertain measure } a_i \\ 0 \text{ with uncertain measure } 1 - a_i \end{cases} \tag{1.148}$$

for $i = 1, 2, \cdots, n$. Then the maximum

$$\xi = \xi_1 \vee \xi_2 \vee \cdots \vee \xi_n \tag{1.149}$$

is a Boolean uncertain variable such that

$$\mathcal{M}\{\xi = 1\} = a_1 \vee a_2 \vee \cdots \vee a_n, \tag{1.150}$$

$$\mathcal{M}\{\xi = 0\} = (1 - a_1) \wedge (1 - a_2) \wedge \cdots \wedge (1 - a_n). \tag{1.151}$$

Proof: Since ξ is the maximum of Boolean uncertain variables, the corresponding Boolean function is

$$f(x_1, x_2, \cdots, x_n) = x_1 \vee x_2 \vee \cdots \vee x_n. \tag{1.152}$$

Without loss of generality, we assume $a_1 \geq a_2 \geq \cdots \geq a_n$. Then we have

$$\sup_{f(x_1, x_2, \cdots, x_n)=1} \min_{1 \leq i \leq n} \nu_i(x_i) = a_1 \wedge \min_{1 < i \leq n} (a_i \vee (1 - a_i)),$$

$$\sup_{f(x_1, x_2, \cdots, x_n)=0} \min_{1 \leq i \leq n} \nu_i(x_i) = \min_{1 \leq i \leq n} \nu_i(0) = 1 - a_1$$

where $\nu_i(x_i)$ are defined by (1.141) for $i = 1, 2, \cdots, n$, respectively. When $a_1 \geq 0.5$, we have

$$\sup_{f(x_1, x_2, \cdots, x_n)=1} \min_{1 \leq i \leq n} \nu_i(x_i) \geq 0.5.$$

It follows from Theorem 1.27 that

$$\mathcal{M}\{\xi = 1\} = 1 - \sup_{f(x_1, x_2, \cdots, x_n)=0} \min_{1 \leq i \leq n} \nu_i(x_i) = 1 - (1 - a_1) = a_1.$$

When $a_1 < 0.5$, we have

$$\sup_{f(x_1, x_2, \cdots, x_n)=1} \min_{1 \leq i \leq n} \nu_i(x_i) = a_1 < 0.5.$$

It follows from Theorem 1.27 that

$$\mathcal{M}\{\xi = 1\} = \sup_{f(x_1, x_2, \cdots, x_n)=1} \min_{1 \leq i \leq n} \nu_i(x_i) = a_1.$$

Thus $\mathcal{M}\{\xi = 1\}$ is always a_1, i.e., the maximum value of $a_1, a_2, \cdots, a_n$. Thus the equation (1.150) is proved. The equation (1.151) may be verified by the self-duality of uncertain measure.

Theorem 1.30. *Assume that $\xi_1, \xi_2, \cdots, \xi_n$ are independent Boolean uncertain variables, i.e.,*

$$\xi_i = \begin{cases} 1 & \text{with uncertain measure } a_i \\ 0 & \text{with uncertain measure } 1 - a_i \end{cases} \tag{1.153}$$

for $i = 1, 2, \cdots, n$. Then (k-out-of-n)

$$\xi = \begin{cases} 1, & \text{if } \xi_1 + \xi_2 + \cdots + \xi_n \geq k \\ 0, & \text{if } \xi_1 + \xi_2 + \cdots + \xi_n < k \end{cases} \tag{1.154}$$

is a Boolean uncertain variable such that

$$\mathcal{M}\{\xi = 1\} = \text{the kth largest value of } a_1, a_2, \cdots, a_n, \tag{1.155}$$

$$\mathcal{M}\{\xi = 0\} = \text{the kth smallest value of } 1 - a_1, 1 - a_2, \cdots, 1 - a_n. \tag{1.156}$$

Proof: This is the so-called k-out-of-n system. The corresponding Boolean function is

$$f(x_1, x_2, \cdots, x_n) = \begin{cases} 1, & \text{if } x_1 + x_2 + \cdots + x_n \geq k \\ 0, & \text{if } x_1 + x_2 + \cdots + x_n < k. \end{cases} \tag{1.157}$$

Without loss of generality, we assume $a_1 \geq a_2 \geq \cdots \geq a_n$. Then we have

$$\sup_{f(x_1, x_2, \cdots, x_n)=1} \min_{1 \leq i \leq n} \nu_i(x_i) = a_k \wedge \min_{k < i \leq n} (a_i \vee (1 - a_i)),$$

$$\sup_{f(x_1, x_2, \cdots, x_n)=0} \min_{1 \leq i \leq n} \nu_i(x_i) = (1 - a_k) \wedge \min_{k < i \leq n} (a_i \vee (1 - a_i))$$

where $\nu_i(x_i)$ are defined by (1.141) for $i = 1, 2, \cdots, n$, respectively. When $a_k \geq 0.5$, we have

$$\sup_{f(x_1, x_2, \cdots, x_n)=1} \min_{1 \leq i \leq n} \nu_i(x_i) \geq 0.5.$$

It follows from Theorem 1.27 that

$$\mathcal{M}\{\xi = 1\} = 1 - \sup_{f(x_1, x_2, \cdots, x_n)=0} \min_{1 \leq i \leq n} \nu_i(x_i) = 1 - (1 - a_k) = a_k.$$

When $a_k < 0.5$, we have

$$\sup_{f(x_1, x_2, \cdots, x_n)=1} \min_{1 \leq i \leq n} \nu_i(x_i) = a_k < 0.5.$$

It follows from Theorem 1.27 that

$$\mathcal{M}\{\xi = 1\} = \sup_{f(x_1, x_2, \cdots, x_n)=1} \min_{1 \leq i \leq n} \nu_i(x_i) = a_k.$$

Thus $\mathcal{M}\{\xi = 1\}$ is always a_k, i.e., the kth largest value of $a_1, a_2, \cdots, a_n$. Thus the equation (1.155) is proved. The equation (1.156) may be verified by the self-duality of uncertain measure.

Operational Law with Joint Uncertainty Distribution

Let $\xi_1, \xi_2, \cdots, \xi_n$ be uncertain variables with joint uncertainty distribution Φ. It is clear that $\Phi^{-1}(\alpha)$ is a set of $\Re^n$ rather than a single point. Assume $f : \Re^n \to \Re$ is an increasing function. It follows from the operational law and maximum uncertainty principle that $f(\xi_1, \xi_2, \cdots, \xi_n)$ is an uncertain variable with inverse uncertainty distribution

$$\Psi^{-1}(\alpha) = \begin{cases} \min_{(x_1, x_2, \cdots, x_n) \in \Phi^{-1}(\alpha)} f(x_1, x_2, \cdots, x_n), & \text{if } \alpha \leq 0.5 \\ \max_{(x_1, x_2, \cdots, x_n) \in \Phi^{-1}(\alpha)} f(x_1, x_2, \cdots, x_n), & \text{if } \alpha > 0.5. \end{cases} \tag{1.158}$$

If $f : \Re^n \to \Re$ is a decreasing function, then $f(\xi_1, \xi_2, \cdots, \xi_n)$ is an uncertain variable with inverse uncertainty distribution

$$\Psi^{-1}(\alpha) = \begin{cases} \min_{(x_1,x_2,\cdots,x_n)\in\Phi^{-1}(1-\alpha)} f(x_1, x_2, \cdots, x_n), & \text{if } \alpha \leq 0.5 \\ \max_{(x_1,x_2,\cdots,x_n)\in\Phi^{-1}(1-\alpha)} f(x_1, x_2, \cdots, x_n), & \text{if } \alpha > 0.5. \end{cases} \qquad (1.159)$$

1.6   Expected Value

Expected value is the average value of uncertain variable in the sense of uncertain measure, and represents the size of uncertain variable.

Definition 1.20 *(Liu [120]). Let ξ be an uncertain variable. Then the expected value of ξ is defined by*

$$E[\xi] = \int_0^{+\infty} \mathcal{M}\{\xi \geq r\}\mathrm{d}r - \int_{-\infty}^0 \mathcal{M}\{\xi \leq r\}\mathrm{d}r \qquad (1.160)$$

provided that at least one of the two integrals is finite.

Theorem 1.31. *Let ξ be an uncertain variable with uncertainty distribution Φ. If the expected value exists, then*

$$E[\xi] = \int_0^{+\infty} (1 - \Phi(x))\mathrm{d}x - \int_{-\infty}^0 \Phi(x)\mathrm{d}x. \qquad (1.161)$$

Proof: It follows from the definitions of expected value operator and uncertainty distribution that

$$\begin{aligned} E[\xi] &= \int_0^{+\infty} \mathcal{M}\{\xi \geq r\}\mathrm{d}r - \int_{-\infty}^0 \mathcal{M}\{\xi \leq r\}\mathrm{d}r \\ &= \int_0^{+\infty} (1 - \Phi(x))\mathrm{d}x - \int_{-\infty}^0 \Phi(x)\mathrm{d}x. \end{aligned}$$

See Figure 1.13. The theorem is proved.

Theorem 1.32. *Let ξ be an uncertain variable with uncertainty distribution Φ. If the expected value exists, then*

$$E[\xi] = \int_0^1 \Phi^{-1}(\alpha)\mathrm{d}\alpha. \qquad (1.162)$$

Proof: It follows from the definitions of expected value operator and uncertainty distribution that

$$\begin{aligned} E[\xi] &= \int_0^{+\infty} \mathcal{M}\{\xi \geq r\}\mathrm{d}r - \int_{-\infty}^0 \mathcal{M}\{\xi \leq r\}\mathrm{d}r \\ &= \int_{\Phi(0)}^1 \Phi^{-1}(\alpha)\mathrm{d}\alpha + \int_0^{\Phi(0)} \Phi^{-1}(\alpha)\mathrm{d}\alpha = \int_0^1 \Phi^{-1}(\alpha)\mathrm{d}\alpha. \end{aligned}$$

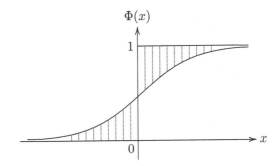

Figure 1.13: $E[\xi] = \int_0^{+\infty} (1 - \Phi(x))\mathrm{d}x - \int_{-\infty}^0 \Phi(x)\mathrm{d}x$

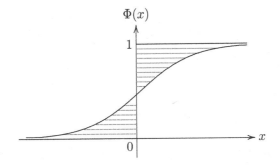

Figure 1.14: $E[\xi] = \int_0^1 \Phi^{-1}(\alpha)\mathrm{d}\alpha$

See Figure 1.14. The theorem is proved.

Theorem 1.33. *Let ξ be an uncertain variable with uncertainty distribution Φ. If the expected value exists, then*

$$E[\xi] = \int_{-\infty}^{+\infty} x\mathrm{d}\Phi(x). \tag{1.163}$$

Proof: It follows from Theorem 1.32 that

$$E[\xi] = \int_0^1 \Phi^{-1}(\alpha)\mathrm{d}\alpha.$$

Now write $\Phi^{-1}(\alpha) = x$. Then we immediately have $\alpha = \Phi(x)$. The change of variable of integral produces (1.163). The theorem is verified.

Example 1.33: Suppose that ξ is a discrete uncertain variable represented by

$$\xi = \begin{array}{|c|c|c|c|} \hline \alpha_1 & \alpha_2 & \cdots & \alpha_m \\ \hline x_1 & x_2 & \cdots & x_m \\ \hline \end{array} \tag{1.164}$$

where $x_1 < x_2 < \cdots < x_m$ and $0 \le \alpha_1 \le \alpha_2 \le \cdots \le \alpha_m = 1$. The uncertainty distribution Φ of ξ is a step function shown in (1.35). Write $\alpha_0 \equiv 0$. If $x_1 \ge 0$, then the expected value is

$$E[\xi] = \int_0^{x_1} 1 \mathrm{d}x + \sum_{i=1}^{m-1} \int_{x_i}^{x_{i+1}} (1 - \alpha_i) \mathrm{d}x + \int_{x_m}^{+\infty} 0 \mathrm{d}x$$

$$= x_1 + \sum_{i=1}^{m-1} (1 - \alpha_i)(x_{i+1} - x_i) + 0$$

$$= \sum_{i=1}^{m} (\alpha_i - \alpha_{i-1}) x_i.$$

If $x_m \le 0$, then the expected value is

$$E[\xi] = -\int_{-\infty}^{x_1} 0 \mathrm{d}x - \sum_{i=1}^{m-1} \int_{x_i}^{x_{i+1}} \alpha_i \mathrm{d}x - \int_{x_m}^{0} 1 \mathrm{d}x$$

$$= 0 - \sum_{i=1}^{m-1} \alpha_i (x_{i+1} - x_i) + x_m$$

$$= \sum_{i=1}^{m} (\alpha_i - \alpha_{i-1}) x_i.$$

If there exists an index k such that $x_k \le 0 \le x_{k+1}$, then the expected value is

$$E[\xi] = \int_0^{x_{k+1}} (1 - \alpha_k) \mathrm{d}x + \sum_{i=k+1}^{m-1} \int_{x_i}^{x_{i+1}} (1 - \alpha_i) \mathrm{d}x$$

$$- \sum_{i=1}^{k-1} \int_{x_i}^{x_{i+1}} \alpha_i \mathrm{d}x - \int_{x_k}^{0} \alpha_k \mathrm{d}x$$

$$= x_{k+1}(1 - \alpha_k) + \sum_{i=k+1}^{m-1} (1 - \alpha_i)(x_{i+1} - x_i)$$

$$- \sum_{i=1}^{k-1} \alpha_i (x_{i+1} - x_i) + x_k \alpha_k$$

$$= \sum_{i=1}^{m} (\alpha_i - \alpha_{i-1}) x_i.$$

Thus we always have the expected value

$$E[\xi] = \sum_{i=1}^{m} (\alpha_i - \alpha_{i-1}) x_i \tag{1.165}$$

where $\alpha_0 \equiv 0$ and $\alpha_m \equiv 1$.

Example 1.34: Let $\xi \sim \mathcal{L}(a, b)$ be a linear uncertain variable. If $a \geq 0$, then the expected value is

$$E[\xi] = \left(\int_0^a 1 \mathrm{d}x + \int_a^b \left(1 - \frac{x-a}{b-a} \right) \mathrm{d}x + \int_b^{+\infty} 0 \mathrm{d}x \right) - \int_{-\infty}^0 0 \mathrm{d}x = \frac{a+b}{2}.$$

If $b \leq 0$, then the expected value is

$$E[\xi] = \int_0^{+\infty} 0 \mathrm{d}x - \left(\int_{-\infty}^a 0 \mathrm{d}x + \int_a^b \frac{x-a}{b-a} \mathrm{d}x + \int_b^0 1 \mathrm{d}x \right) = \frac{a+b}{2}.$$

If $a < 0 < b$, then the expected value is

$$E[\xi] = \int_0^b \left(1 - \frac{x-a}{b-a} \right) \mathrm{d}x - \int_a^0 \frac{x-a}{b-a} \mathrm{d}x = \frac{a+b}{2}.$$

Thus we always have the expected value

$$E[\xi] = \frac{a+b}{2}. \tag{1.166}$$

Example 1.35: The zigzag uncertain variable $\xi \sim \mathcal{Z}(a, b, c)$ has an expected value

$$E[\xi] = \frac{a + 2b + c}{4}. \tag{1.167}$$

Example 1.36: The normal uncertain variable $\xi \sim \mathcal{N}(e, \sigma)$ has an expected value e, i.e.,

$$E[\xi] = e. \tag{1.168}$$

Example 1.37: If $\sigma < \pi/\sqrt{3}$, then the lognormal uncertain variable $\xi \sim \mathcal{LOGN}(e, \sigma)$ has an expected value

$$E[\xi] = \sqrt{3}\sigma \exp(e) \csc(\sqrt{3}\sigma). \tag{1.169}$$

Otherwise, $E[\xi] = +\infty$.

Linearity of Expected Value Operator

Theorem 1.34. *Let ξ and η be independent uncertain variables with finite expected values. Then for any real numbers a and b, we have*

$$E[a\xi + b\eta] = aE[\xi] + bE[\eta]. \tag{1.170}$$

Proof: Suppose that ξ and η have uncertainty distributions Φ and Ψ, respectively.

STEP 1: We first prove $E[a\xi] = aE[\xi]$. If $a = 0$, then the equation holds trivially. If $a > 0$, then the inverse uncertainty distribution of $a\xi$ is

$$\Upsilon^{-1}(\alpha) = a\Phi^{-1}(\alpha).$$

It follows from Theorem 1.32 that

$$E[a\xi] = \int_0^1 a\Phi^{-1}(\alpha)d\alpha = a\int_0^1 \Phi^{-1}(\alpha)d\alpha = aE[\xi].$$

If $a < 0$, then the inverse uncertainty distribution of $a\xi$ is

$$\Upsilon^{-1}(\alpha) = a\Phi^{-1}(1 - \alpha).$$

It follows from Theorem 1.32 that

$$E[a\xi] = \int_0^1 a\Phi^{-1}(1 - \alpha)d\alpha = a\int_0^1 \Phi^{-1}(\alpha)d\alpha = aE[\xi].$$

Thus we always have $E[a\xi] = aE[\xi]$.

STEP 2: We prove $E[\xi + \eta] = E[\xi] + E[\eta]$. The inverse uncertainty distribution of the sum $\xi + \eta$ is

$$\Upsilon^{-1}(\alpha) = \Phi^{-1}(\alpha) + \Psi^{-1}(\alpha).$$

It follows from Theorem 1.32 that

$$E[\xi + \eta] = \int_0^1 \Upsilon^{-1}(\alpha)d\alpha = \int_0^1 \Phi^{-1}(\alpha)d\alpha + \int_0^1 \Psi^{-1}(\alpha)d\alpha = E[\xi] + E[\eta].$$

STEP 3: Finally, for any real numbers a and b, it follows from Steps 1 and 2 that

$$E[a\xi + b\eta] = E[a\xi] + E[b\eta] = aE[\xi] + bE[\eta].$$

The theorem is proved.

Example 1.38: Generally speaking, the expected value operator is not necessarily linear if ξ and η are not independent. For example, take $(\Gamma, \mathcal{L}, \mathcal{M})$ to be $\{\gamma_1, \gamma_2, \gamma_3\}$ with $\mathcal{M}\{\gamma_1\} = 0.7$, $\mathcal{M}\{\gamma_2\} = 0.3$, $\mathcal{M}\{\gamma_3\} = 0.2$, $\mathcal{M}\{\gamma_1, \gamma_2\} = 0.8$, $\mathcal{M}\{\gamma_1, \gamma_3\} = 0.7$, $\mathcal{M}\{\gamma_2, \gamma_3\} = 0.3$. The uncertain variables are defined by

$$\xi_1(\gamma) = \begin{cases} 1, & \text{if } \gamma = \gamma_1 \\ 0, & \text{if } \gamma = \gamma_2 \\ 2, & \text{if } \gamma = \gamma_3, \end{cases} \qquad \xi_2(\gamma) = \begin{cases} 0, & \text{if } \gamma = \gamma_1 \\ 2, & \text{if } \gamma = \gamma_2 \\ 3, & \text{if } \gamma = \gamma_3. \end{cases}$$

Note that ξ_1 and ξ_2 are not independent, and their sum is

$$(\xi_1 + \xi_2)(\gamma) = \begin{cases} 1, & \text{if } \gamma = \gamma_1 \\ 2, & \text{if } \gamma = \gamma_2 \\ 5, & \text{if } \gamma = \gamma_3. \end{cases}$$

Thus $E[\xi_1] = 0.9$, $E[\xi_2] = 0.8$, and $E[\xi_1 + \xi_2] = 1.9$. This fact implies that

$$E[\xi_1 + \xi_2] > E[\xi_1] + E[\xi_2].$$

If the uncertain variables are defined by

$$\eta_1(\gamma) = \begin{cases} 0, & \text{if } \gamma = \gamma_1 \\ 1, & \text{if } \gamma = \gamma_2 \\ 2, & \text{if } \gamma = \gamma_3, \end{cases} \qquad \eta_2(\gamma) = \begin{cases} 0, & \text{if } \gamma = \gamma_1 \\ 3, & \text{if } \gamma = \gamma_2 \\ 1, & \text{if } \gamma = \gamma_3. \end{cases}$$

Then we have

$$(\eta_1 + \eta_2)(\gamma) = \begin{cases} 0, & \text{if } \gamma = \gamma_1 \\ 4, & \text{if } \gamma = \gamma_2 \\ 3, & \text{if } \gamma = \gamma_3. \end{cases}$$

Thus $E[\eta_1] = 0.5$, $E[\eta_2] = 0.9$, and $E[\eta_1 + \eta_2] = 1.2$. This fact implies that

$$E[\eta_1 + \eta_2] < E[\eta_1] + E[\eta_2].$$

Expected Value of Function of Single Uncertain Variable

Let ξ be an uncertain variable, and $f : \Re \to \Re$ a function. Then the expected value of $f(\xi)$ is

$$E[f(\xi)] = \int_0^{+\infty} \mathcal{M}\{f(\xi) \geq r\}\mathrm{d}r - \int_{-\infty}^0 \mathcal{M}\{f(\xi) \leq r\}\mathrm{d}r.$$

For random case, it has been proved that the expected value $E[f(\xi)]$ is the Lebesgue-Stieltjes integral of $f(x)$ with respect to the uncertainty distribution Φ of ξ if the integral exists. However, generally speaking, it is not true for uncertain case.

Example 1.39: We consider an uncertain variable ξ whose first identification function is given by

$$\lambda(x) = \begin{cases} 0.3, & \text{if } -1 \leq x < 0 \\ 0.5, & \text{if } 0 \leq x \leq 1. \end{cases}$$

Then the expected value $E[\xi^2] = 0.5$. However, the uncertainty distribution of ξ is

$$\Phi(x) = \begin{cases} 0, & \text{if } x < -1 \\ 0.3, & \text{if } -1 \leq x < 0 \\ 0.5, & \text{if } 0 \leq x < 1 \\ 1, & \text{if } x \geq 1 \end{cases}$$

and the Lebesgue-Stieltjes integral

$$\int_{-\infty}^{+\infty} x^2 \mathrm{d}\Phi(x) = (-1)^2 \times 0.3 + 0^2 \times 0.2 + 1^2 \times 0.5 = 0.8 \neq E[\xi^2].$$

Theorem 1.35 *(Liu and Ha [132]). Let ξ be an uncertain variable whose uncertainty distribution Φ. If $f(x)$ is a strictly monotone function such that the expected value $E[f(\xi)]$ exists, then*

$$E[f(\xi)] = \int_{-\infty}^{+\infty} f(x)\mathrm{d}\Phi(x). \tag{1.171}$$

Proof: We first suppose that $f(x)$ is a strictly increasing function. Then $f(\xi)$ has an uncertainty distribution $\Phi(f^{-1}(x))$. It follows from the change of variable of integral that

$$E[f(\xi)] = \int_{-\infty}^{+\infty} x\mathrm{d}\Phi(f^{-1}(x)) = \int_{-\infty}^{+\infty} f(x)\mathrm{d}\Phi(x).$$

If $f(x)$ is a strictly decreasing function, then $-f(x)$ is a strictly increasing function. Hence

$$E[f(\xi)] = -E[-f(\xi)] = -\int_{-\infty}^{+\infty} -f(x)\mathrm{d}\Phi(x) = \int_{-\infty}^{+\infty} f(y)\mathrm{d}\Phi(y).$$

The theorem is verified.

Example 1.40: Let ξ be a positive linear uncertain variable $\mathcal{L}(a, b)$. Then its uncertainty distribution is $\Phi(x) = (x - a)/(b - a)$. Thus

$$E[\xi^2] = \int_a^b x^2 \mathrm{d}\Phi(x) = \frac{a^2 + b^2 + ab}{3}.$$

Example 1.41: Let ξ be a positive linear uncertain variable $\mathcal{L}(a, b)$. Then its uncertainty distribution is $\Phi(x) = (x - a)/(b - a)$. Thus

$$E[\exp(\xi)] = \int_a^b \exp(x)\mathrm{d}\Phi(x) = \frac{\exp(b) - \exp(a)}{b - a}.$$

Theorem 1.36 *(Liu and Ha [132]). Assume ξ is an uncertain variable with uncertainty distribution Φ. If $f(x)$ is a strictly monotone function such that the expected value $E[f(\xi)]$ exists, then*

$$E[f(\xi)] = \int_0^1 f(\Phi^{-1}(\alpha))\mathrm{d}\alpha. \tag{1.172}$$

Proof: Suppose that f is a strictly increasing function. It follows from Theorem 1.20 that the inverse uncertainty distribution of $f(\xi)$ is

$$\Psi^{-1}(\alpha) = f(\Phi^{-1}(\alpha)).$$

By using Theorem 1.32, the equation (1.172) is proved. When f is a strictly decreasing function, it follows from Theorem 1.25 that the inverse uncertainty distribution of $f(\xi)$ is

$$\Psi^{-1}(\alpha) = f(\Phi^{-1}(1-\alpha)).$$

By using Theorem 1.32 and the change of variable of integral, we get the equation (1.172). The theorem is verified.

Example 1.42: Let ξ be a nonnegative uncertain variable with uncertainty distribution Φ. Then

$$E[\sqrt{\xi}] = \int_0^1 \sqrt{\Phi^{-1}(\alpha)}\,d\alpha. \tag{1.173}$$

Example 1.43: Let ξ be a positive uncertain variable with uncertainty distribution Φ. Then

$$E\left[\frac{1}{\xi}\right] = \int_0^1 \frac{1}{\Phi^{-1}(1-\alpha)}\,d\alpha = \int_0^1 \frac{1}{\Phi^{-1}(\alpha)}\,d\alpha. \tag{1.174}$$

Expected Value of Function of Multiple Uncertain Variables

Theorem 1.37 *(Liu and Ha [132]). Assume $\xi_1, \xi_2, \cdots, \xi_n$ are independent uncertain variables with uncertainty distributions $\Phi_1, \Phi_2, \cdots, \Phi_n$, respectively. If $f : \Re^n \to \Re$ is a strictly monotone function, then the uncertain variable $\xi = f(\xi_1, \xi_2, \cdots, \xi_n)$ has an expected value*

$$E[\xi] = \int_0^1 f(\Phi_1^{-1}(\alpha), \Phi_2^{-1}(\alpha), \cdots, \Phi_n^{-1}(\alpha))\,d\alpha \tag{1.175}$$

provided that the expected value $E[\xi]$ exists.

Proof: Suppose that f is a strictly increasing function. It follows from Theorem 1.20 that the inverse uncertainty distribution of ξ is

$$\Psi^{-1}(\alpha) = f(\Phi_1^{-1}(\alpha), \Phi_2^{-1}(\alpha), \cdots, \Phi_n^{-1}(\alpha)).$$

By using Theorem 1.32, we obtain (1.175). When f is a strictly decreasing function, it follows from Theorem 1.25 that the inverse uncertainty distribution of ξ is

$$\Psi^{-1}(\alpha) = f(\Phi_1^{-1}(1-\alpha), \Phi_2^{-1}(1-\alpha), \cdots, \Phi_n^{-1}(1-\alpha)).$$

By using Theorem 1.32 and the change of variable of integral, we obtain (1.175). The theorem is proved.

Example 1.44: Let ξ and η be independent and nonnegative uncertain variables with uncertainty distributions Φ and Ψ, respectively. Then

$$E[\xi\eta] = \int_0^1 \Phi^{-1}(\alpha)\Psi^{-1}(\alpha)\mathrm{d}\alpha. \qquad (1.176)$$

Exercise 1.1: What is the expected value of an alternating monotone function of uncertain variables?

Exercise 1.2: Let ξ and η be independent and positive uncertain variables with uncertainty distributions Φ and Ψ, respectively. Prove

$$E\left[\frac{\xi}{\eta}\right] = \int_0^1 \frac{\Phi^{-1}(\alpha)}{\Psi^{-1}(1-\alpha)}\mathrm{d}\alpha. \qquad (1.177)$$

1.7 Variance

The variance of uncertain variable provides a degree of the spread of the distribution around its expected value. A small value of variance indicates that the uncertain variable is tightly concentrated around its expected value; and a large value of variance indicates that the uncertain variable has a wide spread around its expected value.

Definition 1.21 *(Liu [120]). Let ξ be an uncertain variable with finite expected value e. Then the variance of ξ is defined by $V[\xi] = E[(\xi - e)^2]$.*

Let ξ be an uncertain variable with expected value e. If we only know its uncertainty distribution Φ, then the variance

$$\begin{aligned}
V[\xi] &= \int_0^{+\infty} \mathcal{M}\{(\xi - e)^2 \geq r\}\mathrm{d}r \\
&= \int_0^{+\infty} \mathcal{M}\{(\xi \geq e + \sqrt{r}) \cup (\xi \leq e - \sqrt{r})\}\mathrm{d}r \\
&\leq \int_0^{+\infty} (\mathcal{M}\{\xi \geq e + \sqrt{r}\} + \mathcal{M}\{\xi \leq e - \sqrt{r}\})\mathrm{d}r \\
&= \int_0^{+\infty} (1 - \Phi(e + \sqrt{r}) + \Phi(e - \sqrt{r}))\mathrm{d}r \\
&= \int_e^{+\infty} 2(r - e)(1 - \Phi(r) + \Phi(2e - r))\mathrm{d}r.
\end{aligned}$$

For this case, we will stipulate that the variance is

$$V[\xi] = 2\int_e^{+\infty} (r - e)(1 - \Phi(r) + \Phi(2e - r))\mathrm{d}r. \qquad (1.178)$$

Mention that this is a stipulation rather than a precise formula!

Example 1.45: It has been verified that the linear uncertain variable $\xi \sim \mathcal{L}(a, b)$ has an expected value $(a+b)/2$. Note that the uncertainty distribution is $\Phi(x) = (x - a)/(b - a)$ when $a \le x \le b$. It follows from the stipulation (1.178) that the variance is

$$V[\xi] = 2 \int_{(a+b)/2}^{b} \left(r - \frac{a+b}{2} \right) \left(1 - \frac{r-a}{b-a} + \frac{b-r}{b-a} \right) dr = \frac{(b-a)^2}{12}.$$

In fact, a precise conclusion is $(b - a)^2/24 \le V[\xi] \le (b - a)^2/12$.

Example 1.46: It has been verified that the normal uncertain variable $\xi \sim \mathcal{N}(e, \sigma)$ has expected value e. It follows from the stipulation (1.178) that the variance is

$$V[\xi] = \sigma^2. \tag{1.179}$$

In fact, a precise conclusion is $\sigma^2/2 \le V[\xi] \le \sigma^2$.

Theorem 1.38. *If ξ is an uncertain variable with finite expected value, a and b are real numbers, then $V[a\xi + b] = a^2 V[\xi]$.*

Proof: It follows from the definition of variance that

$$V[a\xi + b] = E\left[(a\xi + b - aE[\xi] - b)^2 \right] = a^2 E[(\xi - E[\xi])^2] = a^2 V[\xi].$$

Theorem 1.39. *Let ξ be an uncertain variable with expected value e. Then $V[\xi] = 0$ if and only if $\mathcal{M}\{\xi = e\} = 1$.*

Proof: If $V[\xi] = 0$, then $E[(\xi - e)^2] = 0$. Note that

$$E[(\xi - e)^2] = \int_0^{+\infty} \mathcal{M}\{(\xi - e)^2 \ge r\} dr$$

which implies $\mathcal{M}\{(\xi - e)^2 \ge r\} = 0$ for any $r > 0$. Hence we have

$$\mathcal{M}\{(\xi - e)^2 = 0\} = 1.$$

That is, $\mathcal{M}\{\xi = e\} = 1$. Conversely, if $\mathcal{M}\{\xi = e\} = 1$, then we have $\mathcal{M}\{(\xi - e)^2 = 0\} = 1$ and $\mathcal{M}\{(\xi - e)^2 \ge r\} = 0$ for any $r > 0$. Thus

$$V[\xi] = \int_0^{+\infty} \mathcal{M}\{(\xi - e)^2 \ge r\} dr = 0.$$

The theorem is proved.

Maximum Variance Theorem

Let ξ be an uncertain variable that takes values in $[a, b]$, but whose uncertainty distribution is otherwise arbitrary. If its expected value is given, what is the possible maximum variance? The maximum variance theorem will answer this question, thus playing an important role in treating games against nature.

Theorem 1.40. *Let f be a convex function on $[a, b]$, and ξ an uncertain variable that takes values in $[a, b]$ and has expected value e. Then*

$$E[f(\xi)] \leq \frac{b - e}{b - a} f(a) + \frac{e - a}{b - a} f(b). \tag{1.180}$$

Proof: For each $\gamma \in \Gamma$, we have $a \leq \xi(\gamma) \leq b$ and

$$\xi(\gamma) = \frac{b - \xi(\gamma)}{b - a} a + \frac{\xi(\gamma) - a}{b - a} b.$$

It follows from the convexity of f that

$$f(\xi(\gamma)) \leq \frac{b - \xi(\gamma)}{b - a} f(a) + \frac{\xi(\gamma) - a}{b - a} f(b).$$

Taking expected values on both sides, we obtain the inequality.

Theorem 1.41 *(Maximum Variance Theorem). Let ξ be an uncertain variable that takes values in $[a, b]$ and has expected value e. Then*

$$V[\xi] \leq (e - a)(b - e) \tag{1.181}$$

and equality holds if the uncertain variable ξ is determined by

$$\mathcal{M}\{\xi = x\} = \begin{cases} \dfrac{b - e}{b - a}, & \text{if } x = a \\[2mm] \dfrac{e - a}{b - a}, & \text{if } x = b. \end{cases} \tag{1.182}$$

Proof: It follows from Theorem 1.40 immediately by defining $f(x) = (x - e)^2$. It is also easy to verify that the uncertain variable determined by (1.182) has variance $(e - a)(b - e)$. The theorem is proved.

1.8   Moments

Definition 1.22 *(Liu [120]). Let ξ be an uncertain variable. Then for any positive integer k,*
(a) the expected value $E[\xi^k]$ is called the kth moment;
(b) the expected value $E[|\xi|^k]$ is called the kth absolute moment;
(c) the expected value $E[(\xi - E[\xi])^k]$ is called the kth central moment;
(d) the expected value $E[|\xi - E[\xi]|^k]$ is called the kth absolute central moment.

Note that the first central moment is always 0, the first moment is just the expected value, and the second central moment is just the variance.

Theorem 1.42. *Let ξ be a nonnegative uncertain variable, and k a positive number. Then the k-th moment*

$$E[\xi^k] = k \int_0^{+\infty} r^{k-1} \mathcal{M}\{\xi \geq r\} \mathrm{d}r. \tag{1.183}$$

Proof: It follows from the nonnegativity of ξ that

$$E[\xi^k] = \int_0^\infty \mathcal{M}\{\xi^k \geq x\}dx = \int_0^\infty \mathcal{M}\{\xi \geq r\}dr^k = k\int_0^\infty r^{k-1}\mathcal{M}\{\xi \geq r\}dr.$$

The theorem is proved.

Theorem 1.43. *Let ξ be an uncertain variable, and t a positive number. If $E[|\xi|^t] < \infty$, then*

$$\lim_{x\to\infty} x^t \mathcal{M}\{|\xi| \geq x\} = 0. \tag{1.184}$$

Conversely, if (1.184) holds for some positive number t, then $E[|\xi|^s] < \infty$ for any $0 \leq s < t$.

Proof: It follows from the definition of expected value operator that

$$E[|\xi|^t] = \int_0^{+\infty} \mathcal{M}\{|\xi|^t \geq r\}dr < \infty.$$

Thus we have

$$\lim_{x\to\infty} \int_{x^t/2}^{+\infty} \mathcal{M}\{|\xi|^t \geq r\}dr = 0.$$

The equation (1.184) is proved by the following relation,

$$\int_{x^t/2}^{+\infty} \mathcal{M}\{|\xi|^t \geq r\}dr \geq \int_{x^t/2}^{x^t} \mathcal{M}\{|\xi|^t \geq r\}dr \geq \frac{1}{2}x^t\mathcal{M}\{|\xi| \geq x\}.$$

Conversely, if (1.184) holds, then there exists a number $a > 0$ such that

$$x^t\mathcal{M}\{|\xi| \geq x\} \leq 1, \ \forall x \geq a.$$

Thus we have

$$\begin{aligned}
E[|\xi|^s] &= \int_0^a \mathcal{M}\{|\xi|^s \geq r\}dr + \int_a^{+\infty} \mathcal{M}\{|\xi|^s \geq r\}dr \\
&= \int_0^a \mathcal{M}\{|\xi|^s \geq r\}dr + \int_a^{+\infty} sr^{s-1}\mathcal{M}\{|\xi| \geq r\}dr \\
&\leq \int_0^a \mathcal{M}\{|\xi|^s \geq r\}dr + s\int_a^{+\infty} r^{s-t-1}dr \\
&< +\infty. \quad \left(\text{by } \int_a^{+\infty} r^p dr < \infty \text{ for any } p < -1\right)
\end{aligned}$$

The theorem is proved.

Theorem 1.44. *Let ξ be an uncertain variable that takes values in $[a, b]$ and has expected value e. Then for any positive integer k, the kth absolute moment and kth absolute central moment satisfy the following inequalities,*

$$E[|\xi|^k] \leq \frac{b-e}{b-a}|a|^k + \frac{e-a}{b-a}|b|^k, \tag{1.185}$$

$$E[|\xi - e|^k] \leq \frac{b-e}{b-a}(e-a)^k + \frac{e-a}{b-a}(b-e)^k. \tag{1.186}$$

Proof: It follows from Theorem 1.40 immediately by defining $f(x) = |x|^k$ and $f(x) = |x - e|^k$.

1.9   Critical Values

In order to rank uncertain variables, we may use two critical values: optimistic value and pessimistic value.

Definition 1.23 *(Liu [120]). Let ξ be an uncertain variable, and $\alpha \in (0, 1]$. Then*

$$\xi_{\sup}(\alpha) = \sup\left\{r \mid \mathcal{M}\{\xi \geq r\} \geq \alpha\right\} \tag{1.187}$$

is called the α-optimistic value to ξ, and

$$\xi_{\inf}(\alpha) = \inf\left\{r \mid \mathcal{M}\{\xi \leq r\} \geq \alpha\right\} \tag{1.188}$$

is called the α-pessimistic value to ξ.

This means that the uncertain variable ξ will reach upwards of the α-optimistic value $\xi_{\sup}(\alpha)$ with uncertain measure α, and will be below the α-pessimistic value $\xi_{\inf}(\alpha)$ with uncertain measure α.

Theorem 1.45. *Let ξ be an uncertain variable with uncertainty distribution Φ. Then its α-optimistic value and α-pessimistic value are*

$$\xi_{\sup}(\alpha) = \Phi^{-1}(1 - \alpha), \tag{1.189}$$

$$\xi_{\inf}(\alpha) = \Phi^{-1}(\alpha). \tag{1.190}$$

Proof: It follows from the definition of α-optimistic value and α-pessimistic value immediately.

Example 1.47: Let ξ be a linear uncertain variable $\mathcal{L}(a, b)$. Then its α-optimistic and α-pessimistic values are

$$\xi_{\sup}(\alpha) = \alpha a + (1 - \alpha)b,$$

$$\xi_{\inf}(\alpha) = (1 - \alpha)a + \alpha b.$$

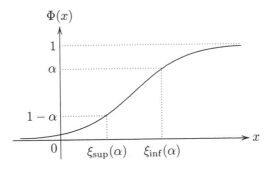

Figure 1.15: Optimistic Value and Pessimistic Value

Example 1.48: Let ξ be a zigzag uncertain variable $\mathcal{Z}(a,b,c)$. Then its α-optimistic and α-pessimistic values are

$$\xi_{\sup}(\alpha) = \begin{cases} 2\alpha b + (1 - 2\alpha)c, & \text{if } \alpha < 0.5 \\ (2\alpha - 1)a + (2 - 2\alpha)b, & \text{if } \alpha \geq 0.5, \end{cases}$$

$$\xi_{\inf}(\alpha) = \begin{cases} (1 - 2\alpha)a + 2\alpha b, & \text{if } \alpha < 0.5 \\ (2 - 2\alpha)b + (2\alpha - 1)c, & \text{if } \alpha \geq 0.5. \end{cases}$$

Example 1.49: Let ξ be a normal uncertain variable $\mathcal{N}(e,\sigma)$. Then its α-optimistic and α-pessimistic values are

$$\xi_{\sup}(\alpha) = e - \frac{\sigma\sqrt{3}}{\pi}\ln\frac{\alpha}{1-\alpha},$$

$$\xi_{\inf}(\alpha) = e + \frac{\sigma\sqrt{3}}{\pi}\ln\frac{\alpha}{1-\alpha}.$$

Example 1.50: Let ξ be a lognormal uncertain variable $\mathcal{LOGN}(e,\sigma)$. Then its α-optimistic and α-pessimistic values are

$$\xi_{\sup}(\alpha) = \exp(e)\left(\frac{1-\alpha}{\alpha}\right)^{\sqrt{3}\sigma/\pi},$$

$$\xi_{\inf}(\alpha) = \exp(e)\left(\frac{\alpha}{1-\alpha}\right)^{\sqrt{3}\sigma/\pi}.$$

Theorem 1.46. *Let ξ be an uncertain variable, and $\alpha \in (0,1]$. Then for any positive number ε, we have*

$$\mathcal{M}\{\xi \leq \xi_{\inf}(\alpha) + \varepsilon\} \geq \alpha, \quad \mathcal{M}\{\xi \geq \xi_{\sup}(\alpha) - \varepsilon\} \geq \alpha. \tag{1.191}$$

Proof: It follows from the definition of α-pessimistic value that there exists a decreasing sequence $\{x_i\}$ such that $\mathcal{M}\{\xi \leq x_i\} \geq \alpha$ and $x_i \downarrow \xi_{\inf}(\alpha)$ as $i \to \infty$. Thus for any positive number ε, there exists an index i such that $x_i < \xi_{\inf}(\alpha) + \varepsilon$. Hence

$$\mathcal{M}\{\xi \leq \xi_{\inf}(\alpha) + \varepsilon\} \geq \mathcal{M}\{\xi \leq x_i\} \geq \alpha.$$

Similarly, there exists an increasing sequence $\{x_i\}$ such that $\mathcal{M}\{\xi \geq x_i\} \geq \alpha$ and $x_i \uparrow \xi_{\sup}(\alpha)$ as $i \to \infty$. Thus for any positive number ε, there exists an index i such that $x_i > \xi_{\sup}(\alpha) - \varepsilon$. Hence

$$\mathcal{M}\{\xi \geq \xi_{\sup}(\alpha) - \varepsilon\} \geq \mathcal{M}\{\xi \geq x_i\} \geq \alpha.$$

The theorem is proved.

Theorem 1.47. *Let ξ be an uncertain variable, and $\alpha \in (0, 1]$. Then we have*
(a) $\xi_{\inf}(\alpha)$ is an increasing and left-continuous function of α;
(b) $\xi_{\sup}(\alpha)$ is a decreasing and left-continuous function of α.

Proof: (a) Let α_1 and α_2 be two numbers with $0 < \alpha_1 < \alpha_2 \leq 1$. Then for any number $r < \xi_{\sup}(\alpha_2)$, we have

$$\mathcal{M}\{\xi \geq r\} \geq \alpha_2 > \alpha_1.$$

Thus, by the definition of optimistic value, we obtain $\xi_{\sup}(\alpha_1) \geq \xi_{\sup}(\alpha_2)$. That is, the value $\xi_{\sup}(\alpha)$ is a decreasing function of α. Next, we prove the left-continuity of $\xi_{\inf}(\alpha)$ with respect to α. Let $\{\alpha_i\}$ be an arbitrary sequence of positive numbers such that $\alpha_i \uparrow \alpha$. Then $\{\xi_{\inf}(\alpha_i)\}$ is an increasing sequence. If the limitation is equal to $\xi_{\inf}(\alpha)$, then the left-continuity is proved. Otherwise, there exists a number z^* such that

$$\lim_{i \to \infty} \xi_{\inf}(\alpha_i) < z^* < \xi_{\inf}(\alpha).$$

Thus $\mathcal{M}\{\xi \leq z^*\} \geq \alpha_i$ for each i. Letting $i \to \infty$, we get $\mathcal{M}\{\xi \leq z^*\} \geq \alpha$. Hence $z^* \geq \xi_{\inf}(\alpha)$. A contradiction proves the left-continuity of $\xi_{\inf}(\alpha)$ with respect to α. The part (b) may be proved similarly.

Theorem 1.48. *Let ξ be an uncertain variable, and $\alpha \in (0, 1]$. Then we have*
(a) if $\alpha > 0.5$, then $\xi_{\inf}(\alpha) \geq \xi_{\sup}(\alpha)$;
(b) if $\alpha \leq 0.5$, then $\xi_{\inf}(\alpha) \leq \xi_{\sup}(\alpha)$.

Proof: Part (a): Write $\bar{\xi}(\alpha) = (\xi_{\inf}(\alpha) + \xi_{\sup}(\alpha))/2$. If $\xi_{\inf}(\alpha) < \xi_{\sup}(\alpha)$, then we have

$$1 \geq \mathcal{M}\{\xi < \bar{\xi}(\alpha)\} + \mathcal{M}\{\xi > \bar{\xi}(\alpha)\} \geq \alpha + \alpha > 1.$$

A contradiction proves $\xi_{\inf}(\alpha) \geq \xi_{\sup}(\alpha)$. Part (b): Assume that $\xi_{\inf}(\alpha) > \xi_{\sup}(\alpha)$. It follows from the definition of $\xi_{\inf}(\alpha)$ that $\mathcal{M}\{\xi \leq \bar{\xi}(\alpha)\} < \alpha$.

Similarly, it follows from the definition of $\xi_{\sup}(\alpha)$ that $\mathcal{M}\{\xi \geq \bar{\xi}(\alpha)\} < \alpha$. Thus

$$1 \leq \mathcal{M}\{\xi \leq \bar{\xi}(\alpha)\} + \mathcal{M}\{\xi \geq \bar{\xi}(\alpha)\} < \alpha + \alpha \leq 1.$$

A contradiction proves $\xi_{\inf}(\alpha) \leq \xi_{\sup}(\alpha)$. The theorem is verified.

Theorem 1.49 *(Zuo [247]). Let $\xi_1, \xi_2, \cdots, \xi_n$ be independent uncertain variables with uncertainty distributions. If $f : \Re^n \to \Re$ is a continuous and strictly increasing function, then $\xi = f(\xi_1, \xi_2, \cdots, \xi_n)$ is an uncertain variable, and*

$$\xi_{\sup}(\alpha) = f(\xi_{1\,\sup}(\alpha), \xi_{2\,\sup}(\alpha), \cdots, \xi_{n\,\sup}(\alpha)), \qquad (1.192)$$

$$\xi_{\inf}(\alpha) = f(\xi_{1\,\inf}(\alpha), \xi_{2\,\inf}(\alpha), \cdots, \xi_{n\,\inf}(\alpha)). \qquad (1.193)$$

Proof: Since f is a strictly increasing function, it follows from Theorem 1.20 that the inverse uncertainty distribution of ξ is

$$\Psi^{-1}(\alpha) = f(\Phi_1^{-1}(\alpha), \Phi_2^{-1}(\alpha), \cdots, \Phi_n^{-1}(\alpha))$$

where $\Phi_1, \Phi_2, \cdots, \Phi_n$ are uncertainty distributions of $\xi_1, \xi_2, \cdots, \xi_n$, respectively. By using Theorem 1.45, we get (1.192) and (1.193). The theorem is proved.

Example 1.51: Let ξ be an uncertain variable, and $\alpha \in (0,1]$. If $c \geq 0$, then

$$(c\xi)_{\sup}(\alpha) = c\xi_{\sup}(\alpha), \quad (c\xi)_{\inf}(\alpha) = c\xi_{\inf}(\alpha).$$

Example 1.52: Suppose that ξ and η are independent uncertain variables, and $\alpha \in (0,1]$. Then we have

$$(\xi + \eta)_{\sup}(\alpha) = \xi_{\sup}(\alpha) + \eta_{\sup}(\alpha), \quad (\xi + \eta)_{\inf}(\alpha) = \xi_{\inf}(\alpha) + \eta_{\inf}(\alpha),$$

$$(\xi \vee \eta)_{\sup}(\alpha) = \xi_{\sup}(\alpha) \vee \eta_{\sup}(\alpha), \quad (\xi \vee \eta)_{\inf}(\alpha) = \xi_{\inf}(\alpha) \vee \eta_{\inf}(\alpha),$$

$$(\xi \wedge \eta)_{\sup}(\alpha) = \xi_{\sup}(\alpha) \wedge \eta_{\sup}(\alpha), \quad (\xi \wedge \eta)_{\inf}(\alpha) = \xi_{\inf}(\alpha) \wedge \eta_{\inf}(\alpha).$$

Example 1.53: Let ξ and η be independent and positive uncertain variables. Since $f(x, y) = xy$ is a strictly increasing function when $x > 0$ and $y > 0$, we immediately have

$$(\xi\eta)_{\sup}(\alpha) = \xi_{\sup}(\alpha)\eta_{\sup}(\alpha), \quad (\xi\eta)_{\inf}(\alpha) = \xi_{\inf}(\alpha)\eta_{\inf}(\alpha). \qquad (1.194)$$

Theorem 1.50 *(Zuo [247]). Let $\xi_1, \xi_2, \cdots, \xi_n$ be independent uncertain variables with uncertainty distributions. If f is a continuous and strictly decreasing function, then*

$$\xi_{\sup}(\alpha) = f(\xi_{1\,\inf}(\alpha), \xi_{2\,\inf}(\alpha), \cdots, \xi_{n\,\inf}(\alpha)), \qquad (1.195)$$

$$\xi_{\inf}(\alpha) = f(\xi_{1\,\sup}(\alpha), \xi_{2\,\sup}(\alpha), \cdots, \xi_{n\,\sup}(\alpha)). \qquad (1.196)$$

Proof: Since f is a strictly decreasing function, it follows from Theorem 1.25 that the inverse uncertainty distribution of ξ is

$$\Psi^{-1}(\alpha) = f(\Phi_1^{-1}(1-\alpha), \Phi_2^{-1}(1-\alpha), \cdots, \Phi_n^{-1}(1-\alpha)).$$

By using Theorem 1.45, we get (1.195) and (1.196). The theorem is proved.

Example 1.54: Let ξ be an uncertain variable, and $\alpha \in (0,1]$. If $c < 0$, then

$$(c\xi)_{\sup}(\alpha) = c\xi_{\inf}(\alpha), \quad (c\xi)_{\inf}(\alpha) = c\xi_{\sup}(\alpha).$$

Exercise 1.3: What are the critical values to an alternating monotone function of uncertain variables?

Exercise 1.4: Let ξ and η be independent and positive uncertain variables. Prove

$$\left(\frac{\xi}{\eta}\right)_{\sup}(\alpha) = \frac{\xi_{\sup}(\alpha)}{\eta_{\inf}(\alpha)}, \qquad \left(\frac{\xi}{\eta}\right)_{\inf}(\alpha) = \frac{\xi_{\inf}(\alpha)}{\eta_{\sup}(\alpha)}. \tag{1.197}$$

1.10　Entropy

This section provides a definition of entropy to characterize the uncertainty of uncertain variables resulting from information deficiency.

Definition 1.24 *(Liu [123]). Suppose that ξ is an uncertain variable with uncertainty distribution Φ. Then its entropy is defined by*

$$H[\xi] = \int_{-\infty}^{+\infty} S(\Phi(x))\mathrm{d}x \tag{1.198}$$

where $S(t) = -t\ln t - (1-t)\ln(1-t)$.

Example 1.55: Let ξ be an uncertain variable with uncertainty distribution

$$\Phi(x) = \begin{cases} 0, & \text{if } x < a \\ 1, & \text{if } x \geq a. \end{cases} \tag{1.199}$$

Essentially, ξ is a constant a. It follows from the definition of entropy that

$$H[\xi] = -\int_{-\infty}^{a}(0\ln 0 + 1\ln 1)\,\mathrm{d}x - \int_{a}^{+\infty}(1\ln 1 + 0\ln 0)\,\mathrm{d}x = 0.$$

This means a constant has no uncertainty.

Example 1.56: Let ξ be a linear uncertain variable $\mathcal{L}(a,b)$. Then its entropy is

$$H[\xi] = -\int_{a}^{b}\left(\frac{x-a}{b-a}\ln\frac{x-a}{b-a} + \frac{b-x}{b-a}\ln\frac{b-x}{b-a}\right)\mathrm{d}x = \frac{b-a}{2}. \tag{1.200}$$

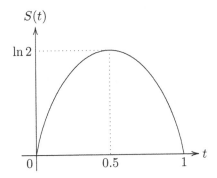

Figure 1.16: Function $S(t) = -t \ln t - (1 - t) \ln(1 - t)$. It is easy to verify that $S(t)$ is a symmetric function about $t = 0.5$, strictly increases on the interval $[0, 0.5]$, strictly decreases on the interval $[0.5, 1]$, and reaches its unique maximum $\ln 2$ at $t = 0.5$.

Example 1.57: Let ξ be a zigzag uncertain variable $\mathcal{Z}(a, b, c)$. Then its entropy is

$$H[\xi] = \frac{c - a}{2}. \tag{1.201}$$

Example 1.58: Let ξ be a normal uncertain variable $\mathcal{N}(e, \sigma)$. Then its entropy is

$$H[\xi] = \frac{\pi \sigma}{\sqrt{3}}. \tag{1.202}$$

Theorem 1.51. *Let ξ be an uncertain variable. Then $H[\xi] \geq 0$ and equality holds if ξ is essentially a constant.*

Proof: The positivity is clear. In addition, when an uncertain variable tends to a constant, its entropy tends to the minimum 0.

Theorem 1.52. *Let ξ be an uncertain variable taking values on the interval $[a, b]$. Then*

$$H[\xi] \leq (b - a) \ln 2 \tag{1.203}$$

and equality holds if ξ has an uncertainty distribution $\Phi(x) = 0.5$ on $[a, b]$.

Proof: The theorem follows from the fact that the function $S(t)$ reaches its maximum $\ln 2$ at $t = 0.5$.

Theorem 1.53. *Let ξ be an uncertain variable, and let c be a real number. Then*

$$H[\xi + c] = H[\xi]. \tag{1.204}$$

That is, the entropy is invariant under arbitrary translations.

Proof: Write the uncertainty distribution of ξ by Φ. Then the uncertain variable $\xi + c$ has an uncertainty distribution $\Phi(x - c)$. It follows from the definition of entropy that

$$H[\xi + c] = \int_{-\infty}^{+\infty} S\left(\Phi(x - c)\right) \mathrm{d}x = \int_{-\infty}^{+\infty} S(\Phi(x)) \mathrm{d}x = H[\xi].$$

The theorem is proved.

Theorem 1.54 *(Dai and Chen [24]). Let ξ be an uncertain variable with uncertainty distribution Φ. Then*

$$H[\xi] = \int_0^1 \Phi^{-1}(\alpha) \ln \frac{\alpha}{1 - \alpha} \mathrm{d}\alpha. \tag{1.205}$$

Proof: It is clear that $S(\alpha)$ is a derivable function with $S'(\alpha) = -\ln \alpha/(1 - \alpha)$. Since

$$S(\Phi(x)) = \int_0^{\Phi(x)} S'(\alpha) \mathrm{d}\alpha = - \int_{\Phi(x)}^1 S'(\alpha) \mathrm{d}\alpha,$$

we have

$$H[\xi] = \int_{-\infty}^{+\infty} S(\Phi(x)) \mathrm{d}x = \int_{-\infty}^0 \int_0^{\Phi(x)} S'(\alpha) \mathrm{d}\alpha \mathrm{d}x - \int_0^{+\infty} \int_{\Phi(x)}^1 S'(\alpha) \mathrm{d}\alpha \mathrm{d}x.$$

It follows from Fubini theorem that

$$H[\xi] = \int_0^{\Phi(0)} \int_{\Phi^{-1}(\alpha)}^0 S'(\alpha) \mathrm{d}x \mathrm{d}\alpha - \int_{\Phi(0)}^1 \int_0^{\Phi^{-1}(\alpha)} S'(\alpha) \mathrm{d}x \mathrm{d}\alpha$$

$$= - \int_0^{\Phi(0)} \Phi^{-1}(\alpha) S'(\alpha) \mathrm{d}\alpha - \int_{\Phi(0)}^1 \Phi^{-1}(\alpha) S'(\alpha) \mathrm{d}\alpha$$

$$= - \int_0^1 \Phi^{-1}(\alpha) S'(\alpha) \mathrm{d}\alpha = \int_0^1 \Phi^{-1}(\alpha) \ln \frac{\alpha}{1 - \alpha} \mathrm{d}\alpha.$$

The theorem is verified.

Theorem 1.55 *(Dai and Chen [24]). Let ξ and η be independent uncertain variables. Then for any real numbers a and b, we have*

$$H[a\xi + b\eta] = |a|H[\xi] + |b|H[\eta]. \tag{1.206}$$

Proof: Suppose that ξ and η have uncertainty distributions Φ and Ψ, respectively.

STEP 1: We prove $H[a\xi] = |a|H[\xi]$. If $a > 0$, then the inverse uncertainty distribution of $a\xi$ is

$$\Upsilon^{-1}(\alpha) = a\Phi^{-1}(\alpha).$$

It follows from Theorem 1.54 that

$$H[a\xi] = \int_0^1 a\Phi^{-1}(\alpha)\ln\frac{\alpha}{1-\alpha}\mathrm{d}\alpha = a\int_0^1 \Phi^{-1}(\alpha)\ln\frac{\alpha}{1-\alpha}\mathrm{d}\alpha = |a|H[\xi].$$

If $a = 0$, we immediately have $H[a\xi] = 0 = |a|H[\xi]$. If $a < 0$, then the inverse uncertainty distribution of $a\xi$ is

$$\Upsilon^{-1}(\alpha) = a\Phi^{-1}(1-\alpha).$$

It follows from Theorem 1.54 that

$$H[a\xi] = \int_0^1 a\Phi^{-1}(1-\alpha)\ln\frac{\alpha}{1-\alpha}\mathrm{d}\alpha = (-a)\int_0^1 \Phi^{-1}(\alpha)\ln\frac{\alpha}{1-\alpha}\mathrm{d}\alpha = |a|H[\xi].$$

Thus we always have $H[a\xi] = |a|H[\xi]$.

STEP 2: We prove $H[\xi + \eta] = H[\xi] + H[\eta]$. Note that the inverse uncertainty distribution of $\xi + \eta$ is

$$\Upsilon^{-1}(\alpha) = \Phi^{-1}(\alpha) + \Psi^{-1}(\alpha).$$

It follows from Theorem 1.54 that

$$H[\xi + \eta] = \int_0^1 (\Phi^{-1}(\alpha) + \Psi^{-1}(\alpha))\ln\frac{\alpha}{1-\alpha}\mathrm{d}\alpha = H[\xi] + H[\eta].$$

STEP 3: Finally, for any real numbers a and b, it follows from Steps 1 and 2 that
$$H[a\xi + b\eta] = H[a\xi] + H[b\eta] = |a|H[\xi] + |b|H[\eta].$$

The theorem is proved.

Entropy of Function of Uncertain Variables

Theorem 1.56 *(Dai and Chen [24]). Let $\xi_1, \xi_2, \cdots, \xi_n$ be independent uncertain variables with uncertainty distributions $\Phi_1, \Phi_2, \cdots, \Phi_n$, respectively. If $f : \Re^n \to \Re$ is a strictly increasing function, then the uncertain variable $\xi = f(\xi_1, \xi_2, \cdots, \xi_n)$ has an entropy*

$$H[\xi] = \int_0^1 f(\Phi_1^{-1}(\alpha), \Phi_2^{-1}(\alpha), \cdots, \Phi_n^{-1}(\alpha))\ln\frac{\alpha}{1-\alpha}\mathrm{d}\alpha. \qquad (1.207)$$

Proof: Since f is a strictly increasing function, it follows from Theorem 1.20 that the inverse uncertainty distribution of ξ is

$$\Psi^{-1}(\alpha) = f(\Phi_1^{-1}(\alpha), \Phi_2^{-1}(\alpha), \cdots, \Phi_n^{-1}(\alpha)).$$

By using Theorem 1.54, we get (1.207). The theorem is thus verified.

Exercise 1.5: Let ξ and η be independent and nonnegative uncertain variables with uncertainty distributions Φ and Ψ, respectively. Then

$$H[\xi\eta] = \int_0^1 \Phi^{-1}(\alpha)\Psi^{-1}(\alpha)\ln\frac{\alpha}{1-\alpha}\mathrm{d}\alpha. \qquad (1.208)$$

Theorem 1.57 *(Dai and Chen [24]). Let $\xi_1, \xi_2, \cdots, \xi_n$ be independent uncertain variables with uncertainty distributions $\Phi_1, \Phi_2, \cdots, \Phi_n$, respectively. If f is a strictly decreasing function, then*

$$H[\xi] = \int_0^1 f(\Phi_1^{-1}(\alpha), \Phi_2^{-1}(\alpha), \cdots, \Phi_n^{-1}(\alpha))\ln\frac{1-\alpha}{\alpha}\mathrm{d}\alpha. \qquad (1.209)$$

Proof: Since f is a strictly decreasing function, it follows from Theorem 1.25 that the inverse uncertainty distribution of ξ is

$$\Psi^{-1}(\alpha) = f(\Phi_1^{-1}(1-\alpha), \Phi_2^{-1}(1-\alpha), \cdots, \Phi_n^{-1}(1-\alpha)).$$

By using Theorem 1.54, we get (1.209). The theorem is thus verified.

Exercise 1.6: What is the entropy of an alternating monotone function of uncertain variables?

Exercise 1.7: Let ξ and η be independent and positive uncertain variables with uncertainty distributions Φ and Ψ, respectively. Prove

$$H\left[\frac{\xi}{\eta}\right] = \int_0^1 \frac{\Phi^{-1}(\alpha)}{\Psi^{-1}(1-\alpha)}\ln\frac{\alpha}{1-\alpha}\mathrm{d}\alpha. \qquad (1.210)$$

Maximum Entropy Principle

Given some constraints, for example, expected value and variance, there are usually multiple compatible uncertainty distributions. Which uncertainty distribution shall we take? The *maximum entropy principle* attempts to select the uncertainty distribution that maximizes the value of entropy and satisfies the prescribed constraints.

Theorem 1.58 *(Chen and Dai [19]). Let ξ be an uncertain variable whose uncertainty distribution is arbitrary but the expected value e and variance σ^2. Then*

$$H[\xi] \leq \frac{\pi\sigma}{\sqrt{3}} \qquad (1.211)$$

and the equality holds if ξ is a normal uncertain variable $\mathcal{N}(e, \sigma)$.

Proof: Let $\Phi(x)$ be the uncertainty distribution of ξ and write $\Psi(x) = \Phi(2e-x)$ for $x \geq e$. It follows from the stipulation (1.178) that the variance is

$$V[\xi] = 2\int_e^{+\infty} (x-e)(1-\Phi(x))\mathrm{d}x + 2\int_e^{+\infty} (x-e)\Psi(x)\mathrm{d}x = \sigma^2.$$

Thus there exists a real number κ such that

$$2 \int_e^{+\infty} (x - e)(1 - \Phi(x))\mathrm{d}x = \kappa\sigma^2,$$

$$2 \int_e^{+\infty} (x - e)\Psi(x)\mathrm{d}x = (1 - \kappa)\sigma^2.$$

The maximum entropy distribution Φ should maximize the entropy

$$H[\xi] = \int_{-\infty}^{+\infty} S(\Phi(x))\mathrm{d}x = \int_e^{+\infty} S(\Phi(x))\mathrm{d}x + \int_e^{+\infty} S(\Psi(x))\mathrm{d}x$$

subject to the above two constraints. The Lagrangian is

$$L = \int_e^{+\infty} S(\Phi(x))\mathrm{d}x + \int_e^{+\infty} S(\Psi(x))\mathrm{d}x$$

$$-\alpha \left(2 \int_e^{+\infty} (x - e)(1 - \Phi(x))\mathrm{d}x - \kappa\sigma^2 \right)$$

$$-\beta \left(2 \int_e^{+\infty} (x - e)\Psi(x)\mathrm{d}x - (1 - \kappa)\sigma^2 \right).$$

The maximum entropy distribution meets Euler-Lagrange equations

$$\ln \Phi(x) - \ln(1 - \Phi(x)) = 2\alpha(x - e),$$

$$\ln \Psi(x) - \ln(1 - \Psi(x)) = 2\beta(e - x).$$

Thus Φ and Ψ have the forms

$$\Phi(x) = (1 + \exp(2\alpha(e - x)))^{-1},$$

$$\Psi(x) = (1 + \exp(2\beta(x - e)))^{-1}.$$

Substituting them into the variance constraints, we get

$$\Phi(x) = \left(1 + \exp\left(\frac{\pi(e - x)}{\sqrt{6\kappa}\sigma} \right) \right)^{-1},$$

$$\Psi(x) = \left(1 + \exp\left(\frac{\pi(x - e)}{\sqrt{6(1 - \kappa)}\sigma} \right) \right)^{-1}.$$

Then the entropy is

$$H[\xi] = \frac{\pi\sigma\sqrt{\kappa}}{\sqrt{6}} + \frac{\pi\sigma\sqrt{1 - \kappa}}{\sqrt{6}}$$

which achieves the maximum when $\kappa = 1/2$. Thus the maximum entropy distribution is just the normal uncertainty distribution $\mathcal{N}(e, \sigma)$.

1.11   Distance

Definition 1.25 *(Liu [120]).   The distance between uncertain variables ξ and η is defined as*

$$d(\xi, \eta) = E[|\xi - \eta|]. \qquad (1.212)$$

Theorem 1.59. *Let ξ, η, τ be uncertain variables, and let $d(\cdot, \cdot)$ be the distance. Then we have*
(a) (Nonnegativity) $d(\xi, \eta) \geq 0$;
(b) (Identification) $d(\xi, \eta) = 0$ if and only if $\xi = \eta$;
(c) (Symmetry) $d(\xi, \eta) = d(\eta, \xi)$;
(d) (Triangle Inequality) $d(\xi, \eta) \leq 2d(\xi, \tau) + 2d(\eta, \tau)$.

Proof: The parts (a), (b) and (c) follow immediately from the definition. Now we prove the part (d). It follows from the countable subadditivity axiom that

$$
\begin{aligned}
d(\xi, \eta) &= \int_0^{+\infty} \mathcal{M}\{|\xi - \eta| \geq r\}\, dr \\
&\leq \int_0^{+\infty} \mathcal{M}\{|\xi - \tau| + |\tau - \eta| \geq r\}\, dr \\
&\leq \int_0^{+\infty} \mathcal{M}\{(|\xi - \tau| \geq r/2) \cup (|\tau - \eta| \geq r/2)\}\, dr \\
&\leq \int_0^{+\infty} (\mathcal{M}\{|\xi - \tau| \geq r/2\} + \mathcal{M}\{|\tau - \eta| \geq r/2\})\, dr \\
&= 2E[|\xi - \tau|] + 2E[|\tau - \eta|] = 2d(\xi, \tau) + 2d(\tau, \eta).
\end{aligned}
$$

Example 1.59: Let $\Gamma = \{\gamma_1, \gamma_2, \gamma_3\}$. Define $\mathcal{M}\{\emptyset\} = 0$, $\mathcal{M}\{\Gamma\} = 1$ and $\mathcal{M}\{\Lambda\} = 1/2$ for any subset Λ (excluding $\emptyset$ and Γ). We set uncertain variables ξ, η and τ as follows,

$$
\xi(\gamma) = \begin{cases} 1, & \text{if } \gamma = \gamma_1 \\ 1, & \text{if } \gamma = \gamma_2 \\ 0, & \text{if } \gamma = \gamma_3, \end{cases} \qquad
\eta(\gamma) = \begin{cases} 0, & \text{if } \gamma = \gamma_1 \\ -1, & \text{if } \gamma = \gamma_2 \\ -1, & \text{if } \gamma = \gamma_3, \end{cases} \qquad
\tau(\gamma) \equiv 0.
$$

It is easy to verify that $d(\xi, \tau) = d(\tau, \eta) = 1/2$ and $d(\xi, \eta) = 3/2$. Thus

$$d(\xi, \eta) = \frac{3}{2}(d(\xi, \tau) + d(\tau, \eta)).$$

A conjecture is $d(\xi, \eta) \leq 1.5(d(\xi, \tau) + d(\tau, \eta))$ for arbitrary uncertain variables ξ, η and τ. This is an open problem.

1.12 Inequalities

Theorem 1.60 *(Liu [120]). Let ξ be an uncertain variable, and f a non-negative function. If f is even and increasing on $[0,\infty)$, then for any given number $t > 0$, we have*

$$\mathcal{M}\{|\xi| \geq t\} \leq \frac{E[f(\xi)]}{f(t)}. \tag{1.213}$$

Proof: It is clear that $\mathcal{M}\{|\xi| \geq f^{-1}(r)\}$ is a monotone decreasing function of r on $[0,\infty)$. It follows from the nonnegativity of $f(\xi)$ that

$$E[f(\xi)] = \int_0^{+\infty} \mathcal{M}\{f(\xi) \geq r\}\mathrm{d}r = \int_0^{+\infty} \mathcal{M}\{|\xi| \geq f^{-1}(r)\}\mathrm{d}r$$

$$\geq \int_0^{f(t)} \mathcal{M}\{|\xi| \geq f^{-1}(r)\}\mathrm{d}r \geq \int_0^{f(t)} \mathrm{d}r \cdot \mathcal{M}\{|\xi| \geq f^{-1}(f(t))\}$$

$$= f(t) \cdot \mathcal{M}\{|\xi| \geq t\}$$

which proves the inequality.

Theorem 1.61 *(Liu [120], Markov Inequality). Let ξ be an uncertain variable. Then for any given numbers $t > 0$ and $p > 0$, we have*

$$\mathcal{M}\{|\xi| \geq t\} \leq \frac{E[|\xi|^p]}{t^p}. \tag{1.214}$$

Proof: It is a special case of Theorem 1.60 when $f(x) = |x|^p$.

Example 1.60: For any given positive number t, we define an uncertain variable as follows,

$$\xi = \begin{cases} 0 \text{ with uncertain measure } 1/2 \\ t \text{ with uncertain measure } 1/2. \end{cases}$$

Then $E[\xi^p] = t^p/2$ and $\mathcal{M}\{\xi \geq t\} = 1/2 = E[\xi^p]/t^p$.

Theorem 1.62 *(Liu [120], Chebyshev Inequality). Let ξ be an uncertain variable whose variance $V[\xi]$ exists. Then for any given number $t > 0$, we have*

$$\mathcal{M}\{|\xi - E[\xi]| \geq t\} \leq \frac{V[\xi]}{t^2}. \tag{1.215}$$

Proof: It is a special case of Theorem 1.60 when the uncertain variable ξ is replaced with $\xi - E[\xi]$, and $f(x) = x^2$.

Example 1.61: For any given positive number t, we define an uncertain variable as follows,

$$\xi = \begin{cases} -t \text{ with uncertain measure } 1/2 \\ t \text{ with uncertain measure } 1/2. \end{cases}$$

Then $V[\xi] = t^2$ and $\mathcal{M}\{|\xi - E[\xi]| \geq t\} = 1 = V[\xi]/t^2$.

Theorem 1.63 *(Liu [120], Hölder's Inequality). Let p and q be positive numbers with $1/p + 1/q = 1$, and let ξ and η be independent uncertain variables with $E[|\xi|^p] < \infty$ and $E[|\eta|^q] < \infty$. Then we have*

$$E[|\xi\eta|] \leq \sqrt[p]{E[|\xi|^p]} \sqrt[q]{E[|\eta|^q]}. \tag{1.216}$$

Proof: The inequality holds trivially if at least one of ξ and η is zero a.s. Now we assume $E[|\xi|^p] > 0$ and $E[|\eta|^q] > 0$. It is easy to prove that the function $f(x, y) = \sqrt[p]{x}\sqrt[q]{y}$ is a concave function on $\{(x, y) : x \geq 0, y \geq 0\}$. Thus for any point (x_0, y_0) with $x_0 > 0$ and $y_0 > 0$, there exist two real numbers a and b such that

$$f(x, y) - f(x_0, y_0) \leq a(x - x_0) + b(y - y_0), \quad \forall x \geq 0, y \geq 0.$$

Letting $x_0 = E[|\xi|^p]$, $y_0 = E[|\eta|^q]$, $x = |\xi|^p$ and $y = |\eta|^q$, we have

$$f(|\xi|^p, |\eta|^q) - f(E[|\xi|^p], E[|\eta|^q]) \leq a(|\xi|^p - E[|\xi|^p]) + b(|\eta|^q - E[|\eta|^q]).$$

Taking the expected values on both sides, we obtain

$$E[f(|\xi|^p, |\eta|^q)] \leq f(E[|\xi|^p], E[|\eta|^q]).$$

Hence the inequality (1.216) holds.

Theorem 1.64 *(Liu [120], Minkowski Inequality). Let p be a real number with $p \geq 1$, and let ξ and η be independent uncertain variables with $E[|\xi|^p] < \infty$ and $E[|\eta|^p] < \infty$. Then we have*

$$\sqrt[p]{E[|\xi + \eta|^p]} \leq \sqrt[p]{E[|\xi|^p]} + \sqrt[p]{E[|\eta|^p]}. \tag{1.217}$$

Proof: The inequality holds trivially if at least one of ξ and η is zero a.s. Now we assume $E[|\xi|^p] > 0$ and $E[|\eta|^p] > 0$. It is easy to prove that the function $f(x, y) = (\sqrt[p]{x} + \sqrt[p]{y})^p$ is a concave function on $\{(x, y) : x \geq 0, y \geq 0\}$. Thus for any point (x_0, y_0) with $x_0 > 0$ and $y_0 > 0$, there exist two real numbers a and b such that

$$f(x, y) - f(x_0, y_0) \leq a(x - x_0) + b(y - y_0), \quad \forall x \geq 0, y \geq 0.$$

Letting $x_0 = E[|\xi|^p]$, $y_0 = E[|\eta|^p]$, $x = |\xi|^p$ and $y = |\eta|^p$, we have

$$f(|\xi|^p, |\eta|^p) - f(E[|\xi|^p], E[|\eta|^p]) \leq a(|\xi|^p - E[|\xi|^p]) + b(|\eta|^p - E[|\eta|^p]).$$

Taking the expected values on both sides, we obtain

$$E[f(|\xi|^p, |\eta|^p)] \leq f(E[|\xi|^p], E[|\eta|^p]).$$

Hence the inequality (1.217) holds.

Theorem 1.65 *(Liu [120], Jensen's Inequality). Let ξ be an uncertain variable, and $f : \Re \to \Re$ a convex function. If $E[\xi]$ and $E[f(\xi)]$ are finite, then*

$$f(E[\xi]) \leq E[f(\xi)]. \tag{1.218}$$

Especially, when $f(x) = |x|^p$ and $p \geq 1$, we have $|E[\xi]|^p \leq E[|\xi|^p]$.

Proof: Since f is a convex function, for each y, there exists a number k such that $f(x) - f(y) \geq k \cdot (x - y)$. Replacing x with ξ and y with $E[\xi]$, we obtain

$$f(\xi) - f(E[\xi]) \geq k \cdot (\xi - E[\xi]).$$

Taking the expected values on both sides, we have

$$E[f(\xi)] - f(E[\xi]) \geq k \cdot (E[\xi] - E[\xi]) = 0$$

which proves the inequality.

1.13   Convergence Concepts

We have the following four convergence concepts of uncertain sequence: convergence almost surely (a.s.), convergence in measure, convergence in mean, and convergence in distribution.

Table 1.1: Relationship among Convergence Concepts

Convergence in Mean		Convergence in Measure		Convergence in Distribution
	$\Rightarrow$		$\Rightarrow$	

Definition 1.26 *(Liu [120]). Suppose that $\xi, \xi_1, \xi_2, \cdots$ are uncertain variables defined on the uncertainty space $(\Gamma, \mathcal{L}, \mathcal{M})$. The sequence $\{\xi_i\}$ is said to be convergent a.s. to ξ if there exists an event Λ with $\mathcal{M}\{\Lambda\} = 1$ such that*

$$\lim_{i \to \infty} |\xi_i(\gamma) - \xi(\gamma)| = 0 \tag{1.219}$$

for every $\gamma \in \Lambda$. In that case we write $\xi_i \to \xi$, a.s.

Definition 1.27 *(Liu [120]). Suppose that $\xi, \xi_1, \xi_2, \cdots$ are uncertain variables. We say that the sequence $\{\xi_i\}$ converges in measure to ξ if*

$$\lim_{i \to \infty} \mathcal{M}\{|\xi_i - \xi| \geq \varepsilon\} = 0 \tag{1.220}$$

for every $\varepsilon > 0$.

Definition 1.28 *(Liu [120]). Suppose that $\xi, \xi_1, \xi_2, \cdots$ are uncertain variables with finite expected values. We say that the sequence $\{\xi_i\}$ converges in mean to ξ if*

$$\lim_{i \to \infty} E[|\xi_i - \xi|] = 0. \tag{1.221}$$

In addition, the sequence $\{\xi_i\}$ is said to converge in mean square to ξ if

$$\lim_{i \to \infty} E[|\xi_i - \xi|^2] = 0. \tag{1.222}$$

Definition 1.29 *(Liu [120]). Suppose that $\Phi, \Phi_1, \Phi_2, \cdots$ are the uncertainty distributions of uncertain variables $\xi, \xi_1, \xi_2, \cdots$, respectively. We say that $\{\xi_i\}$ converges in distribution to ξ if*

$$\lim_{i \to \infty} \Phi_i(x) = \Phi(x) \tag{1.223}$$

at any continuity point x of Φ.

Convergence in Mean vs. Convergence in Measure

Theorem 1.30 *(Liu [120]). Suppose that $\xi, \xi_1, \xi_2, \cdots$ are uncertain variables. If $\{\xi_i\}$ converges in mean to ξ, then $\{\xi_i\}$ converges in measure to ξ.*

Proof: It follows from the Markov inequality that for any given number $\varepsilon > 0$, we have

$$\mathcal{M}\{|\xi_i - \xi| \geq \varepsilon\} \leq \frac{E[|\xi_i - \xi|]}{\varepsilon} \to 0$$

as $i \to \infty$. Thus $\{\xi_i\}$ converges in measure to ξ. The theorem is proved.

Example 1.62: Convergence in measure does not imply convergence in mean. Take an uncertainty space $(\Gamma, \mathcal{L}, \mathcal{M})$ to be $\{\gamma_1, \gamma_2, \cdots\}$ with

$$\mathcal{M}\{\Lambda\} = \begin{cases} \sup_{\gamma_i \in \Lambda} 1/i, & \text{if } \sup_{\gamma_i \in \Lambda} 1/i < 0.5 \\ 1 - \sup_{\gamma_i \notin \Lambda} 1/i, & \text{if } \sup_{\gamma_i \notin \Lambda} 1/i < 0.5 \\ 0.5, & \text{otherwise.} \end{cases}$$

The uncertain variables are defined by

$$\xi_i(\gamma_j) = \begin{cases} i, & \text{if } j = i \\ 0, & \text{otherwise} \end{cases}$$

for $i = 1, 2, \cdots$ and $\xi \equiv 0$. For some small number $\varepsilon > 0$, we have

$$\mathcal{M}\{|\xi_i - \xi| \geq \varepsilon\} = \mathcal{M}\{|\xi_i - \xi| \geq \varepsilon\} = \frac{1}{i} \to 0.$$

That is, the sequence $\{\xi_i\}$ converges in measure to ξ. However, for each i, we have

$$E[|\xi_i - \xi|] = 1.$$

That is, the sequence $\{\xi_i\}$ does not converge in mean to ξ.

Convergence in Measure vs. Convergence in Distribution

Theorem 1.31 *(Liu [120]). Suppose $\xi, \xi_1, \xi_2, \cdots$ are uncertain variables. If $\{\xi_i\}$ converges in measure to ξ, then $\{\xi_i\}$ converges in distribution to ξ.*

Proof: Let x be a given continuity point of the uncertainty distribution Φ. On the one hand, for any $y > x$, we have

$$\{\xi_i \leq x\} = \{\xi_i \leq x, \xi \leq y\} \cup \{\xi_i \leq x, \xi > y\} \subset \{\xi \leq y\} \cup \{|\xi_i - \xi| \geq y - x\}.$$

It follows from the countable subadditivity axiom that

$$\Phi_i(x) \leq \Phi(y) + \mathcal{M}\{|\xi_i - \xi| \geq y - x\}.$$

Since $\{\xi_i\}$ converges in measure to ξ, we have $\mathcal{M}\{|\xi_i - \xi| \geq y - x\} \to 0$ as $i \to \infty$. Thus we obtain $\limsup_{i \to \infty} \Phi_i(x) \leq \Phi(y)$ for any $y > x$. Letting $y \to x$, we get

$$\limsup_{i \to \infty} \Phi_i(x) \leq \Phi(x). \tag{1.224}$$

On the other hand, for any $z < x$, we have

$$\{\xi \leq z\} = \{\xi_i \leq x, \xi \leq z\} \cup \{\xi_i > x, \xi \leq z\} \subset \{\xi_i \leq x\} \cup \{|\xi_i - \xi| \geq x - z\}$$

which implies that

$$\Phi(z) \leq \Phi_i(x) + \mathcal{M}\{|\xi_i - \xi| \geq x - z\}.$$

Since $\mathcal{M}\{|\xi_i - \xi| \geq x - z\} \to 0$, we obtain $\Phi(z) \leq \liminf_{i \to \infty} \Phi_i(x)$ for any $z < x$. Letting $z \to x$, we get

$$\Phi(x) \leq \liminf_{i \to \infty} \Phi_i(x). \tag{1.225}$$

It follows from (1.224) and (1.225) that $\Phi_i(x) \to \Phi(x)$. The theorem is proved.

Example 1.63: Convergence in distribution does not imply convergence in measure. Take an uncertainty space $(\Gamma, \mathcal{L}, \mathcal{M})$ to be $\{\gamma_1, \gamma_2\}$ with $\mathcal{M}\{\gamma_1\} = \mathcal{M}\{\gamma_2\} = 1/2$. We define an uncertain variables as

$$\xi(\gamma) = \begin{cases} -1, & \text{if } \gamma = \gamma_1 \\ 1, & \text{if } \gamma = \gamma_2. \end{cases}$$

We also define $\xi_i = -\xi$ for $i = 1, 2, \cdots$ Then ξ_i and ξ have the same chance distribution. Thus $\{\xi_i\}$ converges in distribution to ξ. However, for some small number $\varepsilon > 0$, we have

$$\mathcal{M}\{|\xi_i - \xi| \geq \varepsilon\} = \mathcal{M}\{|\xi_i - \xi| \geq \varepsilon\} = 1.$$

That is, the sequence $\{\xi_i\}$ does not converge in measure to ξ.

Convergence Almost Surely vs. Convergence in Measure

Example 1.64: Convergence a.s. does not imply convergence in measure. Take an uncertainty space $(\Gamma, \mathcal{L}, \mathcal{M})$ to be $\{\gamma_1, \gamma_2, \cdots\}$ with

$$\mathcal{M}\{\Lambda\} = \begin{cases} \sup_{\gamma_i \in \Lambda} i/(2i+1), & \text{if } \sup_{\gamma_i \in \Lambda} i/(2i+1) < 0.5 \\ 1 - \sup_{\gamma_i \notin \Lambda} i/(2i+1), & \text{if } \sup_{\gamma_i \notin \Lambda} i/(2i+1) < 0.5 \\ 0.5, & \text{otherwise.} \end{cases}$$

Then we define uncertain variables as

$$\xi_i(\gamma_j) = \begin{cases} i, & \text{if } j = i \\ 0, & \text{otherwise} \end{cases}$$

for $i = 1, 2, \cdots$ and $\xi \equiv 0$. The sequence $\{\xi_i\}$ converges a.s. to ξ. However, for some small number $\varepsilon > 0$, we have

$$\mathcal{M}\{|\xi_i - \xi| \geq \varepsilon\} = \mathcal{M}\{|\xi_i - \xi| \geq \varepsilon\} = \frac{i}{2i+1} \to \frac{1}{2}.$$

That is, the sequence $\{\xi_i\}$ does not converge in measure to ξ.

Example 1.65: Convergence in measure does not imply convergence a.s. Take an uncertainty space $(\Gamma, \mathcal{L}, \mathcal{M})$ to be $[0, 1]$ with Borel algebra and Lebesgue measure. For any positive integer i, there is an integer j such that $i = 2^j + k$, where k is an integer between 0 and $2^j - 1$. Then we define uncertain variables as

$$\xi_i(\gamma) = \begin{cases} 1, & \text{if } k/2^j \leq \gamma \leq (k+1)/2^j \\ 0, & \text{otherwise} \end{cases}$$

for $i = 1, 2, \cdots$ and $\xi \equiv 0$. For some small number $\varepsilon > 0$, we have

$$\mathcal{M}\{|\xi_i - \xi| \geq \varepsilon\} = \mathcal{M}\{|\xi_i - \xi| \geq \varepsilon\} = \frac{1}{2^j} \to 0$$

as $i \to \infty$. That is, the sequence $\{\xi_i\}$ converges in measure to ξ. However, for any $\gamma \in [0, 1]$, there is an infinite number of intervals of the form $[k/2^j, (k+1)/2^j]$ containing γ. Thus $\xi_i(\gamma)$ does not converge to 0. In other words, the sequence $\{\xi_i\}$ does not converge a.s. to ξ.

Convergence Almost Surely vs. Convergence in Mean

Example 1.66: Convergence a.s. does not imply convergence in mean. Take an uncertainty space $(\Gamma, \mathcal{L}, \mathcal{M})$ to be $\{\gamma_1, \gamma_2, \cdots\}$ with

$$\mathcal{M}\{\Lambda\} = \sum_{\gamma_i \in \Lambda} \frac{1}{2^i}.$$

The uncertain variables are defined by

$$
\xi_i(\gamma_j) = \begin{cases} 2^i, & \text{if } j = i \\ 0, & \text{otherwise} \end{cases}
$$

for $i = 1, 2, \cdots$ and $\xi \equiv 0$. Then ξ_i converges a.s. to ξ. However, the sequence $\{\xi_i\}$ does not converge in mean to ξ because $E[|\xi_i - \xi|] \equiv 1$.

Example 1.67: Convergence in mean does not imply convergence a.s. Take an uncertainty space $(\Gamma, \mathcal{L}, \mathcal{M})$ to be $[0, 1]$ with Borel algebra and Lebesgue measure. For any positive integer i, there is an integer j such that $i = 2^j + k$, where k is an integer between 0 and $2^j - 1$. The uncertain variables are defined by

$$
\xi_i(\gamma) = \begin{cases} 1, & \text{if } k/2^j \le \gamma \le (k+1)/2^j \\ 0, & \text{otherwise} \end{cases}
$$

for $i = 1, 2, \cdots$ and $\xi \equiv 0$. Then

$$
E[|\xi_i - \xi|] = \frac{1}{2^j} \to 0.
$$

That is, the sequence $\{\xi_i\}$ converges in mean to ξ. However, for any $\gamma \in [0, 1]$, there is an infinite number of intervals of the form $[k/2^j, (k+1)/2^j]$ containing γ. Thus $\xi_i(\gamma)$ does not converge to 0. In other words, the sequence $\{\xi_i\}$ does not converge a.s. to ξ.

Convergence Almost Surely vs. Convergence in Distribution

Example 1.68: Convergence in distribution does not imply convergence a.s. Take an uncertainty space $(\Gamma, \mathcal{L}, \mathcal{M})$ to be $\{\gamma_1, \gamma_2\}$ with $\mathcal{M}\{\gamma_1\} = \mathcal{M}\{\gamma_2\} = 1/2$. We define an uncertain variable ξ as

$$
\xi(\gamma) = \begin{cases} -1, & \text{if } \gamma = \gamma_1 \\ 1, & \text{if } \gamma = \gamma_2. \end{cases}
$$

We also define $\xi_i = -\xi$ for $i = 1, 2, \cdots$ Then ξ_i and ξ have the same uncertainty distribution. Thus $\{\xi_i\}$ converges in distribution to ξ. However, the sequence $\{\xi_i\}$ does not converge a.s. to ξ.

Example 1.69: Convergence a.s. does not imply convergence in distribution. Take an uncertainty space $(\Gamma, \mathcal{L}, \mathcal{M})$ to be $\{\gamma_1, \gamma_2, \cdots\}$ with

$$
\mathcal{M}\{\Lambda\} = \begin{cases} \sup_{\gamma_i \in \Lambda} i/(2i+1), & \text{if } \sup_{\gamma_i \in \Lambda} i/(2i+1) < 0.5 \\ 1 - \sup_{\gamma_i \notin \Lambda} i/(2i+1), & \text{if } \sup_{\gamma_i \notin \Lambda} i/(2i+1) < 0.5 \\ 0.5, & \text{otherwise.} \end{cases}
$$

The uncertain variables are defined by

$$\xi_i(\gamma_j) = \begin{cases} i, & \text{if } j = i \\ 0, & \text{otherwise} \end{cases}$$

for $i = 1, 2, \cdots$ and $\xi \equiv 0$. Then the sequence $\{\xi_i\}$ converges a.s. to ξ. However, the uncertainty distributions of ξ_i are

$$\Phi_i(x) = \begin{cases} 0, & \text{if } x < 0 \\ (i+1)/(2i+1), & \text{if } 0 \le x < i \\ 1, & \text{if } x \ge i \end{cases}$$

for $i = 1, 2, \cdots$, respectively. The uncertainty distribution of ξ is

$$\Phi(x) = \begin{cases} 0, & \text{if } x < 0 \\ 1, & \text{if } x \ge 0. \end{cases}$$

It is clear that $\Phi_i(x)$ does not converge to $\Phi(x)$ at $x > 0$. That is, the sequence $\{\xi_i\}$ does not converge in distribution to ξ.

1.14 Conditional Uncertainty

We consider the uncertain measure of an event A after it has been learned that some other event B has occurred. This new uncertain measure of A is called the conditional uncertain measure of A given B.

In order to define a conditional uncertain measure $\mathcal{M}\{A|B\}$, at first we have to enlarge $\mathcal{M}\{A \cap B\}$ because $\mathcal{M}\{A \cap B\} < 1$ for all events whenever $\mathcal{M}\{B\} < 1$. It seems that we have no alternative but to divide $\mathcal{M}\{A \cap B\}$ by $\mathcal{M}\{B\}$. Unfortunately, $\mathcal{M}\{A \cap B\}/\mathcal{M}\{B\}$ is not always an uncertain measure. However, the value $\mathcal{M}\{A|B\}$ should not be greater than $\mathcal{M}\{A \cap B\}/\mathcal{M}\{B\}$ (otherwise the normality will be lost), i.e.,

$$\mathcal{M}\{A|B\} \le \frac{\mathcal{M}\{A \cap B\}}{\mathcal{M}\{B\}}. \tag{1.226}$$

On the other hand, in order to preserve the self-duality, we should have

$$\mathcal{M}\{A|B\} = 1 - \mathcal{M}\{A^c|B\} \ge 1 - \frac{\mathcal{M}\{A^c \cap B\}}{\mathcal{M}\{B\}}. \tag{1.227}$$

Furthermore, since $(A \cap B) \cup (A^c \cap B) = B$, we have $\mathcal{M}\{B\} \le \mathcal{M}\{A \cap B\} + \mathcal{M}\{A^c \cap B\}$ by using the countable subadditivity axiom. Thus

$$0 \le 1 - \frac{\mathcal{M}\{A^c \cap B\}}{\mathcal{M}\{B\}} \le \frac{\mathcal{M}\{A \cap B\}}{\mathcal{M}\{B\}} \le 1. \tag{1.228}$$

Hence any numbers between $1 - \mathcal{M}\{A^c \cap B\}/\mathcal{M}\{B\}$ and $\mathcal{M}\{A \cap B\}/\mathcal{M}\{B\}$ are reasonable values that the conditional uncertain measure may take. Based on the maximum uncertainty principle, we have the following conditional uncertain measure.

Definition 1.32 *(Liu [120]). Let* $(\Gamma, \mathcal{L}, \mathcal{M})$ *be an uncertainty space, and* A, $B \in \mathcal{L}$. *Then the conditional uncertain measure of* A *given* B *is defined by*

$$
\mathcal{M}\{A|B\} = \begin{cases} \dfrac{\mathcal{M}\{A \cap B\}}{\mathcal{M}\{B\}}, & if \ \dfrac{\mathcal{M}\{A \cap B\}}{\mathcal{M}\{B\}} < 0.5 \\[3mm] 1 - \dfrac{\mathcal{M}\{A^c \cap B\}}{\mathcal{M}\{B\}}, & if \ \dfrac{\mathcal{M}\{A^c \cap B\}}{\mathcal{M}\{B\}} < 0.5 \\[3mm] 0.5, & otherwise \end{cases} \tag{1.229}
$$

provided that $\mathcal{M}\{B\} > 0$.

It follows immediately from the definition of conditional uncertain measure that

$$
1 - \frac{\mathcal{M}\{A^c \cap B\}}{\mathcal{M}\{B\}} \le \mathcal{M}\{A|B\} \le \frac{\mathcal{M}\{A \cap B\}}{\mathcal{M}\{B\}}. \tag{1.230}
$$

Furthermore, the conditional uncertain measure obeys the maximum uncertainty principle, and takes values as close to 0.5 as possible.

Remark 1.6: Assume that we know the *prior* uncertain measures $\mathcal{M}\{B\}$, $\mathcal{M}\{A \cap B\}$ and $\mathcal{M}\{A^c \cap B\}$. Then the conditional uncertain measure $\mathcal{M}\{A|B\}$ yields the *posterior* uncertain measure of A after the occurrence of event B.

Theorem 1.66. *Let* $(\Gamma, \mathcal{L}, \mathcal{M})$ *be an uncertainty space, and* B *an event with* $\mathcal{M}\{B\} > 0$. *Then* $\mathcal{M}\{\cdot|B\}$ *defined by (1.229) is an uncertain measure, and* $(\Gamma, \mathcal{L}, \mathcal{M}\{\cdot|B\})$ *is an uncertainty space.*

Proof: It is sufficient to prove that $\mathcal{M}\{\cdot|B\}$ satisfies the normality, monotonicity, self-duality and countable subadditivity axioms. At first, it satisfies the normality axiom, i.e.,

$$
\mathcal{M}\{\Gamma|B\} = 1 - \frac{\mathcal{M}\{\Gamma^c \cap B\}}{\mathcal{M}\{B\}} = 1 - \frac{\mathcal{M}\{\emptyset\}}{\mathcal{M}\{B\}} = 1.
$$

For any events A_1 and A_2 with $A_1 \subset A_2$, if

$$
\frac{\mathcal{M}\{A_1 \cap B\}}{\mathcal{M}\{B\}} \le \frac{\mathcal{M}\{A_2 \cap B\}}{\mathcal{M}\{B\}} < 0.5,
$$

then

$$
\mathcal{M}\{A_1|B\} = \frac{\mathcal{M}\{A_1 \cap B\}}{\mathcal{M}\{B\}} \le \frac{\mathcal{M}\{A_2 \cap B\}}{\mathcal{M}\{B\}} = \mathcal{M}\{A_2|B\}.
$$

If

$$
\frac{\mathcal{M}\{A_1 \cap B\}}{\mathcal{M}\{B\}} \le 0.5 \le \frac{\mathcal{M}\{A_2 \cap B\}}{\mathcal{M}\{B\}},
$$

then $\mathcal{M}\{A_1|B\} \le 0.5 \le \mathcal{M}\{A_2|B\}$. If

$$
0.5 < \frac{\mathcal{M}\{A_1 \cap B\}}{\mathcal{M}\{B\}} \le \frac{\mathcal{M}\{A_2 \cap B\}}{\mathcal{M}\{B\}},
$$

then we have

$$\mathcal{M}\{A_1|B\} = \left(1 - \frac{\mathcal{M}\{A_1^c \cap B\}}{\mathcal{M}\{B\}}\right) \vee 0.5 \leq \left(1 - \frac{\mathcal{M}\{A_2^c \cap B\}}{\mathcal{M}\{B\}}\right) \vee 0.5 = \mathcal{M}\{A_2|B\}.$$

This means that $\mathcal{M}\{\cdot|B\}$ satisfies the monotonicity axiom. For any event A, if

$$\frac{\mathcal{M}\{A \cap B\}}{\mathcal{M}\{B\}} \geq 0.5, \quad \frac{\mathcal{M}\{A^c \cap B\}}{\mathcal{M}\{B\}} \geq 0.5,$$

then we have $\mathcal{M}\{A|B\} + \mathcal{M}\{A^c|B\} = 0.5 + 0.5 = 1$ immediately. Otherwise, without loss of generality, suppose

$$\frac{\mathcal{M}\{A \cap B\}}{\mathcal{M}\{B\}} < 0.5 < \frac{\mathcal{M}\{A^c \cap B\}}{\mathcal{M}\{B\}},$$

then we have

$$\mathcal{M}\{A|B\} + \mathcal{M}\{A^c|B\} = \frac{\mathcal{M}\{A \cap B\}}{\mathcal{M}\{B\}} + \left(1 - \frac{\mathcal{M}\{A \cap B\}}{\mathcal{M}\{B\}}\right) = 1.$$

That is, $\mathcal{M}\{\cdot|B\}$ satisfies the self-duality axiom. Finally, for any countable sequence $\{A_i\}$ of events, if $\mathcal{M}\{A_i|B\} < 0.5$ for all i, it follows from the countable subadditivity axiom that

$$\mathcal{M}\left\{\bigcup_{i=1}^{\infty} A_i \cap B\right\} \leq \frac{\mathcal{M}\left\{\bigcup_{i=1}^{\infty} A_i \cap B\right\}}{\mathcal{M}\{B\}} \leq \frac{\sum_{i=1}^{\infty}\mathcal{M}\{A_i \cap B\}}{\mathcal{M}\{B\}} = \sum_{i=1}^{\infty}\mathcal{M}\{A_i|B\}.$$

Suppose there is one term greater than 0.5, say

$$\mathcal{M}\{A_1|B\} \geq 0.5, \quad \mathcal{M}\{A_i|B\} < 0.5, \quad i = 2, 3, \cdots$$

If $\mathcal{M}\{\cup_i A_i|B\} = 0.5$, then we immediately have

$$\mathcal{M}\left\{\bigcup_{i=1}^{\infty} A_i \cap B\right\} \leq \sum_{i=1}^{\infty}\mathcal{M}\{A_i|B\}.$$

If $\mathcal{M}\{\cup_i A_i|B\} > 0.5$, we may prove the above inequality by the following facts:

$$A_1^c \cap B \subset \bigcup_{i=2}^{\infty}(A_i \cap B) \cup \left(\bigcap_{i=1}^{\infty} A_i^c \cap B\right),$$

$$\mathcal{M}\{A_1^c \cap B\} \leq \sum_{i=2}^{\infty}\mathcal{M}\{A_i \cap B\} + \mathcal{M}\left\{\bigcap_{i=1}^{\infty} A_i^c \cap B\right\},$$

$$\mathcal{M}\left\{\bigcup_{i=1}^{\infty} A_i|B\right\} = 1 - \frac{\mathcal{M}\left\{\bigcap_{i=1}^{\infty} A_i^c \cap B\right\}}{\mathcal{M}\{B\}},$$

$$\sum_{i=1}^{\infty} \mathcal{M}\{A_i|B\} \geq 1 - \frac{\mathcal{M}\{A_1^c \cap B\}}{\mathcal{M}\{B\}} + \frac{\sum_{i=2}^{\infty} \mathcal{M}\{A_i \cap B\}}{\mathcal{M}\{B\}}.$$

If there are at least two terms greater than 0.5, then the countable subadditivity is clearly true. Thus $\mathcal{M}\{\cdot|B\}$ satisfies the countable subadditivity axiom. Hence $\mathcal{M}\{\cdot|B\}$ is an uncertain measure. Furthermore, $(\Gamma, \mathcal{L}, \mathcal{M}\{\cdot|B\})$ is an uncertainty space.

Definition 1.33 *(Liu [120]). The conditional uncertainty distribution Φ: $\Re \to [0,1]$ of an uncertain variable ξ given B is defined by*

$$\Phi(x|B) = \mathcal{M}\{\xi \leq x|B\} \qquad (1.231)$$

provided that $\mathcal{M}\{B\} > 0$.

Theorem 1.67. *Let ξ be an uncertain variable with uncertainty distribution $\Phi(x)$, and t a real number with $\Phi(t) < 1$. Then the conditional uncertainty distribution of ξ given $\xi > t$ is*

$$\Phi(x|(t,+\infty)) = \begin{cases} 0, & if \ \Phi(x) \leq \Phi(t) \\ \dfrac{\Phi(x)}{1-\Phi(t)} \wedge 0.5, & if \ \Phi(t) < \Phi(x) \leq (1+\Phi(t))/2 \\ \dfrac{\Phi(x)-\Phi(t)}{1-\Phi(t)}, & if \ (1+\Phi(t))/2 \leq \Phi(x). \end{cases}$$

Proof: It follows from $\Phi(x|(t,+\infty)) = \mathcal{M}\{\xi \leq x|\xi > t\}$ and the definition of conditional uncertainty that

$$\Phi(x|(t,+\infty)) = \begin{cases} \dfrac{\mathcal{M}\{(\xi \leq x) \cap (\xi > t)\}}{\mathcal{M}\{\xi > t\}}, & if \ \dfrac{\mathcal{M}\{(\xi \leq x) \cap (\xi > t)\}}{\mathcal{M}\{\xi > t\}} < 0.5 \\ 1 - \dfrac{\mathcal{M}\{(\xi > x) \cap (\xi > t)\}}{\mathcal{M}\{\xi > t\}}, & if \ \dfrac{\mathcal{M}\{(\xi > x) \cap (\xi > t)\}}{\mathcal{M}\{\xi > t\}} < 0.5 \\ 0.5, & otherwise. \end{cases}$$

When $\Phi(x) \leq \Phi(t)$, we have $x \leq t$, and

$$\frac{\mathcal{M}\{(\xi \leq x) \cap (\xi > t)\}}{\mathcal{M}\{\xi > t\}} = \frac{\mathcal{M}\{\emptyset\}}{1-\Phi(t)} = 0 < 0.5.$$

Thus

$$\Phi(x|(t,+\infty)) = \frac{\mathcal{M}\{(\xi \leq x) \cap (\xi > t)\}}{\mathcal{M}\{\xi > t\}} = 0.$$

When $\Phi(t) < \Phi(x) \leq (1+\Phi(t))/2$, we have $x > t$, and

$$\frac{\mathcal{M}\{(\xi > x) \cap (\xi > t)\}}{\mathcal{M}\{\xi > t\}} = \frac{1-\Phi(x)}{1-\Phi(t)} \geq \frac{1-(1+\Phi(t))/2}{1-\Phi(t)} = 0.5$$

and

$$\frac{\mathcal{M}\{(\xi \leq x) \cap (\xi > t)\}}{\mathcal{M}\{\xi > t\}} \leq \frac{\Phi(x)}{1 - \Phi(t)}.$$

It follows from the maximum uncertainty principle that

$$\Phi(x|(t, +\infty)) = \frac{\Phi(x)}{1 - \Phi(t)} \wedge 0.5.$$

When $(1 + \Phi(t))/2 \leq \Phi(x)$, we have $x \geq t$, and

$$\frac{\mathcal{M}\{(\xi > x) \cap (\xi > t)\}}{\mathcal{M}\{\xi > t\}} = \frac{1 - \Phi(x)}{1 - \Phi(t)} \leq \frac{1 - (1 + \Phi(t))/2}{1 - \Phi(t)} \leq 0.5.$$

Thus

$$\Phi(x|(t, +\infty)) = 1 - \frac{\mathcal{M}\{(\xi > x) \cap (\xi > t)\}}{\mathcal{M}\{\xi > t\}} = 1 - \frac{1 - \Phi(x)}{1 - \Phi(t)} = \frac{\Phi(x) - \Phi(t)}{1 - \Phi(t)}.$$

The theorem is proved.

Example 1.70: Let ξ be a linear uncertain variable $\mathcal{L}(a, b)$, and t a real number with $a < t < b$. Then the conditional uncertainty distribution of ξ given $\xi > t$ is

$$\Phi(x|(t, +\infty)) = \begin{cases} 0, & \text{if } x \leq t \\ \dfrac{x - a}{b - t} \wedge 0.5, & \text{if } t < x \leq (b + t)/2 \\ \dfrac{x - t}{b - t} \wedge 1, & \text{if } (b + t)/2 \leq x. \end{cases}$$

Theorem 1.68. *Let ξ be an uncertain variable with uncertainty distribution $\Phi(x)$, and t a real number with $\Phi(t) > 0$. Then the conditional uncertainty distribution of ξ given $\xi \leq t$ is*

$$\Phi(x|(-\infty, t]) = \begin{cases} \dfrac{\Phi(x)}{\Phi(t)}, & \text{if } \Phi(x) \leq \Phi(t)/2 \\ \dfrac{\Phi(x) + \Phi(t) - 1}{\Phi(t)} \vee 0.5, & \text{if } \Phi(t)/2 \leq \Phi(x) < \Phi(t) \\ 1, & \text{if } \Phi(t) \leq \Phi(x). \end{cases}$$

Proof: It follows from $\Phi(x|(-\infty, t]) = \mathcal{M}\{\xi \leq x | \xi \leq t\}$ and the definition of conditional uncertainty that

$$\Phi(x|(-\infty, t]) = \begin{cases} \dfrac{\mathcal{M}\{(\xi \leq x) \cap (\xi \leq t)\}}{\mathcal{M}\{\xi \leq t\}}, & \text{if } \dfrac{\mathcal{M}\{(\xi \leq x) \cap (\xi \leq t)\}}{\mathcal{M}\{\xi \leq t\}} < 0.5 \\ 1 - \dfrac{\mathcal{M}\{(\xi > x) \cap (\xi \leq t)\}}{\mathcal{M}\{\xi \leq t\}}, & \text{if } \dfrac{\mathcal{M}\{(\xi > x) \cap (\xi \leq t)\}}{\mathcal{M}\{\xi \leq t\}} < 0.5 \\ 0.5, & \text{otherwise.} \end{cases}$$

When $\Phi(x) \leq \Phi(t)/2$, we have $x < t$, and

$$\frac{\mathcal{M}\{(\xi \leq x) \cap (\xi \leq t)\}}{\mathcal{M}\{\xi \leq t\}} = \frac{\Phi(x)}{\Phi(t)} \leq \frac{\Phi(t)/2}{\Phi(t)} = 0.5.$$

Thus

$$\Phi(x|(-\infty, t]) = \frac{\mathcal{M}\{(\xi \leq x) \cap (\xi \leq t)\}}{\mathcal{M}\{\xi \leq t\}} = \frac{\Phi(x)}{\Phi(t)}.$$

When $\Phi(t)/2 \leq \Phi(x) < \Phi(t)$, we have $x < t$, and

$$\frac{\mathcal{M}\{(\xi \leq x) \cap (\xi \leq t)\}}{\mathcal{M}\{\xi \leq t\}} = \frac{\Phi(x)}{\Phi(t)} \geq \frac{\Phi(t)/2}{\Phi(t)} = 0.5$$

and

$$\frac{\mathcal{M}\{(\xi > x) \cap (\xi \leq t)\}}{\mathcal{M}\{\xi \leq t\}} \leq \frac{1 - \Phi(x)}{\Phi(t)},$$

i.e.,

$$1 - \frac{\mathcal{M}\{(\xi > x) \cap (\xi \leq t)\}}{\mathcal{M}\{\xi \leq t\}} \geq \frac{\Phi(x) + \Phi(t) - 1}{\Phi(t)}.$$

It follows from the maximum uncertainty principle that

$$\Phi(x|(-\infty, t]) = \frac{\Phi(x) + \Phi(t) - 1}{\Phi(t)} \vee 0.5.$$

When $\Phi(t) \leq \Phi(x)$, we have $x \geq t$, and

$$\frac{\mathcal{M}\{(\xi > x) \cap (\xi \leq t)\}}{\mathcal{M}\{\xi \leq t\}} = \frac{\mathcal{M}\{\emptyset\}}{\Phi(t)} = 0 < 0.5.$$

Thus

$$\Phi(x|(-\infty, t]) = 1 - \frac{\mathcal{M}\{(\xi > x) \cap (\xi \leq t)\}}{\mathcal{M}\{\xi \leq t\}} = 1 - 0 = 1.$$

The theorem is proved.

Example 1.71: Let ξ be a linear uncertain variable $\mathcal{L}(a, b)$, and t a real number with $a < t < b$. Then the conditional uncertainty distribution of ξ given $\xi \leq t$ is

$$\Phi(x|(-\infty, t]) = \begin{cases} \dfrac{x - a}{t - a} \vee 0, & \text{if } x \leq (a + t)/2 \\ \left(1 - \dfrac{b - x}{t - a}\right) \vee 0.5, & \text{if } (a + t)/2 \leq x < t \\ 1, & \text{if } x \leq t. \end{cases}$$

Definition 1.34 *(Liu [120]). Let ξ be an uncertain variable. Then the conditional expected value of ξ given B is defined by*

$$E[\xi|B] = \int_0^{+\infty} \mathcal{M}\{\xi \geq r|B\}\mathrm{d}r - \int_{-\infty}^0 \mathcal{M}\{\xi \leq r|B\}\mathrm{d}r \qquad (1.232)$$

provided that at least one of the two integrals is finite.

Chapter 2

Uncertain Programming

Uncertain programming was founded by Liu [122] in 2009 as a type of mathematical programming involving uncertain variables. This chapter provides a general framework of uncertain programming, including expected value model, chance-constrained programming, dependent-chance programming, uncertain dynamic programming and uncertain multilevel programming. Finally, we present some uncertain programming models for project scheduling problem, vehicle routing problem, and machine scheduling problem.

2.1 Ranking Criteria

Assume that x is a decision vector, $\boldsymbol{\xi}$ is an uncertain vector, $f(x, \boldsymbol{\xi})$ is a return function, and $g_j(x, \boldsymbol{\xi})$ are constraint functions, $j = 1, 2, \cdots, p$. Let us examine

$$\begin{cases} \max f(x, \boldsymbol{\xi}) \\ \text{subject to:} \\ \quad g_j(x, \boldsymbol{\xi}) \leq 0, \quad j = 1, 2, \cdots, p. \end{cases} \tag{2.1}$$

Mention that the model (2.1) is only a conceptual model rather than a mathematical model because there does not exist a natural ordership in an uncertain world.

Thus an important problem appearing in this area is how to rank uncertain variables. Let ξ and η be two uncertain variables. Liu [122] gave four ranking criteria.

Expected Value Criterion: We say $\xi > \eta$ if and only if $E[\xi] > E[\eta]$.

Optimistic Value Criterion: We say $\xi > \eta$ if and only if, for some predetermined confidence level $\alpha \in (0, 1]$, we have $\xi_{\text{sup}}(\alpha) > \eta_{\text{sup}}(\alpha)$, where $\xi_{\text{sup}}(\alpha)$ and $\eta_{\text{sup}}(\alpha)$ are the α-optimistic values of ξ and η, respectively.

Pessimistic Value Criterion: We say $\xi > \eta$ if and only if, for some predetermined confidence level $\alpha \in (0, 1]$, we have $\xi_{\text{inf}}(\alpha) > \eta_{\text{inf}}(\alpha)$, where $\xi_{\text{inf}}(\alpha)$ and $\eta_{\text{inf}}(\alpha)$ are the α-pessimistic values of ξ and η, respectively.

Chance Criterion: We say $\xi > \eta$ if and only if, for some predetermined levels $\bar{r}$, we have $\mathcal{M}\{\xi \geq \bar{r}\} > \mathcal{M}\{\eta \geq \bar{r}\}$.

B. Liu: Uncertainty Theory: A Branch of Mathematics, SCI 300, pp. 81–113.
springerlink.com   © Springer-Verlag Berlin Heidelberg 2010

2.2   Expected Value Model

Assume that we believe the expected value criterion. In order to obtain a decision with maximum expected return subject to expected constraints, we have the following expected value model,

$$\begin{cases} \max E[f(\boldsymbol{x}, \boldsymbol{\xi})] \\ \text{subject to:} \\ \quad E[g_j(\boldsymbol{x}, \boldsymbol{\xi})] \leq 0, \ j = 1, 2, \cdots, p \end{cases} \tag{2.2}$$

where $\boldsymbol{x}$ is a decision vector, $\boldsymbol{\xi}$ is an uncertain vector, f is a return function, and g_j are constraint functions for $j = 1, 2, \cdots, p$.

Definition 2.1. *A solution x is feasible if and only if $E[g_j(\boldsymbol{x}, \boldsymbol{\xi})] \leq 0$ for $j = 1, 2, \cdots, p$. A feasible solution $\boldsymbol{x}^*$ is an optimal solution to the expected value model (2.2) if $E[f(\boldsymbol{x}^*, \boldsymbol{\xi})] \geq E[f(\boldsymbol{x}, \boldsymbol{\xi})]$ for any feasible solution $\boldsymbol{x}$.*

In practice, a decision maker may want to optimize multiple objectives. Thus we have the following expected value multiobjective programming,

$$\begin{cases} \max \left[E[f_1(\boldsymbol{x}, \boldsymbol{\xi})], E[f_2(\boldsymbol{x}, \boldsymbol{\xi})], \cdots, E[f_m(\boldsymbol{x}, \boldsymbol{\xi})] \right] \\ \text{subject to:} \\ \quad E[g_j(\boldsymbol{x}, \boldsymbol{\xi})] \leq 0, \ j = 1, 2, \cdots, p \end{cases} \tag{2.3}$$

where $f_i(\boldsymbol{x}, \boldsymbol{\xi})$ are return functions for $i = 1, 2, \cdots, m$, and $g_j(\boldsymbol{x}, \boldsymbol{\xi})$ are constraint functions for $j = 1, 2, \cdots, p$.

Definition 2.2. *A feasible solution $\boldsymbol{x}^*$ is said to be a Pareto solution to the expected value multiobjective programming (2.3) if there is no feasible solution $\boldsymbol{x}$ such that*

$$E[f_i(\boldsymbol{x}, \boldsymbol{\xi})] \geq E[f_i(\boldsymbol{x}^*, \boldsymbol{\xi})], \quad i = 1, 2, \cdots, m \tag{2.4}$$

and $E[f_j(\boldsymbol{x}, \boldsymbol{\xi})] > E[f_j(\boldsymbol{x}^*, \boldsymbol{\xi})]$ for at least one index j.

In order to balance multiple conflicting objectives, a decision-maker may establish a hierarchy of importance among these incompatible goals so as to satisfy as many goals as possible in the order specified. Thus we have an expected value goal programming,

$$\begin{cases} \min \sum_{j=1}^{l} P_j \sum_{i=1}^{m} (u_{ij} d_i^+ \vee 0 + v_{ij} d_i^- \vee 0) \\ \text{subject to:} \\ \quad E[f_i(\boldsymbol{x}, \boldsymbol{\xi})] - b_i = d_i^+, \quad i = 1, 2, \cdots, m \\ \quad b_i - E[f_i(\boldsymbol{x}, \boldsymbol{\xi})] = d_i^-, \quad i = 1, 2, \cdots, m \\ \quad E[g_j(\boldsymbol{x}, \boldsymbol{\xi})] \leq 0, \quad\quad\quad\ j = 1, 2, \cdots, p \end{cases} \tag{2.5}$$

where P_j is the preemptive priority factor which expresses the relative importance of various goals, $P_j \gg P_{j+1}$, for all j, u_{ij} is the weighting factor corresponding to positive deviation for goal i with priority j assigned, v_{ij} is the weighting factor corresponding to negative deviation for goal i with priority j assigned, $d_i^+ \vee 0$ is the positive deviation from the target of goal i, $d_i^- \vee 0$ is the negative deviation from the target of goal i, f_i is a function in goal constraints, g_j is a function in real constraints, b_i is the target value according to goal i, l is the number of priorities, m is the number of goal constraints, and p is the number of real constraints.

Theorem 2.1. *Assume $f(\boldsymbol{x}, \boldsymbol{\xi}) = h_1(\boldsymbol{x})\xi_1 + h_2(\boldsymbol{x})\xi_2 + \cdots + h_n(\boldsymbol{x})\xi_n + h_0(\boldsymbol{x})$ where $h_1(\boldsymbol{x}), h_2(\boldsymbol{x}), \cdots, h_n(\boldsymbol{x}), h_0(\boldsymbol{x})$ are real-valued functions and $\xi_1, \xi_2, \cdots, \xi_n$ are independent uncertain variables. Then*

$$E[f(\boldsymbol{x}, \boldsymbol{\xi})] = h_1(\boldsymbol{x})E[\xi_1] + h_2(\boldsymbol{x})E[\xi_2] + \cdots + h_n(\boldsymbol{x})E[\xi_n] + h_0(\boldsymbol{x}). \quad (2.6)$$

Proof: It follows from the linearity of expected value operator immediately.

Theorem 2.2. *Assume that $\xi_1, \xi_2, \cdots, \xi_n$ are independent uncertain variables and $h_1(\boldsymbol{x}), h_2(\boldsymbol{x}), \cdots, h_n(\boldsymbol{x}), h_0(\boldsymbol{x})$ are real-valued functions. Then*

$$E[h_1(\boldsymbol{x})\xi_1 + h_2(\boldsymbol{x})\xi_2 + \cdots + h_n(\boldsymbol{x})\xi_n + h_0(\boldsymbol{x})] \leq 0 \quad (2.7)$$

holds if and only if

$$h_1(\boldsymbol{x})E[\xi_1] + h_2(\boldsymbol{x})E[\xi_2] + \cdots + h_n(\boldsymbol{x})E[\xi_n] + h_0(\boldsymbol{x}) \leq 0. \quad (2.8)$$

Proof: It follows from Theorem 2.1 immediately.

2.3　Chance-Constrained Programming

Since the uncertain constraints $g_j(\boldsymbol{x}, \boldsymbol{\xi}) \leq 0, j = 1, 2, \cdots, p$ do not define a deterministic feasible set, it is naturally desired that the uncertain constraints hold with a confidence level α. Then we have a chance constraint as follows,

$$\mathcal{M}\{g_j(\boldsymbol{x}, \boldsymbol{\xi}) \leq 0, j = 1, 2, \cdots, p\} \geq \alpha. \quad (2.9)$$

Maximax Chance-Constrained Programming

Assume that we believe the optimistic value criterion. If we want to maximize the optimistic value to the uncertain return subject to some chance constraints, then we have the following maximax chance-constrained programming,

$$\begin{cases} \max_{\boldsymbol{x}} \max_{\overline{f}} \overline{f} \\ \text{subject to:} \\ \quad \mathcal{M}\{f(\boldsymbol{x}, \boldsymbol{\xi}) \geq \overline{f}\} \geq \beta \\ \quad \mathcal{M}\{g_j(\boldsymbol{x}, \boldsymbol{\xi}) \leq 0\} \geq \alpha_j, \quad j = 1, 2, \cdots, p \end{cases} \quad (2.10)$$

where α_j and β are specified confidence levels for $j = 1, 2, \cdots, p$, and $\max \overline{f}$ is the β-optimistic return.

In practice, it is possible that there exist multiple objectives. We thus have the following maximax chance-constrained multiobjective programming,

$$
\begin{cases}
\displaystyle \max_{\boldsymbol{x}} \left[\max_{\overline{f}_1} \overline{f}_1, \max_{\overline{f}_2} \overline{f}_2, \cdots, \max_{\overline{f}_m} \overline{f}_m \right] \\
\text{subject to:} \\
\quad \mathcal{M}\left\{ f_i(\boldsymbol{x}, \boldsymbol{\xi}) \geq \overline{f}_i \right\} \geq \beta_i, \quad i = 1, 2, \cdots, m \\
\quad \mathcal{M}\left\{ g_j(\boldsymbol{x}, \boldsymbol{\xi}) \leq 0 \right\} \geq \alpha_j, \quad j = 1, 2, \cdots, p
\end{cases}
\tag{2.11}
$$

where β_i are predetermined confidence levels for $i = 1, 2, \cdots, m$, and $\max \overline{f}_i$ are the β-optimistic values to the return functions $f_i(\boldsymbol{x}, \boldsymbol{\xi})$, $i = 1, 2, \cdots, m$, respectively.

If the priority structure and target levels are set by the decision-maker, then we have a minimin chance-constrained goal programming,

$$
\begin{cases}
\displaystyle \min_{\boldsymbol{x}} \sum_{j=1}^{l} P_j \sum_{i=1}^{m} \left(u_{ij} \left(\min_{d_i^+} d_i^+ \vee 0 \right) + v_{ij} \left(\min_{d_i^-} d_i^- \vee 0 \right) \right) \\
\text{subject to:} \\
\quad \mathcal{M}\left\{ f_i(\boldsymbol{x}, \boldsymbol{\xi}) - b_i \leq d_i^+ \right\} \geq \beta_i^+, \quad i = 1, 2, \cdots, m \\
\quad \mathcal{M}\left\{ b_i - f_i(\boldsymbol{x}, \boldsymbol{\xi}) \leq d_i^- \right\} \geq \beta_i^-, \quad i = 1, 2, \cdots, m \\
\quad \mathcal{M}\left\{ g_j(\boldsymbol{x}, \boldsymbol{\xi}) \leq 0 \right\} \geq \alpha_j, \qquad j = 1, 2, \cdots, p
\end{cases}
\tag{2.12}
$$

where P_j is the preemptive priority factor which expresses the relative importance of various goals, $P_j \gg P_{j+1}$, for all j, u_{ij} is the weighting factor corresponding to positive deviation for goal i with priority j assigned, v_{ij} is the weighting factor corresponding to negative deviation for goal i with priority j assigned, $\min d_i^+ \vee 0$ is the β_i^+-optimistic positive deviation from the target of goal i, $\min d_i^- \vee 0$ is the β_i^--optimistic negative deviation from the target of goal i, b_i is the target value according to goal i, and l is the number of priorities.

Minimax Chance-Constrained Programming

Assume that we believe the pessimistic value criterion. If we want to maximize the pessimistic value subject to some chance constraints, then we have the following minimax chance-constrained programming,

$$
\begin{cases}
\displaystyle \max_{\boldsymbol{x}} \min_{\overline{f}} \overline{f} \\
\text{subject to:} \\
\quad \mathcal{M}\left\{ f(\boldsymbol{x}, \boldsymbol{\xi}) \leq \overline{f} \right\} \geq \beta \\
\quad \mathcal{M}\left\{ g_j(\boldsymbol{x}, \boldsymbol{\xi}) \leq 0 \right\} \geq \alpha_j, \quad j = 1, 2, \cdots, p
\end{cases}
\tag{2.13}
$$

where α_j and β are specified confidence levels for $j = 1, 2, \cdots, p$, and $\min \overline{f}$ is the β-pessimistic return.

If there are multiple objectives, then we have the following minimax chance-constrained multiobjective programming,

$$
\begin{cases}
\max\limits_{\boldsymbol{x}} \left[\min\limits_{\overline{f}_1} \overline{f}_1, \; \min\limits_{\overline{f}_2} \overline{f}_2, \; \cdots, \; \min\limits_{\overline{f}_m} \overline{f}_m \right] \\
\text{subject to:} \\
\quad \mathcal{M}\left\{ f_i(\boldsymbol{x}, \boldsymbol{\xi}) \leq \overline{f}_i \right\} \geq \beta_i, \quad i = 1, 2, \cdots, m \\
\quad \mathcal{M}\left\{ g_j(\boldsymbol{x}, \boldsymbol{\xi}) \leq 0 \right\} \geq \alpha_j, \quad j = 1, 2, \cdots, p
\end{cases}
\tag{2.14}
$$

where $\min \overline{f}_i$ are the β_i-pessimistic values to the return functions $f_i(\boldsymbol{x}, \boldsymbol{\xi})$, $i = 1, 2, \cdots, m$, respectively.

We can also formulate an uncertain decision system as a minimax chance-constrained goal programming according to the priority structure and target levels set by the decision-maker:

$$
\begin{cases}
\min\limits_{\boldsymbol{x}} \sum\limits_{j=1}^{l} P_j \sum\limits_{i=1}^{m} \left[u_{ij} \left(\max\limits_{d_i^+} d_i^+ \vee 0 \right) + v_{ij} \left(\max\limits_{d_i^-} d_i^- \vee 0 \right) \right] \\
\text{subject to:} \\
\quad \mathcal{M}\left\{ f_i(\boldsymbol{x}, \boldsymbol{\xi}) - b_i \geq d_i^+ \right\} \geq \beta_i^+, \quad i = 1, 2, \cdots, m \\
\quad \mathcal{M}\left\{ b_i - f_i(\boldsymbol{x}, \boldsymbol{\xi}) \geq d_i^- \right\} \geq \beta_i^-, \quad i = 1, 2, \cdots, m \\
\quad \mathcal{M}\left\{ g_j(\boldsymbol{x}, \boldsymbol{\xi}) \leq 0 \right\} \geq \alpha_j, \quad j = 1, 2, \cdots, p
\end{cases}
\tag{2.15}
$$

where P_j is the preemptive priority factor which expresses the relative importance of various goals, $P_j \gg P_{j+1}$, for all j, u_{ij} is the weighting factor corresponding to positive deviation for goal i with priority j assigned, v_{ij} is the weighting factor corresponding to negative deviation for goal i with priority j assigned, $\max d_i^+ \vee 0$ is the β_i^+-pessimistic positive deviation from the target of goal i, $\max d_i^- \vee 0$ is the β_i^--pessimistic negative deviation from the target of goal i, b_i is the target value according to goal i, and l is the number of priorities.

Theorem 2.3. *Assume that $\xi_1, \xi_2, \cdots, \xi_n$ are independent uncertain variables with uncertainty distributions $\Phi_1, \Phi_2, \cdots, \Phi_n$, respectively, and $h_1(\boldsymbol{x})$, $h_2(\boldsymbol{x}), \cdots, h_n(\boldsymbol{x}), h_0(\boldsymbol{x})$ are real-valued functions. Then*

$$
\mathcal{M}\left\{ \sum_{i=1}^{n} h_i(\boldsymbol{x})\xi_i \leq h_0(\boldsymbol{x}) \right\} \geq \alpha
\tag{2.16}
$$

holds if and only if

$$
\sum_{i=1}^{n} h_i^+(\boldsymbol{x})\Phi_i^{-1}(\alpha) - \sum_{i=1}^{n} h_i^-(\boldsymbol{x})\Phi_i^{-1}(1 - \alpha) \leq h_0(\boldsymbol{x})
\tag{2.17}
$$

where

$$h_i^+(x) = \begin{cases} h_i(x), & \text{if } h_i(x) > 0 \\ 0, & \text{if } h_i(x) \leq 0, \end{cases} \tag{2.18}$$

$$h_i^-(x) = \begin{cases} 0, & \text{if } h_i(x) \geq 0 \\ -h_i(x), & \text{if } h_i(x) < 0 \end{cases} \tag{2.19}$$

for $i = 1, 2, \cdots, n$. *Especially, if* $h_1(x), h_2(x), \cdots, h_n(x)$ *are all nonnegative, then (2.17) becomes*

$$\sum_{i=1}^{n} h_i(x)\Phi_i^{-1}(\alpha) \leq h_0(x); \tag{2.20}$$

if $h_1(x), h_2(x), \cdots, h_n(x)$ *are all nonpositive, then (2.17) becomes*

$$\sum_{i=1}^{n} h_i(x)\Phi_i^{-1}(1-\alpha) \leq h_0(x). \tag{2.21}$$

Proof: For each i, if $h_i(x) > 0$, then $h_i(x)\xi_i$ is an uncertain variable whose uncertainty distribution is described by

$$\Psi_i^{-1}(\alpha) = h_i^+(x)\Phi_i^{-1}(\alpha), \quad 0 < \alpha < 1.$$

If $h_i(x) < 0$, then $h_i(x)\xi_i$ is an uncertain variable whose uncertainty distribution is described by

$$\Psi_i^{-1}(\alpha) = -h_i^-(x)\Phi_i^{-1}(1-\alpha), \quad 0 < \alpha < 1.$$

It follows from the operational law that the uncertainty distribution of the sum $h_1(x)\xi_1 + h_2(x)\xi_2 + \cdots + h_n(x)\xi_n$ is described by

$$\Psi^{-1}(\alpha) = \Psi_1^{-1}(\alpha) + \Psi_2^{-1}(\alpha) + \cdots + \Psi_n^{-1}(\alpha), \quad 0 < \alpha < 1.$$

From which we may derive the result immediately.

Theorem 2.4. *Assume that* $x_1, x_2, \cdots, x_n$ *are nonnegative decision variables, and* $\xi_1, \xi_2, \cdots, \xi_n, \xi$ *are independently linear uncertain variables* $\mathcal{L}(a_1, b_1)$, $\mathcal{L}(a_2, b_2), \cdots, \mathcal{L}(a_n, b_n), \mathcal{L}(a, b)$, *respectively. Then for any confidence level* $\alpha \in (0, 1)$, *the chance constraint*

$$\mathcal{M}\left\{\sum_{i=1}^{n} \xi_i x_i \leq \xi\right\} \geq \alpha \tag{2.22}$$

holds if and only if

$$\sum_{i=1}^{n}((1-\alpha)a_i + \alpha b_i)x_i \leq \alpha a + (1-\alpha)b. \tag{2.23}$$

Proof: Assume that the uncertain variables $\xi_1, \xi_2, \cdots, \xi_n, \xi$ have uncertainty distributions $\Phi_1, \Phi_2, \cdots, \Phi_n, \Phi$, respectively. Then

$$\Phi_i^{-1}(\alpha) = (1-\alpha)a_i + \alpha b_i, \quad i = 1, 2, \cdots, n,$$

$$\Phi^{-1}(1-\alpha) = \alpha a + (1-\alpha)b.$$

Thus the result follows from Theorem 2.3 immediately.

Theorem 2.5. *Assume that $x_1, x_2, \cdots, x_n$ are nonnegative decision variables, and $\xi_1, \xi_2, \cdots, \xi_n, \xi$ are independently zigzag uncertain variables $\mathcal{Z}(a_1, b_1, c_1)$, $\mathcal{Z}(a_2, b_2, c_2), \cdots, \mathcal{Z}(a_n, b_n, c_n), \mathcal{Z}(a, b, c)$, respectively. Then for any confidence level $\alpha \geq 0.5$, the chance constraint*

$$\mathcal{M}\left\{\sum_{i=1}^{n} \xi_i x_i \leq \xi\right\} \geq \alpha \tag{2.24}$$

holds if and only if

$$\sum_{i=1}^{n}((2-2\alpha)b_i + (2\alpha-1)c_i)x_i \leq \alpha(2\alpha-1)a + (2-2\alpha)b. \tag{2.25}$$

Proof: Assume that the uncertain variables $\xi_1, \xi_2, \cdots, \xi_n, \xi$ have uncertainty distributions $\Phi_1, \Phi_2, \cdots, \Phi_n, \Phi$, respectively. Then

$$\Phi_i^{-1}(\alpha) = (2-2\alpha)b_i + (2\alpha-1)c_i, \quad i = 1, 2, \cdots, n,$$

$$\Phi^{-1}(1-\alpha) = (2\alpha-1)a + (2-2\alpha)b.$$

Thus the result follows from Theorem 2.3 immediately.

Theorem 2.6. *Assume that $x_1, x_2, \cdots, x_n$ are nonnegative decision variables, and $\xi_1, \xi_2, \cdots, \xi_n, \xi$ are independently normal uncertain variables $\mathcal{N}(e_1, \sigma_1)$, $\mathcal{N}(e_2, \sigma_2), \cdots, \mathcal{N}(e_n, \sigma_n), \mathcal{N}(e, \sigma)$, respectively. Then for any confidence level $\alpha \in (0, 1)$, the chance constraint*

$$\mathcal{M}\left\{\sum_{i=1}^{n} \xi_i x_i \leq \xi\right\} \geq \alpha \tag{2.26}$$

holds if and only if

$$\sum_{i=1}^{n}\left(e_i + \frac{\sigma_i \sqrt{3}}{\pi} \ln \frac{\alpha}{1-\alpha}\right) x_i \leq e - \frac{\sigma \sqrt{3}}{\pi} \ln \frac{\alpha}{1-\alpha}. \tag{2.27}$$

Proof: Assume that the uncertain variables $\xi_1, \xi_2, \cdots, \xi_n, \xi$ have uncertainty distributions $\Phi_1, \Phi_2, \cdots, \Phi_n, \Phi$, respectively. Then

$$\Phi_i^{-1}(\alpha) = e_i + \frac{\sigma_i \sqrt{3}}{\pi} \ln \frac{\alpha}{1-\alpha}, \quad i = 1, 2, \cdots, n,$$

$$\Phi^{-1}(1-\alpha) = e - \frac{\sigma \sqrt{3}}{\pi} \ln \frac{\alpha}{1-\alpha}.$$

Thus the result follows from Theorem 2.3 immediately.

Theorem 2.7. *Assume $x_1, x_2, \cdots, x_n$ are nonnegative decision variables, and $\xi_1, \xi_2, \cdots, \xi_n, \xi$ are independently lognormal uncertain variables $\mathcal{LOGN}(e_1, \sigma_1)$, $\mathcal{LOGN}(e_2, \sigma_2), \cdots, \mathcal{LOGN}(e_n, \sigma_n), \mathcal{LOGN}(e, \sigma)$, respectively. Then for any confidence level $\alpha \in (0, 1)$, the chance constraint*

$$\mathcal{M}\left\{\sum_{i=1}^{n} \xi_i x_i \leq \xi\right\} \geq \alpha \tag{2.28}$$

holds if and only if

$$\sum_{i=1}^{n} \exp(e_i) \left(\frac{\alpha}{1-\alpha}\right)^{\sqrt{3}\sigma_i/\pi} x_i \leq \exp(e) \left(\frac{1-\alpha}{\alpha}\right)^{\sqrt{3}\sigma/\pi}. \tag{2.29}$$

Proof: Assume that the uncertain variables $\xi_1, \xi_2, \cdots, \xi_n, \xi$ have uncertainty distributions $\Phi_1, \Phi_2, \cdots, \Phi_n, \Phi$, respectively. Then

$$\Phi_i^{-1}(\alpha) = \exp(e_i) \left(\frac{\alpha}{1-\alpha}\right)^{\sqrt{3}\sigma_i/\pi}, \quad i = 1, 2, \cdots, n,$$

$$\Phi^{-1}(1-\alpha) = \exp(e) \left(\frac{1-\alpha}{\alpha}\right)^{\sqrt{3}\sigma/\pi}.$$

Thus the result follows from Theorem 2.3 immediately.

2.4   Dependent-Chance Programming

In practice, there usually exist multiple tasks in a complex uncertain decision system. Sometimes, the decision-maker believes the chance criterion and wishes to maximize the chance of meeting these tasks. In order to model this type of uncertain decision system, Liu [122] provided the third type of uncertain programming, called *dependent-chance programming*, in which the underlying philosophy is based on selecting the decision with maximal chance to meet the task. Dependent-chance programming breaks the concept of feasible set and replaces it with uncertain environment.

Definition 2.3. *By an uncertain environment we mean the uncertain constraints represented by*

$$g_j(\boldsymbol{x}, \boldsymbol{\xi}) \leq 0, \quad j = 1, 2, \cdots, p \tag{2.30}$$

where $\boldsymbol{x}$ is a decision vector, and $\boldsymbol{\xi}$ is an uncertain vector.

Definition 2.4. *By a task we mean an uncertain inequality (or a system of uncertain inequalities) represented by*

$$h(\boldsymbol{x}, \boldsymbol{\xi}) \leq 0 \tag{2.31}$$

where $\boldsymbol{x}$ is a decision vector, and $\boldsymbol{\xi}$ is an uncertain vector.

Definition 2.5. *The chance function of task $\mathcal{E}$ characterized by (2.31) is defined as the uncertain measure that the task $\mathcal{E}$ is met, i.e.,*

$$f(x) = \mathcal{M}\{h(x, \xi) \leq 0\} \qquad (2.32)$$

subject to the uncertain environment (2.30).

How do we compute the chance function in an uncertain environment? In order to answer this question, we first give some basic definitions. Let $r(x_1, x_2, \cdots, x_n)$ be an n-dimensional function. The ith decision variable x_i is said to be degenerate if

$$r(x_1, \cdots, x_{i-1}, x_i', x_{i+1}, \cdots, x_n) = r(x_1, \cdots, x_{i-1}, x_i'', x_{i+1}, \cdots, x_n)$$

for any x_i' and x_i''; otherwise it is nondegenerate. For example,

$$r(x_1, x_2, x_3, x_4, x_5) = (x_1 + x_3)/x_4$$

is a 5-dimensional function. The variables x_1, x_3, x_4 are nondegenerate, but x_2 and x_5 are degenerate.

Definition 2.6. *Let $\mathcal{E}$ be a task $h(x, \xi) \leq 0$. The support of the task $\mathcal{E}$, denoted by $\mathcal{E}^*$, is defined as the set consisting of all nondegenerate decision variables of $h(x, \xi)$.*

Definition 2.7. *The jth constraint $g_j(x, \xi) \leq 0$ is called an active constraint of task $\mathcal{E}$ if the set of nondegenerate decision variables of $g_j(x, \xi)$ and the support $\mathcal{E}^*$ have nonempty intersection; otherwise it is inactive.*

Definition 2.8. *Let $\mathcal{E}$ be a task $h(x, \xi) \leq 0$ in the uncertain environment $g_j(x, \xi) \leq 0$, $j = 1, 2, \cdots, p$. The dependent support of task $\mathcal{E}$, denoted by $\mathcal{E}^{**}$, is defined as the set consisting of all nondegenerate decision variables of $h(x, \xi)$ and $g_j(x, \xi)$ in the active constraints of task $\mathcal{E}$.*

Remark 2.1: It is obvious that $\mathcal{E}^* \subset \mathcal{E}^{**}$ holds.

Definition 2.9. *The jth constraint $g_j(x, \xi) \leq 0$ is called a dependent constraint of task $\mathcal{E}$ if the set of nondegenerate decision variables of $g_j(x, \xi)$ and the dependent support $\mathcal{E}^{**}$ have nonempty intersection; otherwise it is independent.*

Remark 2.2: An active constraint must be a dependent constraint.

Definition 2.10. *Let $\mathcal{E}$ be a task $h(x, \xi) \leq 0$ in the uncertain environment $g_j(x, \xi) \leq 0$, $j = 1, 2, \cdots, p$. For each decision x and realization ξ, the task $\mathcal{E}$ is said to be consistent in the uncertain environment if the following two conditions hold: (i) $h(x, \xi) \leq 0$; and (ii) $g_j(x, \xi) \leq 0$, $j \in J$, where J is the index set of all dependent constraints.*

In order to maximize the chance of some task in an uncertain environment, a dependent-chance programming may be formulated as follows,

$$
\begin{cases}
\max \mathcal{M}\left\{h(\boldsymbol{x}, \boldsymbol{\xi}) \leq 0\right\} \\
\text{subject to:} \\
\quad g_j(\boldsymbol{x}, \boldsymbol{\xi}) \leq 0, \quad j = 1, 2, \cdots, p
\end{cases}
\tag{2.33}
$$

where $\boldsymbol{x}$ is an n-dimensional decision vector, $\boldsymbol{\xi}$ is an uncertain vector, the task $\mathcal{E}$ is characterized by $h(\boldsymbol{x}, \boldsymbol{\xi}) \leq 0$, and the uncertain environment is described by the uncertain constraints $g_j(\boldsymbol{x}, \boldsymbol{\xi}) \leq 0$, $j = 1, 2, \cdots, p$. The model (2.33) is equivalent to

$$
\max \mathcal{M}\left\{h(\boldsymbol{x}, \boldsymbol{\xi}) \leq 0, g_j(\boldsymbol{x}, \boldsymbol{\xi}) \leq 0, j \in J\right\}
\tag{2.34}
$$

where J is the index set of all dependent constraints.

If there are multiple tasks in an uncertain environment, then we have the following dependent-chance multiobjective programming,

$$
\begin{cases}
\max \left[\mathcal{M}\{h_1(\boldsymbol{x}, \boldsymbol{\xi}) \leq 0\}, \cdots, \mathcal{M}\{h_m(\boldsymbol{x}, \boldsymbol{\xi}) \leq 0\}\right] \\
\text{subject to:} \\
\quad g_j(\boldsymbol{x}, \boldsymbol{\xi}) \leq 0, \quad j = 1, 2, \cdots, p
\end{cases}
\tag{2.35}
$$

where tasks $\mathcal{E}_i$ are characterized by $h_i(\boldsymbol{x}, \boldsymbol{\xi}) \leq 0$, $i = 1, 2, \cdots, m$, respectively. The model (2.35) is equivalent to

$$
\begin{cases}
\max \mathcal{M}\left\{h_1(\boldsymbol{x}, \boldsymbol{\xi}) \leq 0, g_j(\boldsymbol{x}, \boldsymbol{\xi}) \leq 0, j \in J_1\right\} \\
\max \mathcal{M}\left\{h_2(\boldsymbol{x}, \boldsymbol{\xi}) \leq 0, g_j(\boldsymbol{x}, \boldsymbol{\xi}) \leq 0, j \in J_2\right\} \\
\quad \cdots \\
\max \mathcal{M}\left\{h_m(\boldsymbol{x}, \boldsymbol{\xi}) \leq 0, g_j(\boldsymbol{x}, \boldsymbol{\xi}) \leq 0, j \in J_m\right\}
\end{cases}
\tag{2.36}
$$

where J_i are the index sets of all dependent constraints of tasks $\mathcal{E}_i$, $i = 1, 2, \cdots, m$, respectively.

Dependent-chance goal programming is employed to formulate uncertain decision systems according to the priority structure and target levels set by the decision-maker,

$$
\begin{cases}
\min \sum\limits_{j=1}^{l} P_j \sum\limits_{i=1}^{m} (u_{ij} d_i^+ \vee 0 + v_{ij} d_i^- \vee 0) \\
\text{subject to:} \\
\quad \mathcal{M}\left\{h_i(\boldsymbol{x}, \boldsymbol{\xi}) \leq 0\right\} - b_i = d_i^+, \quad i = 1, 2, \cdots, m \\
\quad b_i - \mathcal{M}\left\{h_i(\boldsymbol{x}, \boldsymbol{\xi}) \leq 0\right\} = d_i^-, \quad i = 1, 2, \cdots, m \\
\quad g_j(\boldsymbol{x}, \boldsymbol{\xi}) \leq 0, \qquad\qquad\qquad j = 1, 2, \cdots, p
\end{cases}
$$

where P_j is the preemptive priority factor which expresses the relative importance of various goals, $P_j \gg P_{j+1}$, for all j, u_{ij} is the weighting factor corresponding to positive deviation for goal i with priority j assigned, v_{ij} is the weighting factor corresponding to negative deviation for goal i with priority j assigned, $d_i^+ \vee 0$ is the positive deviation from the target of goal i, $d_i^- \vee 0$ is the negative deviation from the target of goal i, g_j is a function in system constraints, b_i is the target value according to goal i, l is the number of priorities, m is the number of goal constraints, and p is the number of system constraints.

Theorem 2.8. *Assume $x_1, x_2, \cdots, x_n$ are nonnegative decision variables, and $\xi_1, \xi_2, \cdots, \xi_n$ are independently linear uncertain variables $\mathcal{L}(a_1, b_1)$, $\mathcal{L}(a_2, b_2)$, $\cdots, \mathcal{L}(a_n, b_n)$, respectively. When*

$$t \in \left[\sum_{i=1}^{n} a_i x_i, \sum_{i=1}^{n} b_i x_i \right], \tag{2.37}$$

we have

$$\mathcal{M} \left\{ \sum_{i=1}^{n} \xi_i x_i \leq t \right\} = \frac{t - \displaystyle\sum_{i=1}^{n} a_i x_i}{\displaystyle\sum_{i=1}^{n} (b_i - a_i) x_i}. \tag{2.38}$$

Otherwise, the measure will be 0 if t is on the left-hand side of interval (2.37) or 1 if t is on the right-hand side.

Proof: Since $\xi_1, \xi_2, \cdots, \xi_n$ are independently linear uncertain variables, their weighted sum $\xi_1 x_1 + \xi_2 x_2 + \cdots + \xi_n x_n$ is also a linear uncertain variable

$$\mathcal{L} \left(\sum_{i=1}^{n} a_i x_i, \sum_{i=1}^{n} b_i x_i \right).$$

From this fact we may derive the result immediately.

Theorem 2.9. *Assume that $x_1, x_2, \cdots, x_n$ are nonnegative decision variables, and $\xi_1, \xi_2, \cdots, \xi_n$ are independently zigzag uncertain variables $\mathcal{Z}(a_1, b_1, c_1)$, $\mathcal{Z}(a_2, b_2, c_2), \cdots, \mathcal{Z}(a_n, b_n, c_n)$, respectively. When*

$$t \in \left[\sum_{i=1}^{n} a_i x_i, \sum_{i=1}^{n} b_i x_i \right], \tag{2.39}$$

we have

$$\mathcal{M} \left\{ \sum_{i=1}^{n} \xi_i x_i \leq t \right\} = \frac{t - \displaystyle\sum_{i=1}^{n} a_i x_i}{2 \displaystyle\sum_{i=1}^{n} (b_i - a_i) x_i}. \tag{2.40}$$

When

$$t \in \left[\sum_{i=1}^{n} b_i x_i, \sum_{i=1}^{n} c_i x_i \right], \tag{2.41}$$

we have

$$\mathcal{M} \left\{ \sum_{i=1}^{n} \xi_i x_i \leq t \right\} = \frac{t + \sum_{i=1}^{n} (c_i - 2b_i) x_i}{2 \sum_{i=1}^{n} (c_i - b_i) x_i}. \tag{2.42}$$

Otherwise, the measure will be 0 if t is on the left-hand side of interval (2.39) or 1 if t is on the right-hand side of interval (2.41).

Proof: Since $\xi_1, \xi_2, \cdots, \xi_n$ are independently zigzag uncertain variables, their weighted sum $\xi_1 x_1 + \xi_2 x_2 + \cdots + \xi_n x_n$ is also a zigzag uncertain variable

$$\mathcal{Z} \left(\sum_{i=1}^{n} a_i x_i, \sum_{i=1}^{n} b_i x_i, \sum_{i=1}^{n} c_i x_i \right).$$

From this fact we may derive the result immediately.

Theorem 2.10. *Assume* $x_1, x_2, \cdots, x_n$ *are nonnegative decision variables, and* $\xi_1, \xi_2, \cdots, \xi_n$ *are independently normal uncertain variables* $\mathcal{N}(e_1, \sigma_1)$, $\mathcal{N}(e_2, \sigma_2), \cdots, \mathcal{N}(e_n, \sigma_n)$, *respectively. Then*

$$\mathcal{M} \left\{ \sum_{i=1}^{n} \xi_i x_i \leq t \right\} = \left(1 + \exp \left(\frac{\pi \left(\sum_{i=1}^{n} e_i x_i - t \right)}{\sqrt{3} \sum_{i=1}^{n} \sigma_i x_i} \right) \right)^{-1}. \tag{2.43}$$

Proof: Since $\xi_1, \xi_2, \cdots, \xi_n$ are independently normal uncertain variables, their weighted sum $\xi_1 x_1 + \xi_2 x_2 + \cdots + \xi_n x_n$ is also a normal uncertain variable

$$\mathcal{N} \left(\sum_{i=1}^{n} e_i x_i, \sum_{i=1}^{n} \sigma_i x_i \right).$$

From this fact we may derive the result immediately.

Theorem 2.11. *Assume* $x_1, x_2, \cdots, x_n$ *are nonnegative decision variables,* $\xi_1, \xi_2, \cdots, \xi_n$ *are independently lognormal uncertain variables* $\mathcal{LOGN}(e_1, \sigma_1)$, $\mathcal{LOGN}(e_2, \sigma_2), \cdots, \mathcal{LOGN}(e_n, \sigma_n)$, *respectively. Then*

$$\mathcal{M} \left\{ \sum_{i=1}^{n} \xi_i x_i \leq t \right\} = \Psi(t) \tag{2.44}$$

where Ψ is determined by

$$\Psi^{-1}(\alpha) = \sum_{i=1}^{n} \exp(e_i) \left(\frac{\alpha}{1-\alpha} \right)^{\sqrt{3}\sigma_i/\pi} x_i. \qquad (2.45)$$

Proof: Since $\xi_1, \xi_2, \cdots, \xi_n$ are independently lognormal uncertain variables, the uncertainty distribution Ψ of $\xi_1 x_1 + \xi_2 x_2 + \cdots + \xi_n x_n$ is just determined by (2.45). From this fact we may derive the result immediately.

2.5   Uncertain Dynamic Programming

In order to model uncertain decision processes, Liu [122] proposed a general framework of uncertain dynamic programming, including expected value dynamic programming, chance-constrained dynamic programming as well as dependent-chance dynamic programming.

Expected Value Dynamic Programming

Consider an N-stage decision system in which $(a_1, a_2, \cdots, a_N)$ represents the state vector, $(x_1, x_2, \cdots, x_N)$ the decision vector, $(\xi_1, \xi_2, \cdots, \xi_N)$ the uncertain vector. We also assume that the state transition function is

$$a_{n+1} = T(a_n, x_n, \xi_n), \quad n = 1, 2, \cdots, N-1. \qquad (2.46)$$

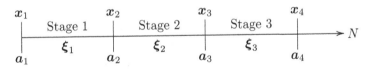

Figure 2.1: A Multistage Decision System

In order to maximize the expected return over the horizon, we may use the following expected value dynamic programming,

$$\begin{cases} f_N(a) = \max_{E[g_N(a,x,\xi_N)] \le 0} E[r_N(a, x, \xi_N)] \\ f_n(a) = \max_{E[g_n(a,x,\xi_n)] \le 0} E[r_n(a, x, \xi_n) + f_{n+1}(T(a, x, \xi_n))] \\ \quad n \le N-1 \end{cases} \qquad (2.47)$$

where r_n are the return functions at the nth stages, $n = 1, 2, \cdots, N$, respectively.

Chance-Constrained Dynamic Programming

In order to maximize the optimistic return over the horizon, we may use the following chance-constrained dynamic programming,

$$\begin{cases} f_N(\boldsymbol{a}) = \max\limits_{\mathcal{M}\{g_N(\boldsymbol{a},\boldsymbol{x},\boldsymbol{\xi}_N)\le 0\}\ge\alpha} \overline{r}_N(\boldsymbol{a},\boldsymbol{x},\boldsymbol{\xi}_N) \\ f_n(\boldsymbol{a}) = \max\limits_{\mathcal{M}\{g_n(\boldsymbol{a},\boldsymbol{x},\boldsymbol{\xi}_n)\le 0\}\ge\alpha} \overline{r}_n(\boldsymbol{a},\boldsymbol{x},\boldsymbol{\xi}_n) + f_{n+1}(T(\boldsymbol{a},\boldsymbol{x},\boldsymbol{\xi}_n)) \\ n \le N-1 \end{cases} \quad (2.48)$$

where the functions $\overline{r}_n$ are defined by

$$\overline{r}_n(\boldsymbol{a},\boldsymbol{x},\boldsymbol{\xi}_n) = \sup\left\{\overline{r} \mid \mathcal{M}\{r_n(\boldsymbol{a},\boldsymbol{x},\boldsymbol{\xi}_n) \ge \overline{r}\} \ge \beta\right\} \quad (2.49)$$

for $n = 1, 2, \cdots, N$. If we want to maximize the pessimistic return over the horizon, then we must define the functions $\overline{r}_n$ as

$$\overline{r}_n(\boldsymbol{a},\boldsymbol{x},\boldsymbol{\xi}_n) = \inf\left\{\overline{r} \mid \mathcal{M}\{r_n(\boldsymbol{a},\boldsymbol{x},\boldsymbol{\xi}_n) \le \overline{r}\} \ge \beta\right\} \quad (2.50)$$

for $n = 1, 2, \cdots, N$.

Dependent-Chance Dynamic Programming

In order to maximize the chance over the horizon, we may employ the following dependent-chance dynamic programming,

$$\begin{cases} f_N(\boldsymbol{a}) = \max\limits_{g_N(\boldsymbol{a},\boldsymbol{x},\boldsymbol{\xi}_N)\le 0} \mathcal{M}\{h_N(\boldsymbol{a},\boldsymbol{x},\boldsymbol{\xi}_N) \le 0\} \\ f_n(\boldsymbol{a}) = \max\limits_{g_n(\boldsymbol{a},\boldsymbol{x},\boldsymbol{\xi}_n)\le 0} \mathcal{M}\{h_n(\boldsymbol{a},\boldsymbol{x},\boldsymbol{\xi}_n) \le 0\} + f_{n+1}(T(\boldsymbol{a},\boldsymbol{x},\boldsymbol{\xi}_n)) \\ n \le N-1 \end{cases}$$

where $h_n(\boldsymbol{a},\boldsymbol{x},\boldsymbol{\xi}_n) \le 0$ are the events, and $g_n(\boldsymbol{a},\boldsymbol{x},\boldsymbol{\xi}_n) \le 0$ are the uncertain environments at the nth stages, $n = 1, 2, \cdots, N$, respectively.

2.6   Uncertain Multilevel Programming

In order to model uncertain decentralized decision systems, Liu [122] presented three types of uncertain multilevel programming, including expected value multilevel programming, chance-constrained multilevel programming and dependent-chance multilevel programming, and provided the concept of Stackelberg-Nash equilibrium to uncertain multilevel programming.

Expected Value Multilevel Programming

Assume that in a decentralized two-level decision system there is one leader and m followers. Let $\boldsymbol{x}$ and $\boldsymbol{y}_i$ be the control vectors of the leader and the ith followers, $i = 1, 2, \cdots, m$, respectively. We also assume that the objective functions of the leader and ith followers are $F(\boldsymbol{x}, \boldsymbol{y}_1, \cdots, \boldsymbol{y}_m, \boldsymbol{\xi})$ and $f_i(\boldsymbol{x}, \boldsymbol{y}_1, \cdots, \boldsymbol{y}_m, \boldsymbol{\xi})$, $i = 1, 2, \cdots, m$, respectively, where $\boldsymbol{\xi}$ is an uncertain vector.

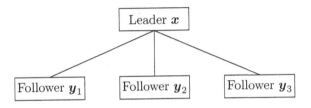

Figure 2.2: A Decentralized Decision System

Let the feasible set of control vector x of the leader be defined by the expected constraint

$$E[G(x, \xi)] \leq 0 \qquad (2.51)$$

where G is a vector-valued function and 0 is a zero vector. Then for each decision x chosen by the leader, the feasibility of control vectors y_i of the ith followers should be dependent on not only x but also $y_1, \cdots, y_{i-1}, y_{i+1}, \cdots, y_m$, and generally represented by the expected constraints,

$$E[g_i(x, y_1, y_2, \cdots, y_m, \xi)] \leq 0 \qquad (2.52)$$

where g_i are vector-valued functions, $i = 1, 2, \cdots, m$, respectively.

Assume that the leader first chooses his control vector x, and the followers determine their control array $(y_1, y_2, \cdots, y_m)$ after that. In order to maximize the expected objective of the leader, we have the following expected value bilevel programming,

$$\begin{cases} \max_{x} E[F(x, y_1^*, y_2^*, \cdots, y_m^*, \xi)] \\ \text{subject to:} \\ \quad E[G(x, \xi)] \leq 0 \\ \quad (y_1^*, y_2^*, \cdots, y_m^*) \text{ solves problems } (i = 1, 2, \cdots, m) \qquad (2.53) \\ \quad \begin{cases} \max_{y_i} E[f_i(x, y_1, y_2, \cdots, y_m, \xi)] \\ \text{subject to:} \\ \quad E[g_i(x, y_1, y_2, \cdots, y_m, \xi)] \leq 0. \end{cases} \end{cases}$$

Definition 2.11. *Let x be a feasible control vector of the leader. A Nash equilibrium of followers is the feasible array $(y_1^*, y_2^*, \cdots, y_m^*)$ with respect to x if*

$$E[f_i(x, y_1^*, \cdots, y_{i-1}^*, y_i, y_{i+1}^*, \cdots, y_m^*, \xi)] \qquad (2.54)$$
$$\leq E[f_i(x, y_1^*, \cdots, y_{i-1}^*, y_i^*, y_{i+1}^*, \cdots, y_m^*, \xi)]$$

for any feasible array $(y_1^*, \cdots, y_{i-1}^*, y_i, y_{i+1}^*, \cdots, y_m^*)$ and $i = 1, 2, \cdots, m$.

Definition 2.12. *Suppose that x^* is a feasible control vector of the leader and $(y_1^*, y_2^*, \cdots, y_m^*)$ is a Nash equilibrium of followers with respect to x^*.*

We call the array $(x^*, y_1^*, y_2^*, \cdots, y_m^*)$ a Stackelberg-Nash equilibrium to the expected value bilevel programming (2.53) if

$$E[F(\overline{x}, \overline{y}_1, \overline{y}_2, \cdots, \overline{y}_m, \xi)] \leq E[F(x^*, y_1^*, y_2^*, \cdots, y_m^*, \xi)] \qquad (2.55)$$

for any feasible control vector $\overline{x}$ and the Nash equilibrium $(\overline{y}_1, \overline{y}_2, \cdots, \overline{y}_m)$ with respect to $\overline{x}$.

Chance-Constrained Multilevel Programming

In order to maximize the optimistic return subject to the chance constraint, we may use the following chance-constrained bilevel programming,

$$
\begin{cases}
\max\limits_{x} \max\limits_{\overline{F}} \overline{F} \\
\text{subject to:} \\
\quad \mathcal{M}\{F(x, y_1^*, y_2^*, \cdots, y_m^*, \xi) \geq \overline{F}\} \geq \beta \\
\quad \mathcal{M}\{G(x, \xi) \leq 0\} \geq \alpha \\
\quad (y_1^*, y_2^*, \cdots, y_m^*) \text{ solves problems } (i = 1, 2, \cdots, m) \\
\quad \begin{cases}
\max\limits_{y_i} \max\limits_{\overline{f}_i} \overline{f}_i \\
\text{subject to:} \\
\quad \mathcal{M}\{f_i(x, y_1, y_2, \cdots, y_m, \xi) \geq \overline{f}_i\} \geq \beta_i \\
\quad \mathcal{M}\{g_i(x, y_1, y_2, \cdots, y_m, \xi) \leq 0\} \geq \alpha_i
\end{cases}
\end{cases} \qquad (2.56)
$$

where $\alpha, \beta, \alpha_i, \beta_i, i = 1, 2, \cdots, m$ are predetermined confidence levels.

Definition 2.13. Let x be a feasible control vector of the leader. A Nash equilibrium of followers is the feasible array $(y_1^*, y_2^*, \cdots, y_m^*)$ with respect to x if

$$
\begin{aligned}
\overline{f}_i(x, y_1^*, &\cdots, y_{i-1}^*, y_i, y_{i+1}^*, \cdots, y_m^*) \\
&\leq \overline{f}_i(x, y_1^*, \cdots, y_{i-1}^*, y_i^*, y_{i+1}^*, \cdots, y_m^*)
\end{aligned} \qquad (2.57)
$$

for any feasible array $(y_1^*, \cdots, y_{i-1}^*, y_i, y_{i+1}^*, \cdots, y_m^*)$ and $i = 1, 2, \cdots, m$.

Definition 2.14. Suppose that x^* is a feasible control vector of the leader and $(y_1^*, y_2^*, \cdots, y_m^*)$ is a Nash equilibrium of followers with respect to x^*. The array $(x^*, y_1^*, y_2^*, \cdots, y_m^*)$ is called a Stackelberg-Nash equilibrium to the chance-constrained bilevel programming (2.56) if

$$\overline{F}(\overline{x}, \overline{y}_1, \overline{y}_2, \cdots, \overline{y}_m) \leq \overline{F}(x^*, y_1^*, y_2^*, \cdots, y_m^*) \qquad (2.58)$$

for any feasible control vector $\overline{x}$ and the Nash equilibrium $(\overline{y}_1, \overline{y}_2, \cdots, \overline{y}_m)$ with respect to $\overline{x}$.

In order to maximize the pessimistic return, we have the following minimax chance-constrained bilevel programming,

$$
\begin{cases}
\max\limits_{\boldsymbol{x}} \min\limits_{\overline{F}} \overline{F} \\
\text{subject to:} \\
\quad \mathcal{M}\{F(\boldsymbol{x}, \boldsymbol{y}_1^*, \boldsymbol{y}_2^*, \cdots, \boldsymbol{y}_m^*, \boldsymbol{\xi}) \leq \overline{F}\} \geq \beta \\
\quad \mathcal{M}\{G(\boldsymbol{x}, \boldsymbol{\xi}) \leq 0\} \geq \alpha \\
\quad (\boldsymbol{y}_1^*, \boldsymbol{y}_2^*, \cdots, \boldsymbol{y}_m^*) \text{ solves problems } (i = 1, 2, \cdots, m) \\
\quad \begin{cases}
\max\limits_{\boldsymbol{y}_i} \min\limits_{\overline{f}_i} \overline{f}_i \\
\text{subject to:} \\
\quad \mathcal{M}\{f_i(\boldsymbol{x}, \boldsymbol{y}_1, \boldsymbol{y}_2, \cdots, \boldsymbol{y}_m, \boldsymbol{\xi}) \leq \overline{f}_i\} \geq \beta_i \\
\quad \mathcal{M}\{g_i(\boldsymbol{x}, \boldsymbol{y}_1, \boldsymbol{y}_2, \cdots, \boldsymbol{y}_m, \boldsymbol{\xi}) \leq 0\} \geq \alpha_i
\end{cases}
\end{cases} \tag{2.59}
$$

where $\alpha, \beta, \alpha_i, \beta_i$, $i = 1, 2, \cdots, m$ are predetermined confidence levels.

Dependent-Chance Multilevel Programming

Let $H(\boldsymbol{x}, \boldsymbol{y}_1, \boldsymbol{y}_2, \cdots, \boldsymbol{y}_m, \boldsymbol{\xi}) \leq 0$ and $h_i(\boldsymbol{x}, \boldsymbol{y}_1, \boldsymbol{y}_2, \cdots, \boldsymbol{y}_m, \boldsymbol{\xi}) \leq 0$ be the tasks of the leader and ith followers, $i = 1, 2, \cdots, m$, respectively. In order to maximize the chance functions of the leader and followers, we have the following dependent-chance bilevel programming,

$$
\begin{cases}
\max\limits_{\boldsymbol{x}} \mathcal{M}\{H(\boldsymbol{x}, \boldsymbol{y}_1^*, \boldsymbol{y}_2^*, \cdots, \boldsymbol{y}_m^*, \boldsymbol{\xi}) \leq 0\} \\
\text{subject to:} \\
\quad G(\boldsymbol{x}, \boldsymbol{\xi}) \leq 0 \\
\quad (\boldsymbol{y}_1^*, \boldsymbol{y}_2^*, \cdots, \boldsymbol{y}_m^*) \text{ solves problems } (i = 1, 2, \cdots, m) \\
\quad \begin{cases}
\max\limits_{\boldsymbol{y}_i} \mathcal{M}\{h_i(\boldsymbol{x}, \boldsymbol{y}_1, \boldsymbol{y}_2, \cdots, \boldsymbol{y}_m, \boldsymbol{\xi}) \leq 0\} \\
\text{subject to:} \\
\quad g_i(\boldsymbol{x}, \boldsymbol{y}_1, \boldsymbol{y}_2, \cdots, \boldsymbol{y}_m, \boldsymbol{\xi}) \leq 0.
\end{cases}
\end{cases} \tag{2.60}
$$

Definition 2.15. *Let $\boldsymbol{x}$ be a control vector of the leader. We call the array $(\boldsymbol{y}_1^*, \boldsymbol{y}_2^*, \cdots, \boldsymbol{y}_m^*)$ a Nash equilibrium of followers with respect to $\boldsymbol{x}$ if*

$$
\begin{aligned}
&\mathcal{M}\{h_i(\boldsymbol{x}, \boldsymbol{y}_1^*, \cdots, \boldsymbol{y}_{i-1}^*, \boldsymbol{y}_i, \boldsymbol{y}_{i+1}^*, \cdots, \boldsymbol{y}_m^*, \boldsymbol{\xi}) \leq 0\} \\
&\leq \mathcal{M}\{h_i(\boldsymbol{x}, \boldsymbol{y}_1^*, \cdots, \boldsymbol{y}_{i-1}^*, \boldsymbol{y}_i^*, \boldsymbol{y}_{i+1}^*, \cdots, \boldsymbol{y}_m^*, \boldsymbol{\xi}) \leq 0\}
\end{aligned} \tag{2.61}
$$

subject to the uncertain environment $g_i(\boldsymbol{x}, \boldsymbol{y}_1, \boldsymbol{y}_2, \cdots, \boldsymbol{y}_m, \boldsymbol{\xi}) \leq 0, i = 1, 2, \cdots, m$ for any array $(\boldsymbol{y}_1^*, \cdots, \boldsymbol{y}_{i-1}^*, \boldsymbol{y}_i, \boldsymbol{y}_{i+1}^*, \cdots, \boldsymbol{y}_m^*)$ and $i = 1, 2, \cdots, m$.

Definition 2.16. *Let $\boldsymbol{x}^*$ be a control vector of the leader, and $(\boldsymbol{y}_1^*, \boldsymbol{y}_2^*, \cdots, \boldsymbol{y}_m^*)$ a Nash equilibrium of followers with respect to $\boldsymbol{x}^*$. Then $(\boldsymbol{x}^*, \boldsymbol{y}_1^*, \boldsymbol{y}_2^*, \cdots, \boldsymbol{y}_m^*)$ is called a Stackelberg-Nash equilibrium to the dependent-chance bilevel programming (2.60) if*

$$\mathcal{M}\{H(\overline{\boldsymbol{x}},\overline{\boldsymbol{y}}_1,\overline{\boldsymbol{y}}_2,\cdots,\overline{\boldsymbol{y}}_m,\boldsymbol{\xi})\leq 0\}\leq \mathcal{M}\{H(\boldsymbol{x}^*,\boldsymbol{y}_1^*,\boldsymbol{y}_2^*,\cdots,\boldsymbol{y}_m^*,\boldsymbol{\xi})\leq 0\}$$

subject to the uncertain environment $G(\boldsymbol{x},\boldsymbol{\xi})\leq 0$ *for any control vector* $\overline{\boldsymbol{x}}$ *and the Nash equilibrium* $(\overline{\boldsymbol{y}}_1,\overline{\boldsymbol{y}}_2,\cdots,\overline{\boldsymbol{y}}_m)$ *with respect to* $\overline{\boldsymbol{x}}$.

2.7   Hybrid Intelligent Algorithm

From the mathematical viewpoint, there is no difference between determin-istic mathematical programming and uncertain programming except for the fact that there exist uncertain functions in the latter. Essentially, there are three types of uncertain functions in uncertain programming,

$$\begin{aligned}
U_1 &: \boldsymbol{x} \to E[f(\boldsymbol{x},\boldsymbol{\xi})], \\
U_2 &: \boldsymbol{x} \to \mathcal{M}\{f(\boldsymbol{x},\boldsymbol{\xi})\leq 0\}, \\
U_3 &: \boldsymbol{x} \to \max\{\overline{f}\mid \mathcal{M}\{f(\boldsymbol{x},\boldsymbol{\xi})\geq \overline{f}\}\geq \alpha\}.
\end{aligned} \qquad (2.62)$$

Note that those uncertain functions may be calculated by the 99-method if the function f is monotone. Otherwise, I give up! It is fortunate for us that almost all functions in practical problems are indeed monotone.

In order to solve uncertain programming models, we must find a numerical method for solving deterministic mathematical programming, for example, genetic algorithm, particle swarm optimization, neural networks, tabu search, or any classical algorithms.

Then, for example, we may integrate the 99-method and the genetic al-gorithm to produce a hybrid intelligent algorithm for solving uncertain pro-gramming models:

Step 1. Initialize chromosomes whose feasibility may be checked by the 99-method.

Step 2. Update the chromosomes by the crossover operation in which the 99-method may be employed to check the feasibility of offsprings.

Step 3. Update the chromosomes by the mutation operation in which the 99-method may be employed to check the feasibility of offsprings.

Step 4. Calculate the objective values for all chromosomes by the 99-method.

Step 5. Compute the fitness of each chromosome based on the objective values.

Step 6. Select the chromosomes by spinning the roulette wheel.

Step 7. Repeat the second to sixth steps a given number of cycles.

Step 8. Report the best chromosome as the optimal solution.

Please visit the website at http://orsc.edu.cn/liu/resources.htm for com-puter source files of hybrid intelligent algorithm and numerical examples.

2.8   Ψ Graph

Any types of uncertain programming (including stochastic programming, fuzzy programming and hybrid programming) may be represented by a Ψ graph

(**Philosophy, Structure, Information**)

which is essentially a coordinate system in which, for example, the plane

"**P** = CCP"

represents the class of chance-constrained programming; the plane

"**S** = MOP"

represents the class of multiobjective programming; the plane

"**I** = Uncertain"

represents the class of uncertain programming; and the point

"(**P,S,I**) = (DCP, GP, Uncertain)"

represents the uncertain dependent-chance goal programming.

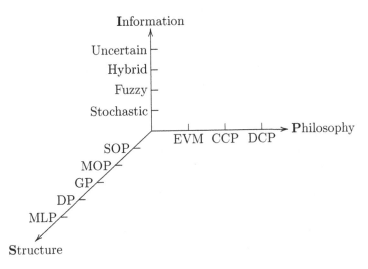

Figure 2.3: Ψ Graph for Uncertain Programming Classifications (Liu [112])

2.9   Project Scheduling Problem

Project scheduling problem is to determine the schedule of allocating re-
sources so as to balance the total cost and the completion time. The study
of project scheduling problem with uncertain factors was started by Liu [122]
in 2009. This section presents an uncertain programming model for project
scheduling problem in which the duration times are assumed to be uncertain
variables with known uncertainty distributions.

Project scheduling is usually represented by a directed acyclic graph where
nodes correspond to milestones, and arcs to activities which are basically
characterized by the times and costs consumed.

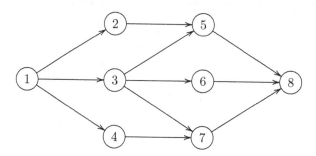

Figure 2.4: A Project with 8 Milestones and 11 Activities

Let $(\mathcal{V}, \mathcal{A})$ be a directed acyclic graph, where $\mathcal{V} = \{1, 2, \cdots, n, n+1\}$ is
the set of nodes, $\mathcal{A}$ is the set of arcs, $(i, j) \in \mathcal{A}$ is the arc of the graph $(\mathcal{V}, \mathcal{A})$
from nodes i to j. It is well-known that we can rearrange the indexes of the
nodes in $\mathcal{V}$ such that $i < j$ for all $(i, j) \in \mathcal{A}$.

Before we begin to study project scheduling problem with uncertain activity duration times, we first make some assumptions: (a) all of the costs
needed are obtained via loans with some given interest rate; and (b) each
activity can be processed only if the loan needed is allocated and all the
foregoing activities are finished.

In order to model the project scheduling problem, we introduce the following indices and parameters:

ξ_{ij}: uncertain duration time of activity (i, j) in $\mathcal{A}$;

Φ_{ij}: uncertainty distribution of ξ_{ij};

c_{ij}: cost of activity (i, j) in $\mathcal{A}$;

r: interest rate;

x_i: integer decision variable representing the allocating time of all loans
needed for all activities (i, j) in $\mathcal{A}$.

Starting Times

For simplicity, we write $\boldsymbol{\xi} = \{\xi_{ij} : (i, j) \in \mathcal{A}\}$ and $\boldsymbol{x} = (x_1, x_2, \cdots, x_n)$.
Assume each uncertain duration time ξ_{ij} is represented by a 99-table,

0.01	0.02	0.03	$\cdots$	0.99
t_{ij}^1	t_{ij}^2	t_{ij}^3	$\cdots$	t_{ij}^{99}

$$(2.63)$$

Let $T_i(\boldsymbol{x}, \boldsymbol{\xi})$ denote the starting time of all activities (i, j) in $\mathcal{A}$. According to the assumptions, the starting time of the total project (i.e., the starting time of of all activities $(1, j)$ in $\mathcal{A}$) should be

$$T_1(\boldsymbol{x}, \boldsymbol{\xi}) = x_1 \qquad (2.64)$$

whose inverse uncertainty distribution may be written as

$$\Psi_1^{-1}(\alpha) = x_1 \qquad (2.65)$$

and has a 99-table,

0.01	0.02	0.03	$\cdots$	0.99
x_1	x_1	x_1	$\cdots$	x_1

$$(2.66)$$

Generally, suppose that the starting time $T_k(\boldsymbol{x}, \boldsymbol{\xi})$ of all activities (k, i) in $\mathcal{A}$ has an inverse uncertainty distribution $\Psi_k^{-1}(\alpha)$ and has a 99-table,

0.01	0.02	0.03	$\cdots$	0.99
y_k^1	y_k^2	y_k^3	$\cdots$	y_k^{99}

$$(2.67)$$

Then the starting time $T_i(\boldsymbol{x}, \boldsymbol{\xi})$ of all activities (i, j) in $\mathcal{A}$ should be

$$T_i(\boldsymbol{x}, \boldsymbol{\xi}) = x_i \vee \max_{(k,i)\in\mathcal{A}} (T_k(\boldsymbol{x}, \boldsymbol{\xi}) + \xi_{ki}) \qquad (2.68)$$

whose inverse uncertainty distribution is

$$\Psi_i^{-1}(\alpha) = x_i \vee \max_{(k,i)\in\mathcal{A}} \left(\Psi_k^{-1}(\alpha) + \Phi_{ki}^{-1}(\alpha)\right) \qquad (2.69)$$

and has a 99-table,

0.01	$\cdots$	0.99
$x_i \vee \max\limits_{(k,i)\in\mathcal{A}} (y_k^1 + t_{ki}^1)$	$\cdots$	$x_i \vee \max\limits_{(k,i)\in\mathcal{A}} (y_k^{99} + t_{ki}^{99})$

$$(2.70)$$

where $y_k^1, y_k^2, \cdots, y_k^{99}$ are determined by (2.67). This recursive process may produce all starting times of activities.

Completion Time

The completion time $T(\boldsymbol{x}, \boldsymbol{\xi})$ of the total project (i.e, the finish time of all activities $(k, n + 1)$ in $\mathcal{A}$) is

$$T(\boldsymbol{x}, \boldsymbol{\xi}) = \max_{(k,n+1)\in\mathcal{A}} (T_k(\boldsymbol{x}, \boldsymbol{\xi}) + \xi_{k,n+1}) \qquad (2.71)$$

whose inverse uncertainty distribution is

$$\Psi^{-1}(\alpha) = \max_{(k,n+1)\in\mathcal{A}} \left(\Psi_k^{-1}(\alpha) + \Phi_{k,n+1}^{-1}(\alpha)\right) \qquad (2.72)$$

and has a 99-table,

0.01	$\cdots$	0.99
$\max\limits_{(k,n+1)\in\mathcal{A}}\left(y_k^1+t_{k,n+1}^1\right)$	$\cdots$	$\max\limits_{(k,n+1)\in\mathcal{A}}\left(y_k^{99}+t_{k,n+1}^{99}\right)$

$$(2.73)$$

where $y_k^1, y_k^2, \cdots, y_k^{99}$ are determined by (2.67).

Total Cost

Based on the completion time $T(\boldsymbol{x},\boldsymbol{\xi})$, the total cost of the project can be written as

$$C(\boldsymbol{x},\boldsymbol{\xi}) = \sum_{(i,j)\in\mathcal{A}} c_{ij}\left(1+r\right)^{\lceil T(\boldsymbol{x},\boldsymbol{\xi})-x_i\rceil} \qquad (2.74)$$

where $\lceil a \rceil$ represents the minimal integer greater than or equal to a. Note that $C(\boldsymbol{x},\boldsymbol{\xi})$ is a discrete uncertain variable whose inverse uncertainty distribution is

$$\Upsilon^{-1}(\boldsymbol{x};\alpha) = \sum_{(i,j)\in\mathcal{A}} c_{ij}\left(1+r\right)^{\lceil \Psi^{-1}(\boldsymbol{x};\alpha)-x_i\rceil} \qquad (2.75)$$

for $0 < \alpha < 1$. Since $T(\boldsymbol{x},\boldsymbol{\xi})$ is obtained by the recursive process and represented by a 99-table,

0.01	0.02	0.03	$\cdots$	0.99
s_1	s_2	s_3	$\cdots$	s_{99}

$$(2.76)$$

the total cost $C(\boldsymbol{x},\boldsymbol{\xi})$ has a 99-table,

0.01	$\cdots$	0.99
$\sum\limits_{(i,j)\in\mathcal{A}} c_{ij}\left(1+r\right)^{\lceil s_1-x_i\rceil}$	$\cdots$	$\sum\limits_{(i,j)\in\mathcal{A}} c_{ij}\left(1+r\right)^{\lceil s_{99}-x_i\rceil}$

$$(2.77)$$

Project Scheduling Model

If we want to minimize the expected cost of the project under the completion time constraint, we may construct the following project scheduling model,

$$\begin{cases} \min E[C(\boldsymbol{x},\boldsymbol{\xi})] \\ \text{subject to:} \\ \quad \mathcal{M}\{T(\boldsymbol{x},\boldsymbol{\xi}) \le T^0\} \ge \alpha \\ \quad \boldsymbol{x} \ge 0, \text{ integer vector} \end{cases} \qquad (2.78)$$

where T^0 is a due date of the project, α is a predetermined confidence level, $T(\boldsymbol{x},\boldsymbol{\xi})$ is the completion time defined by (2.71), and $C(\boldsymbol{x},\boldsymbol{\xi})$ is the total cost defined by (2.74). This model is equivalent to

$$\begin{cases} \min \displaystyle\int_0^{+\infty} (1 - \Upsilon(\boldsymbol{x};z))\mathrm{d}z \\ \text{subject to:} \\ \quad \Psi(\boldsymbol{x};T^0) \ge \alpha \\ \quad \boldsymbol{x} \ge 0, \text{ integer vector} \end{cases} \qquad (2.79)$$

where Ψ is determined by (2.72) and Υ is determined by (2.75). Note that the completion time $T(\boldsymbol{x}, \boldsymbol{\xi})$ and total cost $C(\boldsymbol{x}, \boldsymbol{\xi})$ are obtained by the recursive process and are respectively represented by 99-tables,

0.01	0.02	0.03	$\cdots$	0.99
s_1	s_2	s_3	$\cdots$	s_{99}

0.01	0.02	0.03	$\cdots$	0.99
c_1	c_2	c_3	$\cdots$	c_{99}

$$(2.80)$$

Thus the project scheduling model is simplified as follows,

$$\begin{cases} \min \ (c_1 + c_2 + \cdots + c_{99})/99 \\ \text{subject to:} \\ \quad k/100 \geq \alpha \ \text{if} \ s_k \geq T^0 \\ \quad \boldsymbol{x} \geq 0, \ \text{integer vector.} \end{cases} \qquad (2.81)$$

Numerical Experiment

Consider a project scheduling problem shown by Figure 2.4 in which there are 8 milestones and 11 activities. Assume that all duration times of activities are linear uncertain variables,

$$\xi_{ij} \sim \mathcal{L}(3i, 3j), \quad \forall (i, j) \in \mathcal{A}$$

and assume that the costs of activities are

$$c_{ij} = i + j, \quad \forall (i, j) \in \mathcal{A}.$$

In addition, we also suppose that the interest rate is $r = 0.02$, the due date is $T^0 = 60$, and the confidence level is $\alpha = 0.85$. In order to find an optimal project schedule, we integrate the 99-method and a genetic algorithm to produce a hybrid intelligent algorithm. A run of the computer program (http://orsc.edu.cn/liu/resources.htm) shows that the optimal allocating times of all loans needed for all activities are

Date	7	11	13	23	26	29
Node	1	4	3	2, 7	6	5
Loan	12	11	27	22	14	13

whose expected total cost is 166.8, and $\mathcal{M}\{T(\boldsymbol{x}^*, \boldsymbol{\xi}) \leq 60\} = 0.89$.

2.10 Vehicle Routing Problem

Vehicle routing problem (VRP) is concerned with finding efficient routes, beginning and ending at a central depot, for a fleet of vehicles to serve a number of customers.

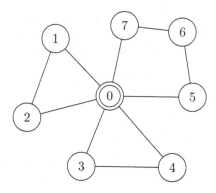

Figure 2.5: A Vehicle Routing Plan with Single Depot and 7 Customers

Due to its wide applicability and economic importance, vehicle routing problem has been extensively studied. Liu [122] first introduced uncertainty theory into the research area of vehicle routing problem in 2009. In this section, vehicle routing problem will be modelled by uncertain programming in which the travel times are assumed to be uncertain variables with known uncertainty distributions.

We assume that (a) a vehicle will be assigned for only one route on which there may be more than one customer; (b) a customer will be visited by one and only one vehicle; (c) each route begins and ends at the depot; and (d) each customer specifies its time window within which the delivery is permitted or preferred to start.

Let us first introduce the following indices and model parameters:

$i = 0$: depot;

$i = 1, 2, \cdots, n$: customers;

$k = 1, 2, \cdots, m$: vehicles;

D_{ij}: travel distance from customers i to j, $i, j = 0, 1, 2, \cdots, n$;

T_{ij}: uncertain travel time from customers i to j, $i, j = 0, 1, 2, \cdots, n$;

Φ_{ij}: uncertainty distribution of T_{ij}, $i, j = 0, 1, 2, \cdots, n$;

$[a_i, b_i]$: time window of customer i, $i = 1, 2, \cdots, n$.

Operational Plan

In this book, the operational plan is represented by the formulation (Liu [112]) via three decision vectors $\boldsymbol{x}$, $\boldsymbol{y}$ and $\boldsymbol{t}$, where

$\boldsymbol{x} = (x_1, x_2, \cdots, x_n)$: integer decision vector representing n customers with $1 \leq x_i \leq n$ and $x_i \neq x_j$ for all $i \neq j$, $i, j = 1, 2, \cdots, n$. That is, the sequence $\{x_1, x_2, \cdots, x_n\}$ is a rearrangement of $\{1, 2, \cdots, n\}$;

$\boldsymbol{y} = (y_1, y_2, \cdots, y_{m-1})$: integer decision vector with $y_0 \equiv 0 \leq y_1 \leq y_2 \leq \cdots \leq y_{m-1} \leq n \equiv y_m$;

$\boldsymbol{t} = (t_1, t_2, \cdots, t_m)$: each t_k represents the starting time of vehicle k at the depot, $k = 1, 2, \cdots, m$.

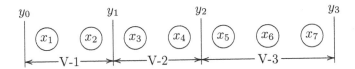

Figure 2.6: Formulation of Operational Plan in which Vehicle 1 Visits Customers x_1, x_2, Vehicle 2 Visits Customers x_3, x_4 and Vehicle 3 Visits Customers x_5, x_6, x_7.

We note that the operational plan is fully determined by the decision vectors $\boldsymbol{x}$, $\boldsymbol{y}$ and $\boldsymbol{t}$ in the following way. For each k ($1 \leq k \leq m$), if $y_k = y_{k-1}$, then vehicle k is not used; if $y_k > y_{k-1}$, then vehicle k is used and starts from the depot at time t_k, and the tour of vehicle k is $0 \rightarrow x_{y_{k-1}+1} \rightarrow x_{y_{k-1}+2} \rightarrow \cdots \rightarrow x_{y_k} \rightarrow 0$. Thus the tours of all vehicles are as follows:

$$\text{Vehicle 1: } 0 \rightarrow x_{y_0+1} \rightarrow x_{y_0+2} \rightarrow \cdots \rightarrow x_{y_1} \rightarrow 0;$$
$$\text{Vehicle 2: } 0 \rightarrow x_{y_1+1} \rightarrow x_{y_1+2} \rightarrow \cdots \rightarrow x_{y_2} \rightarrow 0;$$

$$\cdots$$

$$\text{Vehicle } m\text{: } 0 \rightarrow x_{y_{m-1}+1} \rightarrow x_{y_{m-1}+2} \rightarrow \cdots \rightarrow x_{y_m} \rightarrow 0.$$

It is clear that this type of representation is intuitive, and the total number of decision variables is $n + 2m - 1$. We also note that the above decision variables $\boldsymbol{x}$, $\boldsymbol{y}$ and $\boldsymbol{t}$ ensure that: (a) each vehicle will be used at most one time; (b) all tours begin and end at the depot; (c) each customer will be visited by one and only one vehicle; and (d) there is no subtour.

Arrival Times

Let $f_i(\boldsymbol{x}, \boldsymbol{y}, \boldsymbol{t})$ be the arrival time function of some vehicles at customers i for $i = 1, 2, \cdots, n$. We remind readers that $f_i(\boldsymbol{x}, \boldsymbol{y}, \boldsymbol{t})$ are determined by the decision variables $\boldsymbol{x}$, $\boldsymbol{y}$ and $\boldsymbol{t}$, $i = 1, 2, \cdots, n$. Since unloading can start either immediately, or later, when a vehicle arrives at a customer, the calculation of $f_i(\boldsymbol{x}, \boldsymbol{y}, \boldsymbol{t})$ is heavily dependent on the operational strategy. Here we assume that the customer does not permit a delivery earlier than the time window. That is, the vehicle will wait to unload until the beginning of the time window if it arrives before the time window. If a vehicle arrives at a customer after the beginning of the time window, unloading will start immediately. For each k with $1 \leq k \leq m$, if vehicle k is used (i.e., $y_k > y_{k-1}$), then we have

$$f_{x_{y_{k-1}+1}}(\boldsymbol{x}, \boldsymbol{y}, \boldsymbol{t}) = t_k + T_{0x_{y_{k-1}+1}} \tag{2.82}$$

and

$$f_{x_{y_{k-1}+j}}(\boldsymbol{x}, \boldsymbol{y}, \boldsymbol{t}) = f_{x_{y_{k-1}+j-1}}(\boldsymbol{x}, \boldsymbol{y}, \boldsymbol{t}) \vee a_{x_{y_{k-1}+j-1}} + T_{x_{y_{k-1}+j-1}x_{y_{k-1}+j}} \tag{2.83}$$

for $2 \leq j \leq y_k - y_{k-1}$. It follows from the uncertainty of travel times T_{ij}'s that the arrival times $f_i(\boldsymbol{x}, \boldsymbol{y}, \boldsymbol{t})$, $i = 1, 2, \cdots, n$ are uncertain variables fully determined by (2.82) and (2.83).

Assume that each travel time T_{ij} from customers i to j is represented by a 99-table,

0.01	0.02	0.03	$\cdots$	0.99
t_{ij}^1	t_{ij}^2	t_{ij}^3	$\cdots$	t_{ij}^{99}

(2.84)

If the vehicle k is used, i.e., $y_k > y_{k-1}$, then the arrival time $f_{x_{y_{k-1}+1}}(x, y, t)$ at the customer $x_{y_{k-1}+1}$ is an uncertain variable whose inverse uncertainty distribution is

$$\Psi_{x_{y_{k-1}+1}}^{-1}(\alpha) = t_k + \Phi_{0x_{y_{k-1}+1}}^{-1}(\alpha) \qquad (2.85)$$

and has a 99-table,

0.01	0.02	$\cdots$	0.99
$t_k + t_{0x_{y_{k-1}+1}}^1$	$t_k + t_{0x_{y_{k-1}+1}}^2$	$\cdots$	$t_k + t_{0x_{y_{k-1}+1}}^{99}$

(2.86)

Generally, suppose that the arrival time $f_{x_{y_{k-1}+j-1}}(x, y, t)$ has an inverse uncertainty distribution $\Psi_{x_{y_{k-1}+j-1}}^{-1}(\alpha)$, and has a 99-table,

0.01	0.02	$\cdots$	0.99
$s_{x_{y_{k-1}+j-1}}^1$	$s_{x_{y_{k-1}+j-1}}^2$	$\cdots$	$s_{x_{y_{k-1}+j-1}}^{99}$

(2.87)

Since the arrival time $f_{x_{y_{k-1}+j}}(x, y, t)$ at the customer $x_{y_{k-1}+j}$ has an inverse uncertainty distribution

$$\Psi_{x_{y_{k-1}+j}}^{-1}(\alpha) = \Psi_{x_{y_{k-1}+j-1}}^{-1}(\alpha) \vee a_{x_{y_{k-1}+j-1}} + \Phi_{x_{y_{k-1}+j-1}x_{y_{k-1}+j}}^{-1}(\alpha) \quad (2.88)$$

for $2 \le j \le y_k - y_{k-1}$, the arrival time $f_{x_{y_{k-1}+j}}(x, y, t)$ has a 99-table,

0.01	$\cdots$	0.99
$s_{x_{y_{k-1}+j-1}}^1 \vee a_{x_{y_{k-1}+j-1}}$ $+t_{x_{y_{k-1}+j-1}x_{y_{k-1}+j}}^1$	$\cdots$	$s_{x_{y_{k-1}+j-1}}^{99} \vee a_{x_{y_{k-1}+j-1}}$ $+t_{x_{y_{k-1}+j-1}x_{y_{k-1}+j}}^{99}$

(2.89)

where $s_{x_{y_{k-1}+j-1}}^1, s_{x_{y_{k-1}+j-1}}^2, \cdots, s_{x_{y_{k-1}+j-1}}^{99}$ are determined by (2.87). This recursive process may produce all arrival times at customers.

Travel Distance

Let $g(x, y)$ be the total travel distance of all vehicles. Then we have

$$g(x, y) = \sum_{k=1}^{m} g_k(x, y) \qquad (2.90)$$

where

$$g_k(x, y) = \begin{cases} D_{0x_{y_{k-1}+1}} + \sum_{j=y_{k-1}+1}^{y_k-1} D_{x_j x_{j+1}} + D_{x_{y_k}0}, & \text{if } y_k > y_{k-1} \\ 0, & \text{if } y_k = y_{k-1} \end{cases}$$

for $k = 1, 2, \cdots, m$.

Vehicle Routing Model

If we hope that each customer i ($1 \le i \le n$) is visited within its time window $[a_i, b_i]$ with confidence level α_i (i.e., the vehicle arrives at customer i before time b_i), then we have the following chance constraint,

$$\mathcal{M}\{f_i(\boldsymbol{x}, \boldsymbol{y}, \boldsymbol{t}) \le b_i\} \ge \alpha_i. \tag{2.91}$$

If we want to minimize the total travel distance of all vehicles subject to the time window constraint, then we have the following vehicle routing model,

$$\begin{cases} \min g(\boldsymbol{x}, \boldsymbol{y}) \\ \text{subject to:} \\ \quad \mathcal{M}\{f_i(\boldsymbol{x}, \boldsymbol{y}, \boldsymbol{t}) \le b_i\} \ge \alpha_i, \quad i = 1, 2, \cdots, n \\ \quad 1 \le x_i \le n, \quad i = 1, 2, \cdots, n \\ \quad x_i \ne x_j, \quad i \ne j, \ i, j = 1, 2, \cdots, n \\ \quad 0 \le y_1 \le y_2 \le \cdots \le y_{m-1} \le n \\ \quad x_i, y_j, \quad i = 1, 2, \cdots, n, \quad j = 1, 2, \cdots, m-1, \quad \text{integers} \end{cases} \tag{2.92}$$

which is equivalent to

$$\begin{cases} \min g(\boldsymbol{x}, \boldsymbol{y}) \\ \text{subject to:} \\ \quad \Psi_i(\boldsymbol{x}, \boldsymbol{y}, \boldsymbol{t}; b_i) \ge \alpha_i, \quad i = 1, 2, \cdots, n \\ \quad 1 \le x_i \le n, \quad i = 1, 2, \cdots, n \\ \quad x_i \ne x_j, \quad i \ne j, \ i, j = 1, 2, \cdots, n \\ \quad 0 \le y_1 \le y_2 \le \cdots \le y_{m-1} \le n \\ \quad x_i, y_j, \quad i = 1, 2, \cdots, n, \quad j = 1, 2, \cdots, m-1, \quad \text{integers} \end{cases} \tag{2.93}$$

where Ψ_i are uncertainty distributions determined by (2.85) and (2.88) for $i = 1, 2, \cdots, n$. Note that all arrival times $f_i(\boldsymbol{x}, \boldsymbol{y}, \boldsymbol{t})$, $i = 1, 2, \cdots, n$ are obtained by the 99-method and are respectively represented by 99-tables,

0.01	0.02	0.03	$\cdots$	0.99
s_1^1	s_1^2	s_1^3	$\cdots$	s_1^{99}

0.01	0.02	0.03	$\cdots$	0.99
s_2^1	s_2^2	s_2^3	$\cdots$	s_2^{99}

$$\vdots$$

0.01	0.02	0.03	$\cdots$	0.99
s_n^1	s_n^2	s_n^3	$\cdots$	s_n^{99}

$$\tag{2.94}$$

Thus the vehicle routing model is simplified as follows,

$$\begin{cases} \min g(\boldsymbol{x}, \boldsymbol{y}) \\ \text{subject to:} \\ \quad k/100 \geq \alpha_i \text{ if } s_i^k \geq b_i, \quad i = 1, 2, \cdots, n \\ \quad 1 \leq x_i \leq n, \quad i = 1, 2, \cdots, n \\ \quad x_i \neq x_j, \quad i \neq j, \; i, j = 1, 2, \cdots, n \\ \quad 0 \leq y_1 \leq y_2 \leq \cdots \leq y_{m-1} \leq n \\ \quad x_i, y_j, \quad i = 1, 2, \cdots, n, \quad j = 1, 2, \cdots, m-1, \quad \text{integers.} \end{cases} \qquad (2.95)$$

Numerical Experiment

Assume that there are 3 vehicles and 7 customers with the following time windows,

Node	Window	Node	Window
1	$[7:00, 9:00]$	5	$[15:00, 17:00]$
2	$[7:00, 9:00]$	6	$[19:00, 21:00]$
3	$[15:00, 17:00]$	7	$[19:00, 21:00]$
4	$[15:00, 17:00]$		

and each customer is visited within time windows with confidence level 0.90. We also assume that the distances are

$$D_{ij} = |i - j|, \quad i, j = 0, 1, 2, \cdots, 7$$

and travel times are normal uncertain variables

$$T_{ij} \sim \mathcal{N}(2|i - j|, 1), \quad i, j = 0, 1, 2, \cdots, 7.$$

In order to find an optimal operational plan, we integrate the 99-method and a genetic algorithm to produce a hybrid intelligent algorithm. A run of the computer program (http://orsc.edu.cn/liu/resources.htm) shows that the optimal operational plan is

Vehicle 1: depot→ 1 → 3 →depot, starting time: 6:18
Vehicle 2: deport→ 2 → 5 → 7 →depot, starting time: 4:18
Vehicle 3: depot→ 4 → 6 →depot, starting time: 8:18

whose total travel distance is 32.

2.11   Machine Scheduling Problem

Machine scheduling problem is concerned with finding an efficient schedule during an uninterrupted period of time for a set of machines to process a set of jobs. A lot of research work has been done on this type of problem. The study of machine scheduling problem with uncertain processing times was started by Liu [122] in 2009.

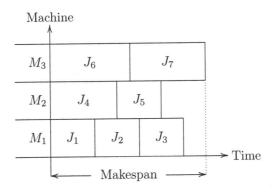

Figure 2.7: A Machine Schedule with 3 Machines and 7 Jobs

In a machine scheduling problem, we assume that (a) each job can be processed on any machine without interruption; (b) each machine can process only one job at a time; and (c) the processing times are uncertain variables with known uncertainty distributions. We also use the following indices and parameters:

$i = 1, 2, \cdots, n$: jobs;

$k = 1, 2, \cdots, m$: machines;

ξ_{ik}: uncertain processing time of job i on machine k;

Φ_{ik}: uncertainty distribution of ξ_{ik}.

How to Represent a Schedule?

The schedule is represented by the formulation (Liu [112]) via two decision vectors $\boldsymbol{x}$ and $\boldsymbol{y}$, where

$\boldsymbol{x} = (x_1, x_2, \cdots, x_n)$: integer decision vector representing n jobs with $1 \leq x_i \leq n$ and $x_i \neq x_j$ for all $i \neq j$, $i, j = 1, 2, \cdots, n$. That is, the sequence $\{x_1, x_2, \cdots, x_n\}$ is a rearrangement of $\{1, 2, \cdots, n\}$;

$\boldsymbol{y} = (y_1, y_2, \cdots, y_{m-1})$: integer decision vector with $y_0 \equiv 0 \leq y_1 \leq y_2 \leq \cdots \leq y_{m-1} \leq n \equiv y_m$.

We note that the schedule is fully determined by the decision vectors $\boldsymbol{x}$ and $\boldsymbol{y}$ in the following way. For each k ($1 \leq k \leq m$), if $y_k = y_{k-1}$, then the machine k is not used; if $y_k > y_{k-1}$, then the machine k is used and processes jobs $x_{y_{k-1}+1}, x_{y_{k-1}+2}, \cdots, x_{y_k}$ in turn. Thus the schedule of all machines is as follows,

$$\text{Machine 1: } x_{y_0+1} \to x_{y_0+2} \to \cdots \to x_{y_1};$$
$$\text{Machine 2: } x_{y_1+1} \to x_{y_1+2} \to \cdots \to x_{y_2}; \qquad (2.96)$$
$$\cdots$$
$$\text{Machine } m: x_{y_{m-1}+1} \to x_{y_{m-1}+2} \to \cdots \to x_{y_m}.$$

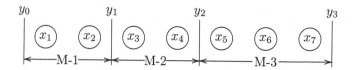

Figure 2.8: Formulation of Schedule in which Machine 1 Processes Jobs x_1, x_2, Machine 2 Processes Jobs x_3, x_4 and Machine 3 Processes Jobs x_5, x_6, x_7.

Completion Times

Let $C_i(\boldsymbol{x}, \boldsymbol{y}, \boldsymbol{\xi})$ be the completion times of jobs i, $i = 1, 2, \cdots, n$, respectively. For each k with $1 \leq k \leq m$, if the machine k is used (i.e., $y_k > y_{k-1}$), then we have

$$C_{x_{y_{k-1}+1}}(\boldsymbol{x}, \boldsymbol{y}, \boldsymbol{\xi}) = \xi_{x_{y_{k-1}+1}k} \qquad (2.97)$$

and

$$C_{x_{y_{k-1}+j}}(\boldsymbol{x}, \boldsymbol{y}, \boldsymbol{\xi}) = C_{x_{y_{k-1}+j-1}}(\boldsymbol{x}, \boldsymbol{y}, \boldsymbol{\xi}) + \xi_{x_{y_{k-1}+j}k} \qquad (2.98)$$

for $2 \leq j \leq y_k - y_{k-1}$.

Assume that each uncertain processing time ξ_{ik} of job i on machine k is represented by a 99-table,

0.01	0.02	0.03	$\cdots$	0.99
t_{ik}^1	t_{ik}^2	t_{ik}^3	$\cdots$	t_{ik}^{99}

(2.99)

If the machine k is used, then the completion time $C_{x_{y_{k-1}+1}}(\boldsymbol{x}, \boldsymbol{y}, \boldsymbol{\xi})$ of job $x_{y_{k-1}+1}$ is an uncertain variable whose inverse uncertainty distribution is

$$\Psi_{x_{y_{k-1}+1}}^{-1}(\alpha) = \Phi_{x_{y_{k-1}+1}k}^{-1}(\alpha) \qquad (2.100)$$

and has a 99-table,

0.01	0.02	$\cdots$	0.99
$t_{x_{y_{k-1}+1}}^1$	$t_{x_{y_{k-1}+1}}^2$	$\cdots$	$t_{x_{y_{k-1}+1}}^{99}$

(2.101)

Generally, suppose the completion time $C_{x_{y_{k-1}+j-1}}(\boldsymbol{x}, \boldsymbol{y}, \boldsymbol{\xi})$ has an inverse uncertainty distribution $\Psi_{x_{y_{k-1}+j-1}}^{-1}(\alpha)$ and is represented by a 99-table,

0.01	0.02	$\cdots$	0.99
$s_{x_{y_{k-1}+j-1}}^1$	$s_{x_{y_{k-1}+j-1}}^2$	$\cdots$	$s_{x_{y_{k-1}+j-1}}^{99}$

(2.102)

Then the completion time $C_{x_{y_{k-1}+j}}(\boldsymbol{x}, \boldsymbol{y}, \boldsymbol{\xi})$ has an inverse uncertainty distribution

$$\Psi_{x_{y_{k-1}+j}}^{-1}(\alpha) = \Psi_{x_{y_{k-1}+j-1}}^{-1}(\alpha) + \Phi_{x_{y_{k-1}+j}k}^{-1}(\alpha) \qquad (2.103)$$

and has a 99-table,

0.01	$\cdots$	0.99
$s^1_{x_{y_{k-1}+j-1}} + t^1_{x_{y_{k-1}+j}k}$	$\cdots$	$s^{99}_{x_{y_{k-1}+j-1}} + t^{99}_{x_{y_{k-1}+j}k}$

where $s^1_{x_{y_{k-1}+j-1}}, s^2_{x_{y_{k-1}+j-1}}, \cdots, s^{99}_{x_{y_{k-1}+j-1}}$ are determined by (2.102), and $t^1_{x_{y_{k-1}+j}k}, t^2_{x_{y_{k-1}+j}k}, \cdots, t^{99}_{x_{y_{k-1}+j}k}$ are determined by (2.99). This recursive process may produce all completion times of jobs.

Makespan

Note that, for each k ($1 \le k \le m$), the value $C_{x_{y_k}}(\boldsymbol{x}, \boldsymbol{y}, \boldsymbol{\xi})$ is just the time that the machine k finishes all jobs assigned to it, and has a 99-table,

0.01	0.02	$\cdots$	0.99
$s^1_{x_{y_k}}$	$s^2_{x_{y_k}}$	$\cdots$	$s^{99}_{x_{y_k}}$

(2.104)

Thus the makespan of the schedule $(\boldsymbol{x}, \boldsymbol{y})$ is determined by

$$f(\boldsymbol{x}, \boldsymbol{y}, \boldsymbol{\xi}) = \max_{1 \le k \le m} C_{x_{y_k}}(\boldsymbol{x}, \boldsymbol{y}, \boldsymbol{\xi}) \tag{2.105}$$

whose inverse uncertainty distribution is

$$\Upsilon^{-1}(\alpha) = \max_{1 \le k \le m} \Psi^{-1}_{x_{y_k}}(\alpha) \tag{2.106}$$

and has a 99-table,

0.01	0.02	$\cdots$	0.99
$\bigvee\limits_{k=1}^{m} s^1_{x_{y_k}}$	$\bigvee\limits_{k=1}^{m} s^2_{x_{y_k}}$	$\cdots$	$\bigvee\limits_{k=1}^{m} s^{99}_{x_{y_k}}$

(2.107)

Machine Scheduling Model

In order to minimize the expected makespan $E[f(\boldsymbol{x}, \boldsymbol{y}, \boldsymbol{\xi})]$, we have the following machine scheduling model,

$$\begin{cases} \min E[f(\boldsymbol{x}, \boldsymbol{y}, \boldsymbol{\xi})] \\ \text{subject to:} \\ \quad 1 \le x_i \le n, \quad i = 1, 2, \cdots, n \\ \quad x_i \ne x_j, \quad i \ne j, \ i, j = 1, 2, \cdots, n \\ \quad 0 \le y_1 \le y_2 \cdots \le y_{m-1} \le n \\ \quad x_i, y_j, \quad i = 1, 2, \cdots, n, \quad j = 1, 2, \cdots, m-1, \quad \text{integers.} \end{cases} \tag{2.108}$$

By using (2.107), the machine scheduling model is simplified as follows,

$$
\begin{cases}
\min \left(\displaystyle\bigvee_{k=1}^{m} s^1_{x_{y_k}} + \bigvee_{k=1}^{m} s^2_{x_{y_k}} + \cdots + \bigvee_{k=1}^{m} s^{99}_{x_{y_k}} \right) /99 \\
\text{subject to:} \\
\quad 1 \le x_i \le n, \quad i = 1, 2, \cdots, n \\
\quad x_i \ne x_j, \quad i \ne j, \ i,j = 1, 2, \cdots, n \\
\quad 0 \le y_1 \le y_2 \cdots \le y_{m-1} \le n \\
\quad x_i, y_j, \quad i = 1, 2, \cdots, n, \quad j = 1, 2, \cdots, m-1, \quad \text{integers.}
\end{cases} \tag{2.109}
$$

Numerical Experiment

Assume that there are 3 machines and 7 jobs with the following linear uncertain processing times

$$
\xi_{ik} \sim \mathcal{L}(i, i+k), \quad i = 1, 2, \cdots, 7, \ k = 1, 2, 3
$$

where i is the index of jobs and k is the index of machines. In order to find an optimal machine schedule, we integrate the 99-method and a genetic algorithm to produce a hybrid intelligent algorithm. A run of the computer program (http://orsc.edu.cn/liu/resources.htm) shows that the optimal machine schedule is

Machine 1: $1 \to 4 \to 5$
Machine 2: $3 \to 7$
Machine 3: $2 \to 6$

whose expected makespan is 12.

2.12   Exercises

In order to enhance your ability in modeling, this section provides some exercises.

Exercise 2.1: One approach to improve system reliability is to provide redundancy for components in a system. There are two ways to provide component redundancy: parallel redundancy and standby redundancy. In parallel redundancy, all redundant elements are required to operate simultaneously. This method is usually used when element replacements are not permitted during the system operation. In standby redundancy, one of the redundant elements begins to work only when the active element fails. This method is usually employed when the replacement is allowable and can be finished immediately. The system reliability design is to determine the optimal number of redundant elements for balancing system performance and total cost. Assume the element lifetimes are uncertain variables with known uncertainty distributions. Please construct an uncertain programming model for the system reliability design.

Exercise 2.2: The facility location problem is to find locations for new facilities such that the conveying cost from facilities to customers is minimized. In practice, some factors such as demands, allocations, even locations of customers and facilities are changing and then are assumed to be uncertain variables with known uncertainty distributions. Please construct an uncertain programming model for the facility location problem.

Exercise 2.3: The inventory problem (or supply chain) is concerned with the issues of *when to order* and *how much to order* of some goods. The purpose is to obtain the right goods in the right place, at the right time, and at low cost. Assume the demands and prices are uncertain variables with known uncertainty distributions. Please construct an uncertain programming model to determine the optimal order quantity.

Exercise 2.4: The capital budgeting problem (or portfolio selection) is concerned with maximizing the total profit subject to budget constraint by selecting appropriate combination of projects. Assume the future returns are uncertain variables with known uncertainty distributions. Please construct an uncertain programming model to determine the optimal investment plan.

Exercise 2.5: One of the basic network optimization problems is the shortest path problem which is to find the shortest path between two given nodes in a network, where the arc lengths are assumed to be uncertain variables. Please construct an uncertain programming model to find the shortest path.

Exercise 2.6: The maximal flow problem is related to maximizing the flow of some commodity through the arcs of a network from a given origin to a given destination, where each arc has an uncertain capacity of flow. Please construct an uncertain programming model to discover the maximum flow.

Exercise 2.7: The transportation problem is to determine the optimal transportation plan of some goods from suppliers to customers such that the total transportation cost is minimum. Assume the unit transportation cost of each route is an uncertain variable. Please construct an uncertain programming model to solve the transportation problem.

Chapter 3

Uncertain Risk Analysis

The term *risk* has been used in different ways in literature. Here the risk is defined as the "accidental loss" plus "uncertain measure of such loss". Uncertain risk analysis was proposed by Liu [126] in 2010 as a tool to quantify risk via uncertainty theory. One main feature of this topic is to model events that almost never occur. This chapter will introduce a definition of risk index and provide some useful formulas for calculating risk index.

3.1 Risk Index

A system usually contains uncertain factors, for example, lifetime, demand, production rate, cost, profit, and resource. Risk index is defined as the uncertain measure that some specified loss occurs. Note that the loss is problem-dependent.

Definition 3.1 *(Liu [126]). Assume a system contains uncertain variables* $\xi_1, \xi_2, \cdots, \xi_n$, *and there is a loss function* L *such that some specified loss occurs if and only if* $L(\xi_1, \xi_2, \cdots, \xi_n) \leq 0$. *Then the risk index is*

$$Risk = \mathcal{M}\{L(\xi_1, \xi_2, \cdots, \xi_n) \leq 0\}. \tag{3.1}$$

Example 3.1: Consider a series system in which there are n elements whose lifetimes are independent uncertain variables $\xi_1, \xi_2, \cdots, \xi_n$ with uncertainty distributions $\Phi_1, \Phi_2, \cdots, \Phi_n$, respectively. Such a system fails if any one element does not work. Thus the system lifetime

$$\xi = \xi_1 \wedge \xi_2 \wedge \cdots \wedge \xi_n \tag{3.2}$$

is an uncertain variable with uncertainty distribution

$$\Psi(x) = \Phi_1(x) \vee \Phi_2(x) \vee \cdots \vee \Phi_n(x). \tag{3.3}$$

If the loss is understood as the case that the system fails before time T, then the risk index is

$$Risk = \mathcal{M}\{\xi \leq T\} = \Phi_1(T) \vee \Phi_2(T) \vee \cdots \vee \Phi_n(T). \tag{3.4}$$

Example 3.2: Consider a parallel system in which there are n elements whose lifetimes are independent uncertain variables $\xi_1, \xi_2, \cdots, \xi_n$ with

B. Liu: Uncertainty Theory: A Branch of Mathematics, SCI 300, pp. 115–123.
springerlink.com © Springer-Verlag Berlin Heidelberg 2010

Figure 3.1: A Series System

uncertainty distributions $\Phi_1, \Phi_2, \cdots, \Phi_n$, respectively. Such a system fails if all elements do not work. Thus the system lifetime

$$\xi = \xi_1 \vee \xi_2 \vee \cdots \vee \xi_n \qquad (3.5)$$

is an uncertain variable with uncertainty distribution

$$\Psi(x) = \Phi_1(x) \wedge \Phi_2(x) \wedge \cdots \wedge \Phi_n(x). \qquad (3.6)$$

If the loss is understood as the case that the system fails before time T, then the risk index is

$$Risk = \mathcal{M}\{\xi \leq T\} = \Phi_1(T) \wedge \Phi_2(T) \wedge \cdots \wedge \Phi_n(T). \qquad (3.7)$$

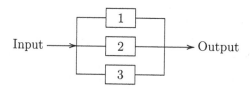

Figure 3.2: A Parallel System

Theorem 3.1 *(Liu [126], Risk Index Theorem). Assume that $\xi_1, \xi_2, \cdots, \xi_n$ are independent uncertain variables with uncertainty distributions $\Phi_1, \Phi_2, \cdots, \Phi_n$, respectively, and L is a strictly increasing function. If some specified loss occurs if and only if $L(\xi_1, \xi_2, \cdots, \xi_n) \leq 0$, then the risk index is*

$$Risk = \alpha \qquad (3.8)$$

where α is the root of

$$L(\Phi_1^{-1}(\alpha), \Phi_2^{-1}(\alpha), \cdots, \Phi_n^{-1}(\alpha)) = 0. \qquad (3.9)$$

Proof: It follows from Theorem 1.20 that $L(\xi_1, \xi_2, \cdots, \xi_n)$ is an uncertain variable whose inverse uncertainty distribution is

$$\Psi^{-1}(\alpha) = L(\Phi_1^{-1}(\alpha), \Phi_2^{-1}(\alpha), \cdots, \Phi_n^{-1}(\alpha)).$$

Since $Risk = \mathcal{M}\{L(\xi_1, \xi_2, \cdots, \xi_n) \leq 0\} = \Psi(0)$, we get (3.8).

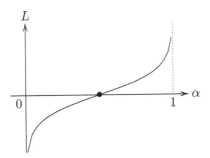

Figure 3.3: The Equation $L(\Phi_1^{-1}(\alpha), \Phi_2^{-1}(\alpha), \cdots, \Phi_n^{-1}(\alpha)) = 0$ whose root α may be estimated by the bisection method since L is a monotone function with respect to α.

Example 3.3: Assume that an investor has n projects whose returns are uncertain variables $\xi_1, \xi_2, \cdots, \xi_n$. If the loss is understood as case that the total return $\xi_1 + \xi_2 + \cdots + \xi_n$ is negative, then the risk index is

$$Risk = \mathcal{M}\{\xi_1 + \xi_2 + \cdots + \xi_n \le 0\}. \tag{3.10}$$

If $\xi_1, \xi_2, \cdots, \xi_n$ are uncertain variables with uncertainty distributions $\Phi_1, \Phi_2, \cdots, \Phi_n$, respectively, then the risk index is just the root α of

$$\Phi_1^{-1}(\alpha) + \Phi_2^{-1}(\alpha) + \cdots + \Phi_n^{-1}(\alpha) = 0. \tag{3.11}$$

Theorem 3.2 *(Liu [126], Risk Index Theorem). Assume that $\xi_1, \xi_2, \cdots, \xi_n$ are independent uncertain variables with uncertainty distributions $\Phi_1, \Phi_2, \cdots, \Phi_n$, respectively, and L is a strictly decreasing function. If some specified loss occurs if and only if $L(\xi_1, \xi_2, \cdots, \xi_n) \le 0$, then the risk index is*

$$Risk = \alpha \tag{3.12}$$

where α is the root of

$$L(\Phi_1^{-1}(1-\alpha), \Phi_2^{-1}(1-\alpha), \cdots, \Phi_n^{-1}(1-\alpha)) = 0. \tag{3.13}$$

Proof: It follows from Theorem 1.25 that $L(\xi_1, \xi_2, \cdots, \xi_n)$ is an uncertain variable whose inverse uncertainty distribution is

$$\Psi^{-1}(\alpha) = L(\Phi_1^{-1}(1-\alpha), \Phi_2^{-1}(1-\alpha), \cdots, \Phi_n^{-1}(1-\alpha)).$$

Since $Risk = \mathcal{M}\{L(\xi_1, \xi_2, \cdots, \xi_n) \le 0\} = \Psi(0)$, we get (3.12).

Theorem 3.3 *(Liu [126], Risk Index Theorem). Assume that $\xi_1, \xi_2, \cdots, \xi_n$ are independent uncertain variables with uncertainty distributions $\Phi_1, \Phi_2, \cdots, \Phi_n$, respectively, and the function $L(x_1, x_2, \cdots, x_n)$ is strictly increasing with* respect to $x_1, x_2, \cdots, x_m$ and strictly decreasing with respect to x_{m+1}, x_{m+2}, $\cdots, x_n$. If some specified loss occurs if and only if $L(\xi_1, \xi_2, \cdots, \xi_n) \leq 0$, then the risk index is

$$Risk = \alpha \qquad (3.14)$$

where α is the root of

$$L(\Phi_1^{-1}(\alpha), \cdots, \Phi_m^{-1}(\alpha), \Phi_{m+1}^{-1}(1-\alpha) \cdots, \Phi_n^{-1}(1-\alpha)) = 0. \qquad (3.15)$$

Proof: It follows from Theorem 1.26 that $L(\xi_1, \xi_2, \cdots, \xi_n)$ is an uncertain variable whose inverse uncertainty distribution is

$$\Psi^{-1}(\alpha) = L(\Phi_1^{-1}(\alpha), \cdots, \Phi_m^{-1}(\alpha), \Phi_{m+1}^{-1}(1-\alpha), \cdots, \Phi_n^{-1}(1-\alpha)).$$

Since $Risk = \mathcal{M}\{L(\xi_1, \xi_2, \cdots, \xi_n) \leq 0\} = \Psi(0)$, we get (3.14).

Example 3.4: Consider a structural system in which ξ is the strength variable and η is the load variable. The system failure occurs whenever the load variable η exceeds the strength variable ξ. If the loss is understood as the system failure, then the risk index is

$$Risk = \mathcal{M}\{\xi \leq \eta\}. \qquad (3.16)$$

If ξ and η are uncertain variables with uncertainty distributions Φ and Ψ, respectively, then the risk index is just the root α of

$$\Phi(\alpha) = \Psi(1 - \alpha). \qquad (3.17)$$

3.2 Hazard Distribution

Suppose that ξ is the lifetime of some system/element. Here it is assumed to be an uncertain variable with a prior uncertainty distribution. At some time t, it is observed that the system/element is working. What is the residual lifetime of the system/element? The following definition answers this question.

Definition 3.2 *(Liu [126]). Let ξ be a nonnegative uncertain variable representing lifetime of some system/element. If ξ has a prior uncertainty distribution Φ, then the hazard distribution (or failure distribution) at time t is*

$$\Phi(x|t) = \begin{cases} 0, & if\ \Phi(x) \leq \Phi(t) \\[2mm] \dfrac{\Phi(x)}{1 - \Phi(t)} \wedge 0.5, & if\ \Phi(t) < \Phi(x) \leq (1 + \Phi(t))/2 \\[2mm] \dfrac{\Phi(x) - \Phi(t)}{1 - \Phi(t)}, & if\ (1 + \Phi(t))/2 \leq \Phi(x) \end{cases} \qquad (3.18)$$

that is just the conditional uncertainty distribution of ξ given $\xi > t$.

The hazard distribution is essentially the posterior uncertainty distribution just after time t given that it is working at time t.

Example 3.5: Let ξ be a linear uncertain variable $\mathcal{L}(a, b)$, and t a real number with $a < t < b$. Then the hazard distribution at time t is

$$\Phi(x|t) = \begin{cases} 0, & \text{if } x \leq t \\[2mm] \dfrac{x-a}{b-t} \wedge 0.5, & \text{if } t < x \leq (b+t)/2 \\[2mm] \dfrac{x-t}{b-t} \wedge 1, & \text{if } (b+t)/2 \leq x. \end{cases}$$

Theorem 3.4 *(Liu [126], Conditional Risk Index Theorem). Consider a system that contains n elements whose uncertain lifetimes $\xi_1, \xi_2, \cdots, \xi_n$ are independent and have uncertainty distributions $\Phi_1, \Phi_2, \cdots, \Phi_n$, respectively. Assume L is a strictly increasing function, and some specified loss occurs if and only if $L(\xi_1, \xi_2, \cdots, \xi_n) \leq 0$. If it is observed that all elements are working at some time t, then the risk index is*

$$Risk = \alpha \tag{3.19}$$

where α is the root of

$$L(\Phi_1^{-1}(\alpha|t), \Phi_2^{-1}(\alpha|t), \cdots, \Phi_n^{-1}(\alpha|t)) = 0 \tag{3.20}$$

where $\Phi_i(x|t)$ are hazard distributions determined by

$$\Phi_i(x|t) = \begin{cases} 0, & \text{if } \Phi_i(x) \leq \Phi_i(t) \\[2mm] \dfrac{\Phi_i(x)}{1 - \Phi_i(t)} \wedge 0.5, & \text{if } \Phi_i(t) < \Phi_i(x) \leq (1 + \Phi_i(t))/2 \\[2mm] \dfrac{\Phi_i(x) - \Phi_i(t)}{1 - \Phi_i(t)}, & \text{if } (1 + \Phi_i(t))/2 \leq \Phi_i(x) \end{cases} \tag{3.21}$$

for $i = 1, 2, \cdots, n$.

Proof: It follows from Definition 3.2 that each hazard distribution of element is determined by (3.21). Thus the conditional risk index is obtained by Theorem 3.1 immediately.

Theorem 3.5 *(Liu [126], Conditional Risk Index Theorem). Consider a system that contains n elements whose uncertain lifetimes $\xi_1, \xi_2, \cdots, \xi_n$ are independent and have uncertainty distributions $\Phi_1, \Phi_2, \cdots, \Phi_n$, respectively. Assume L is a strictly decreasing function, and some specified loss occurs if and only if $L(\xi_1, \xi_2, \cdots, \xi_n) \leq 0$. If it is observed that all elements are working at some time t, then the risk index is*

$$Risk = \alpha \tag{3.22}$$

where α is the root of

$$L(\Phi_1^{-1}(1 - \alpha|t), \Phi_2^{-1}(1 - \alpha|t), \cdots, \Phi_n^{-1}(1 - \alpha|t)) = 0 \qquad (3.23)$$

where $\Phi_i(x|t)$ are hazard distributions determined by (3.21) for $i = 1, 2, \cdots, n$.

Proof: It follows from Definition 3.2 that each hazard distribution of element is determined by (3.21). Thus the conditional risk index is obtained by Theorem 3.2 immediately.

Theorem 3.6 *(Liu [126], Conditional Risk Index Theorem). Consider a system that contains n elements whose uncertain lifetimes $\xi_1, \xi_2, \cdots, \xi_n$ are independent and have uncertainty distributions $\Phi_1, \Phi_2, \cdots, \Phi_n$, respectively. Assume $L(x_1, x_2, \cdots, x_n)$ is strictly increasing with respect to $x_1, x_2, \cdots, x_m$ and strictly decreasing with respect to $x_{m+1}, x_{m+2}, \cdots, x_n$, and some specified loss occurs if and only if $L(\xi_1, \xi_2, \cdots, \xi_n) \leq 0$. If it is observed that all elements are working at some time t, then the risk index is*

$$Risk = \alpha \qquad (3.24)$$

where α is the root of

$$L(\Phi_1^{-1}(\alpha|t), \cdots, \Phi_m^{-1}(\alpha|t), \Phi_{m+1}^{-1}(1 - \alpha|t), \cdots, \Phi_n^{-1}(1 - \alpha|t)) = 0 \quad (3.25)$$

where $\Phi_i(x|t)$ are hazard distributions determined by (3.21) for $i = 1, 2, \cdots, n$.

Proof: It follows from Definition 3.2 that each hazard distribution of element is determined by (3.21). Thus the conditional risk index is obtained by Theorem 3.3 immediately.

3.3 Boolean System

Many real systems may be simplified to a Boolean system in which each element (including the system itself) has two states: working and failure. This section provides a risk index theorem for such a system.

We use ξ to express an element and use a to express its reliability in uncertain measure. Then the element ξ is essentially an uncertain variable

$$\xi = \begin{cases} 1 \text{ with uncertain measure } a \\ 0 \text{ with uncertain measure } 1 - a \end{cases} \qquad (3.26)$$

where $\xi = 1$ means the element is in working state and $\xi = 0$ means ξ is in failure state.

Assume that X is a Boolean system containing elements $\xi_1, \xi_2, \cdots, \xi_n$. Usually there is a function $f : \{0, 1\}^n \to \{0, 1\}$ such that

$$X = 0 \text{ if and only if } f(\xi_1, \xi_2, \cdots, \xi_n) = 0, \qquad (3.27)$$

$$X = 1 \text{ if and only if } f(\xi_1, \xi_2, \cdots, \xi_n) = 1. \tag{3.28}$$

Such a Boolean function f is called the *truth function* of X.

Example 3.6: For a series system, the truth function is a mapping from $\{0,1\}^n$ to $\{0,1\}$, i.e.,

$$f(x_1, x_2, \cdots, x_n) = x_1 \wedge x_2 \wedge \cdots \wedge x_n. \tag{3.29}$$

Example 3.7: For a parallel system, the truth function is a mapping from $\{0,1\}^n$ to $\{0,1\}$, i.e.,

$$f(x_1, x_2, \cdots, x_n) = x_1 \vee x_2 \vee \cdots \vee x_n. \tag{3.30}$$

Example 3.8: For a k-out-of-n system, the truth function is a mapping from $\{0,1\}^n$ to $\{0,1\}$, i.e.,

$$f(x_1, x_2, \cdots, x_n) = \begin{cases} 1, & \text{if } x_1 + x_2 + \cdots + x_n \geq k \\ 0, & \text{if } x_1 + x_2 + \cdots + x_n < k. \end{cases} \tag{3.31}$$

For any system with truth function f, if the loss is understood as the system failure, i.e., $X = 0$, then the risk index is

$$Risk = \mathcal{M}\{f(\xi_1, \xi_2, \cdots, \xi_n) = 0\}. \tag{3.32}$$

Theorem 3.7 *(Liu [126], Risk Index Theorem for Boolean System). Assume that $\xi_1, \xi_2, \cdots, \xi_n$ are independent elements with reliabilities $a_1, a_2, \cdots, a_n$, respectively. If a system contains $\xi_1, \xi_2, \cdots, \xi_n$ and has truth function f, then the risk index is*

$$Risk = \begin{cases} \displaystyle\sup_{f(x_1, x_2, \cdots, x_n)=0} \min_{1 \leq i \leq n} \nu_i(x_i), \\ \quad \text{if } \displaystyle\sup_{f(x_1, x_2, \cdots, x_n)=0} \min_{1 \leq i \leq n} \nu_i(x_i) < 0.5 \\ 1 - \displaystyle\sup_{f(x_1, x_2, \cdots, x_n)=1} \min_{1 \leq i \leq n} \nu_i(x_i), \\ \quad \text{if } \displaystyle\sup_{f(x_1, x_2, \cdots, x_n)=0} \min_{1 \leq i \leq n} \nu_i(x_i) \geq 0.5 \end{cases} \tag{3.33}$$

where x_i take values either 0 or 1, and ν_i are defined by

$$\nu_i(x_i) = \begin{cases} a_i, & \text{if } x_i = 1 \\ 1 - a_i, & \text{if } x_i = 0 \end{cases} \tag{3.34}$$

for $i = 1, 2, \cdots, n$, respectively.

Proof: Since $\xi_1, \xi_2, \cdots, \xi_n$ are Boolean uncertain variables and f is a Boolean function, the equation (3.33) follows from $Risk = \mathcal{M}\{f(\xi_1, \xi_2, \cdots, \xi_n) = 0\}$ and Theorem 1.27 immediately.

Example 3.9: Consider a series system having uncertain elements $\xi_1, \xi_2, \cdots,$ ξ_n with reliabilities $a_1, a_2, \cdots, a_n$, respectively. Note that the truth function is

$$f(x_1, x_2, \cdots, x_n) = x_1 \wedge x_2 \wedge \cdots \wedge x_n. \tag{3.35}$$

It follows from the risk index theorem or Theorem 1.28 that the risk index is

$$Risk = (1 - a_1) \vee (1 - a_2) \vee \cdots \vee (1 - a_n). \tag{3.36}$$

Example 3.10: Consider a parallel system having uncertain elements $\xi_1, \xi_2,$ $\cdots, \xi_n$ with reliabilities $a_1, a_2, \cdots, a_n$, respectively. Note that the truth function is

$$f(x_1, x_2, \cdots, x_n) = x_1 \vee x_2 \vee \cdots \vee x_n. \tag{3.37}$$

It follows from the risk index theorem or Theorem 1.29 that the risk index is

$$Risk = (1 - a_1) \wedge (1 - a_2) \wedge \cdots \wedge (1 - a_n). \tag{3.38}$$

Example 3.11: Consider a k-out-of-n system having uncertain elements $\xi_1, \xi_2, \cdots, \xi_n$ with reliabilities $a_1, a_2, \cdots, a_n$, respectively. Note that the truth function is

$$f(x_1, x_2, \cdots, x_n) = \begin{cases} 1, & \text{if } x_1 + x_2 + \cdots + x_n \geq k \\ 0, & \text{if } x_1 + x_2 + \cdots + x_n < k. \end{cases} \tag{3.39}$$

It follows from the risk index theorem or Theorem 1.30 that the risk index is

$$Risk = \text{"the } k\text{th smallest value of } 1 - a_1, 1 - a_2, \cdots, 1 - a_n\text{"}. \tag{3.40}$$

3.4   Risk Index Calculator

Risk Index Calculator is a software for calculating the risk index of Boolean system, and is available at http://orsc.edu.cn/liu/resources.htm.

Example 3.12: Consider a bridge system shown in Figure 3.4 that consists of 5 elements whose states are denoted by x_1, x_2, x_3, x_4, x_5. It is obvious that there are 4 paths from the input of the system to the output:

$$\text{Path 1: input}-1-4-\text{output},$$
$$\text{Path 2: input}-2-5-\text{output},$$
$$\text{Path 3: input}-1-3-5-\text{output},$$
$$\text{Path 4: input}-2-3-4-\text{output}.$$

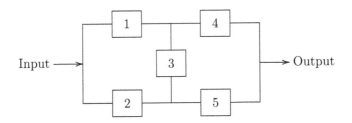

Figure 3.4: A Bridge System

Assume each path works if and only if all elements on which are working. Then the states of the 4 paths are

$$x_1 \wedge x_4, \quad x_2 \wedge x_5, \quad x_1 \wedge x_3 \wedge x_5, \quad x_2 \wedge x_3 \wedge x_4.$$

Assume the system works if and only if there is a path of working elements. Then the truth function of the bridge system is

$$f(x_1, x_2, x_3, x_4, x_5) = (x_1 \wedge x_4) \vee (x_2 \wedge x_5) \vee (x_1 \wedge x_3 \wedge x_5) \vee (x_2 \wedge x_3 \wedge x_4).$$

Assume the 5 elements have reliabilities 0.91, 0.92, 0.93, 0.94, 0.95 in uncertain measure. When the loss is understood as the bridge system failure, a run of Risk Index Calculator shows that the risk index is 0.08 in uncertain measure.

Chapter 4

Uncertain Reliability Analysis

Uncertain reliability analysis was proposed by Liu [126] in 2010 as a tool to deal with system reliability via uncertainty theory. Note that uncertain reliability analysis and uncertain risk analysis have the same root in mathematics. They are separately treated for application convenience in practice rather than theoretical demand.

This chapter will introduce a definition of reliability index and provide some useful formulas for calculating reliability index.

4.1 Reliability Index

Reliability index is defined as the uncertain measure that some system is working.

Definition 4.1 *(Liu [126]). Assume a system contains uncertain variables $\xi_1, \xi_2, \cdots, \xi_n$, and there is a function R such that the system is working if and only if $R(\xi_1, \xi_2, \cdots, \xi_n) \geq 0$. Then the reliability index is*

$$Reliability = \mathcal{M}\{R(\xi_1, \xi_2, \cdots, \xi_n) \geq 0\}. \tag{4.1}$$

Example 4.1: Consider a series system in which there are n elements whose lifetimes are independent uncertain variables $\xi_1, \xi_2, \cdots, \xi_n$ with uncertainty distributions $\Phi_1, \Phi_2, \cdots, \Phi_n$, respectively. Such a system works if all elements are working, and the system lifetime ξ has an uncertainty distribution $\Psi(x) = \Phi_1(x) \vee \Phi_2(x) \vee \cdots \vee \Phi_n(x)$. If we hope the system is working until time T, then the reliability index is

$$Reliability = \mathcal{M}\{\xi \geq T\} = 1 - \Phi_1(T) \vee \Phi_2(T) \vee \cdots \vee \Phi_n(T). \tag{4.2}$$

Example 4.2: Consider a parallel system in which there are n elements whose lifetimes are independent uncertain variables $\xi_1, \xi_2, \cdots, \xi_n$ with uncertainty distributions $\Phi_1, \Phi_2, \cdots, \Phi_n$, respectively. Such a system works if there is at least one working element. Thus the system lifetime ξ has an uncertainty distribution $\Psi(x) = \Phi_1(x) \wedge \Phi_2(x) \wedge \cdots \wedge \Phi_n(x)$. If we hope the system is working until time T, then the reliability index is

B. Liu: Uncertainty Theory: A Branch of Mathematics, SCI 300, pp. 125–130.
springerlink.com   © Springer-Verlag Berlin Heidelberg 2010

$$Reliability = \mathcal{M}\{\xi \geq T\} = 1 - \Phi_1(T) \wedge \Phi_2(T) \wedge \cdots \wedge \Phi_n(T). \qquad (4.3)$$

Theorem 4.1 *(Liu [126], Reliability Index Theorem). Assume $\xi_1, \xi_2, \cdots, \xi_n$ are independent uncertain variables with uncertainty distributions $\Phi_1, \Phi_2, \cdots, \Phi_n$, respectively, and R is a strictly increasing function. If some system is working if and only if $R(\xi_1, \xi_2, \cdots, \xi_n) \geq 0$, then the reliability index is*

$$Reliability = \alpha \qquad (4.4)$$

where α is the root of

$$R(\Phi_1^{-1}(1 - \alpha), \Phi_2^{-1}(1 - \alpha), \cdots, \Phi_n^{-1}(1 - \alpha)) = 0. \qquad (4.5)$$

Proof: It follows from Theorem 1.20 that $R(\xi_1, \xi_2, \cdots, \xi_n)$ is an uncertain variable whose inverse uncertainty distribution is

$$\Psi^{-1}(\alpha) = R(\Phi_1^{-1}(\alpha), \Phi_2^{-1}(\alpha), \cdots, \Phi_n^{-1}(\alpha)).$$

Since $Reliability = \mathcal{M}\{R(\xi_1, \xi_2, \cdots, \xi_n) \geq 0\} = 1 - \Psi(0)$, we get (4.4).

Theorem 4.2 *(Liu [126], Reliability Index Theorem). Assume $\xi_1, \xi_2, \cdots, \xi_n$ are independent uncertain variables with uncertainty distributions $\Phi_1, \Phi_2, \cdots, \Phi_n$, respectively, and R is a strictly decreasing function. If some system is working if and only if $R(\xi_1, \xi_2, \cdots, \xi_n) \geq 0$, then the reliability index is*

$$Reliability = \alpha \qquad (4.6)$$

where α is the root of

$$R(\Phi_1^{-1}(\alpha), \Phi_2^{-1}(\alpha), \cdots, \Phi_n^{-1}(\alpha)) = 0. \qquad (4.7)$$

Proof: It follows from Theorem 1.25 that $R(\xi_1, \xi_2, \cdots, \xi_n)$ is an uncertain variable whose inverse uncertainty distribution is

$$\Psi^{-1}(\alpha) = R(\Phi_1^{-1}(1 - \alpha), \Phi_2^{-1}(1 - \alpha), \cdots, \Phi_n^{-1}(1 - \alpha)).$$

Since $Reliability = \mathcal{M}\{R(\xi_1, \xi_2, \cdots, \xi_n) \geq 0\} = 1 - \Psi(0)$, we get (4.6).

Theorem 4.3 *(Liu [126], Reliability Index Theorem). Assume $\xi_1, \xi_2, \cdots, \xi_n$ are independent uncertain variables with uncertainty distributions $\Phi_1, \Phi_2, \cdots, \Phi_n$, respectively, and the function $R(x_1, x_2, \cdots, x_n)$ is strictly increasing with respect to $x_1, x_2, \cdots, x_m$ and strictly decreasing with respect to $x_{m+1}, x_{m+2}, \cdots, x_n$. If some system is working if and only if $R(\xi_1, \xi_2, \cdots, \xi_n) \geq 0$, then the reliability index is*

$$Reliability = \alpha \qquad (4.8)$$

where α is the root of

$$R(\Phi_1^{-1}(1 - \alpha), \cdots, \Phi_m^{-1}(1 - \alpha), \Phi_{m+1}^{-1}(\alpha) \cdots, \Phi_n^{-1}(\alpha)) = 0. \qquad (4.9)$$

Proof: It follows from Theorem 1.26 that $R(\xi_1, \xi_2, \cdots, \xi_n)$ is an uncertain variable whose inverse uncertainty distribution is

$$\Psi^{-1}(\alpha) = R(\Phi_1^{-1}(\alpha), \cdots, \Phi_m^{-1}(\alpha), \Phi_{m+1}^{-1}(1-\alpha), \cdots, \Phi_n^{-1}(1-\alpha)).$$

Since $Reliability = \mathcal{M}\{R(\xi_1, \xi_2, \cdots, \xi_n) \geq 0\} = 1 - \Psi(0)$, we get (4.8).

Example 4.3: Consider a structural system in which ξ is the strength variable and η is the load variable. The system works whenever the load variable η does not exceed the strength variable ξ. Then the reliability index is

$$Reliability = \mathcal{M}\{\xi \geq \eta\}. \tag{4.10}$$

If ξ and η are uncertain variables with uncertainty distributions Φ and Ψ, respectively, then the reliability index is just the root α of

$$\Phi(1-\alpha) = \Psi(\alpha). \tag{4.11}$$

4.2   Conditional Reliability

This section provides some conditional reliability index theorems given that all elements are working at time t.

Theorem 4.4 *(Liu [126], Conditional Reliability Index Theorem). Consider a system that contains n elements whose uncertain lifetimes $\xi_1, \xi_2, \cdots, \xi_n$ are independent and have uncertainty distributions $\Phi_1, \Phi_2, \cdots, \Phi_n$, respectively. Assume R is a strictly increasing function, and some system is working if and only if $R(\xi_1, \xi_2, \cdots, \xi_n) \geq 0$. If it is observed that all elements are working at some time t, then the reliability index is*

$$Reliability = \alpha \tag{4.12}$$

where α is the root of

$$R(\Phi_1^{-1}(1-\alpha|t), \Phi_2^{-1}(1-\alpha|t), \cdots, \Phi_n^{-1}(1-\alpha|t)) = 0 \tag{4.13}$$

where $\Phi_i(x|t)$ are hazard distributions determined by

$$\Phi_i(x|t) = \begin{cases} 0, & \text{if } \Phi_i(x) \leq \Phi_i(t) \\ \dfrac{\Phi_i(x)}{1 - \Phi_i(t)} \wedge 0.5, & \text{if } \Phi_i(t) < \Phi_i(x) \leq (1 + \Phi_i(t))/2 \\ \dfrac{\Phi_i(x) - \Phi_i(t)}{1 - \Phi_i(t)}, & \text{if } (1 + \Phi_i(t))/2 \leq \Phi_i(x) \end{cases} \tag{4.14}$$

for $i = 1, 2, \cdots, n$.

Proof: Since each hazard distribution of element is determined by (4.14), the conditional reliability index is obtained by Theorem 4.1 immediately.

Theorem 4.5 *(Conditional Reliability Index Theorem). Consider a system that contains n elements whose uncertain lifetimes $\xi_1, \xi_2, \cdots, \xi_n$ are independent and have uncertainty distributions $\Phi_1, \Phi_2, \cdots, \Phi_n$, respectively. Assume R is a strictly decreasing function, and some system is working if and only if $R(\xi_1, \xi_2, \cdots, \xi_n) \geq 0$. If it is observed that all elements are working at some time t, then the reliability index is*

$$Reliability = \alpha \tag{4.15}$$

where α is the root of

$$R(\Phi_1^{-1}(\alpha|t), \Phi_2^{-1}(\alpha|t), \cdots, \Phi_n^{-1}(\alpha|t)) = 0 \tag{4.16}$$

where $\Phi_i(x|t)$ are hazard distributions determined by (4.14) for $i = 1, 2, \cdots, n$.

Proof: Since each hazard distribution of element is determined by (4.14), the conditional reliability index is obtained by Theorem 4.2 immediately.

Theorem 4.6 *(Conditional Reliability Index Theorem). Consider a system that contains n elements whose uncertain lifetimes $\xi_1, \xi_2, \cdots, \xi_n$ are independent and have uncertainty distributions $\Phi_1, \Phi_2, \cdots, \Phi_n$, respectively. Assume $R(x_1, x_2, \cdots, x_n)$ is strictly increasing with respect to $x_1, x_2, \cdots, x_m$ and strictly decreasing with respect to $x_{m+1}, x_{m+2}, \cdots, x_n$, and some system is working if and only if $R(\xi_1, \xi_2, \cdots, \xi_n) \geq 0$. If it is observed that all elements are working at some time t, then the reliability index is*

$$Reliability = \alpha \tag{4.17}$$

where α is the root of

$$R(\Phi_1^{-1}(1-\alpha|t), \cdots, \Phi_m^{-1}(1-\alpha|t), \Phi_{m+1}^{-1}(\alpha|t), \cdots, \Phi_n^{-1}(\alpha|t)) = 0 \tag{4.18}$$

where $\Phi_i(x|t)$ are hazard distributions determined by (4.14) for $i = 1, 2, \cdots, n$.

Proof: Since each hazard distribution of element is determined by (4.14), the conditional reliability index is obtained by Theorem 4.3 immediately.

4.3   Boolean System

Consider a Boolean system with n elements $\xi_1, \xi_2, \cdots, \xi_n$ and a truth function f. Since the system is working if and only if $f(\xi_1, \xi_2, \cdots, \xi_n) = 1$, the reliability index is

$$Reliability = \mathcal{M}\{f(\xi_1, \xi_2, \cdots, \xi_n) = 1\}. \tag{4.19}$$

Theorem 4.7 *(Liu [126], Reliability Index Theorem for Boolean System). Assume $\xi_1, \xi_2, \cdots, \xi_n$ are independent elements with reliabilities $a_1, a_2, \cdots, a_n$,* *respectively. If a system contains $\xi_1, \xi_2, \cdots, \xi_n$ and has truth function f, then the reliability index is*

$$Reliability = \begin{cases} \sup\limits_{f(x_1,x_2,\cdots,x_n)=1} \min\limits_{1\leq i\leq n} \nu_i(x_i), \\ \qquad if \quad \sup\limits_{f(x_1,x_2,\cdots,x_n)=1} \min\limits_{1\leq i\leq n} \nu_i(x_i) < 0.5 \\ 1 - \sup\limits_{f(x_1,x_2,\cdots,x_n)=0} \min\limits_{1\leq i\leq n} \nu_i(x_i), \\ \qquad if \quad \sup\limits_{f(x_1,x_2,\cdots,x_n)=1} \min\limits_{1\leq i\leq n} \nu_i(x_i) \geq 0.5 \end{cases} \qquad (4.20)$$

where x_i take values either 0 or 1, and ν_i are defined by

$$\nu_i(x_i) = \begin{cases} a_i, & if \ x_i = 1 \\ 1 - a_i, & if \ x_i = 0 \end{cases} \qquad (4.21)$$

for $i = 1, 2, \cdots, n$, respectively.

Proof: Since $\xi_1, \xi_2, \cdots, \xi_n$ are Boolean uncertain variables and f is a Boolean function, the equation (4.20) follows from $Reliability = \mathcal{M}\{f(\xi_1, \xi_2, \cdots, \xi_n) = 1\}$ and Theorem 1.27 immediately.

Example 4.4: Consider a series system having uncertain elements $\xi_1, \xi_2, \cdots,$ ξ_n with reliabilities $a_1, a_2, \cdots, a_n$, respectively. Note that the truth function is a Boolean function,

$$f(x_1, x_2, \cdots, x_n) = x_1 \wedge x_2 \wedge \cdots \wedge x_n. \qquad (4.22)$$

It follows from the reliability index theorem or Theorem 1.28 that the reliability index is

$$Reliability = a_1 \wedge a_2 \wedge \cdots \wedge a_n. \qquad (4.23)$$

Example 4.5: Consider a parallel system having uncertain elements $\xi_1, \xi_2,$ $\cdots, \xi_n$ with reliabilities $a_1, a_2, \cdots, a_n$, respectively. Note that the truth function is a Boolean function,

$$f(x_1, x_2, \cdots, x_n) = x_1 \vee x_2 \vee \cdots \vee x_n. \qquad (4.24)$$

It follows from the reliability index theorem or Theorem 1.29 that the reliability index is

$$Reliability = a_1 \vee a_2 \vee \cdots \vee a_n. \qquad (4.25)$$

Example 4.6: Consider a k-out-of-n system having uncertain elements $\xi_1, \xi_2, \cdots, \xi_n$ with reliabilities $a_1, a_2, \cdots, a_n$, respectively. Note that the truth function is a Boolean function,

$$f(x_1, x_2, \cdots, x_n) = \begin{cases} 1, & if \ x_1 + x_2 + \cdots + x_n \geq k \\ 0, & if \ x_1 + x_2 + \cdots + x_n < k. \end{cases} \qquad (4.26)$$

It follows from the reliability index theorem or Theorem 1.30 that the reliability index is

$$Reliability = \text{the } k\text{th largest value of } a_1, a_2, \cdots, a_n. \tag{4.27}$$

4.4 Reliability Index Calculator

Reliability Index Calculator is a software for calculating the reliability index of Boolean system, and is available at http://orsc.edu.cn/liu/resources.htm.

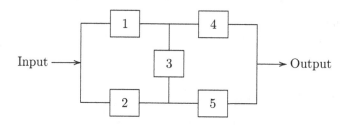

Figure 4.1: A Bridge System

Example 4.7: Consider a bridge system shown in Figure 4.1 that consists of 5 elements whose states are denoted by x_1, x_2, x_3, x_4, x_5. Assume each path works if and only if all elements on which are working and the system works if and only if there is a path of working elements. Then the truth function of the bridge system is

$$f(x_1, x_2, x_3, x_4, x_5) = (x_1 \wedge x_4) \vee (x_2 \wedge x_5) \vee (x_1 \wedge x_3 \wedge x_5) \vee (x_2 \wedge x_3 \wedge x_4).$$

When the 5 elements have reliabilities 0.91, 0.92, 0.93, 0.94, 0.95 in uncertain measure, a run of Reliability Index Calculator shows that the system reliability index is 0.92 in uncertain measure.

Chapter 5

Uncertain Process

An uncertain process is essentially a sequence of uncertain variables indexed by time or space. The study of uncertain process was started by Liu [121] in 2008. This chapter introduces the basic concepts of uncertain process, including renewal process, martingale, Markov process and stationary process.

5.1 Uncertain Process

Definition 5.1 *(Liu [121]). Let T be an index set and let $(\Gamma, \mathcal{L}, \mathcal{M})$ be an uncertainty space. An uncertain process is a measurable function from $T \times (\Gamma, \mathcal{L}, \mathcal{M})$ to the set of real numbers, i.e., for each $t \in T$ and any Borel set B of real numbers, the set*

$$\{X_t \in B\} = \{\gamma \in \Gamma \mid X_t(\gamma) \in B\} \tag{5.1}$$

is an event.

That is, an uncertain process $X_t(\gamma)$ is a function of two variables such that the function $X_{t^*}(\gamma)$ is an uncertain variable for each t^*.

Definition 5.2. *For each fixed γ^*, the function $X_t(\gamma^*)$ is called a sample path of the uncertain process X_t.*

Definition 5.3. *An uncertain process X_t is said to be sample-continuous if almost all sample paths are continuous with respect to t.*

Definition 5.4. *An uncertain process X_t is said to have independent increments if*

$$X_{t_1} - X_{t_0}, \; X_{t_2} - X_{t_1}, \; \cdots, \; X_{t_k} - X_{t_{k-1}} \tag{5.2}$$

are independent uncertain variables for any times $t_0 < t_1 < \cdots < t_k$.

Definition 5.5. *An uncertain process X_t is said to have stationary increments if, for any given $t > 0$, the increments $X_{s+t} - X_s$ are identically distributed uncertain variables for all $s > 0$.*

Definition 5.6. *For any partition of closed interval $[0, t]$ with $0 = t_1 < t_2 < \cdots < t_{k+1} = t$, the mesh is written as*

$$\Delta = \max_{1 \le i \le k} |t_{i+1} - t_i|.$$

B. Liu: Uncertainty Theory: A Branch of Mathematics, SCI 300, pp. 131–138.
springerlink.com   © Springer-Verlag Berlin Heidelberg 2010

Let $m > 0$ be a real number. Then the m-variation of uncertain process X_t is

$$\|X\|_t^m = \lim_{\Delta \to 0} \sum_{i=1}^{k} |X_{t_{i+1}} - X_{t_i}|^m \tag{5.3}$$

provided that the limit exists almost surely and is an uncertain process. Especially,

$$\|X\|_t = \lim_{\Delta \to 0} \sum_{i=1}^{k} |X_{t_{i+1}} - X_{t_i}| \tag{5.4}$$

is called total variation, and

$$\|X\|_t^2 = \lim_{\Delta \to 0} \sum_{i=1}^{k} |X_{t_{i+1}} - X_{t_i}|^2 \tag{5.5}$$

is called the squared variation of uncertain process X_t.

5.2   Renewal Process

Definition 5.7 *(Liu [121]). Let $\xi_1, \xi_2, \cdots$ be iid positive uncertain variables. Define $S_0 = 0$ and $S_n = \xi_1 + \xi_2 + \cdots + \xi_n$ for $n \geq 1$. Then the uncertain process*

$$N_t = \max_{n \geq 0} \left\{ n \mid S_n \leq t \right\} \tag{5.6}$$

is called a renewal process.

If $\xi_1, \xi_2, \cdots$ denote the interarrival times of successive events. Then S_n can be regarded as the waiting time until the occurrence of the nth event, and N_t is the number of renewals in $(0, t]$. The renewal process N_t is not sample-continuous. But each sample path of N_t is a right-continuous and increasing step function taking only nonnegative integer values. Furthermore, the size

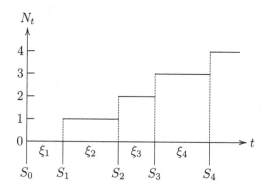

Figure 5.1: A Sample Path of Renewal Process

of each jump of N_t is always 1. In other words, N_t has at most one renewal at each time. In particular, N_t does not jump at time 0. Since $N_t \geq n$ is equivalent to $S_n \leq t$, we immediately have

$$\mathcal{M}\{N_t \geq n\} = \mathcal{M}\{S_n \leq t\}. \tag{5.7}$$

Since $N_t \leq n$ is equivalent to $S_{n+1} > t$, by using the self-duality axiom, we immediately have

$$\mathcal{M}\{N_t \leq n\} = 1 - \mathcal{M}\{S_{n+1} \leq t\}. \tag{5.8}$$

Theorem 5.1. *Let N_t be a renewal process with uncertain interarrival times $\xi_1, \xi_2, \cdots$ If those interarrival times have a common uncertainty distribution Φ, then N_t has an uncertainty distribution*

$$\Upsilon_t(x) = 1 - \Phi\left(\frac{t}{\lfloor x \rfloor + 1}\right), \quad \forall x \geq 0 \tag{5.9}$$

where $\lfloor x \rfloor$ represents the maximal integer less than or equal to x.

Proof: Note that S_{n+1} has an uncertainty distribution $\Phi(x/(n+1))$. It follows from (5.8) that

$$\mathcal{M}\{N_t \leq n\} = 1 - \mathcal{M}\{S_{n+1} \leq t\} = 1 - \Phi\left(\frac{t}{n+1}\right).$$

Since N_t takes integer values, for any $x \geq 0$, we have

$$\Upsilon_t(x) = \mathcal{M}\{N_t \leq x\} = \mathcal{M}\{N_t \leq \lfloor x \rfloor\} = 1 - \Phi\left(\frac{t}{\lfloor x \rfloor + 1}\right).$$

The theorem is verified.

Theorem 5.2. *Let N_t be a renewal process with uncertain interarrival times $\xi_1, \xi_2, \cdots$ If those interarrival times have a common uncertainty distribution Φ, then*

$$E[N_t] = \sum_{n=1}^{\infty} \Phi\left(\frac{t}{n}\right). \tag{5.10}$$

Proof: Since N_t takes only nonnegative integer values, it follows from the definition of expected value and Theorem 5.1 that

$$E[N_t] = \int_0^{\infty} (1 - \Upsilon_t(x)) \mathrm{d}x = \sum_{n=0}^{\infty} (1 - \Upsilon_t(n))$$

$$= \sum_{n=0}^{\infty} \Phi\left(\frac{t}{n+1}\right) = \sum_{n=1}^{\infty} \Phi\left(\frac{t}{n}\right).$$

Thus the theorem is verified.

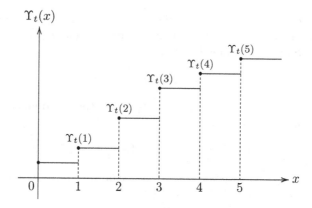

Figure 5.2: Uncertainty Distribution Υ_t of Renewal Process N_t

Theorem 5.3 *(Renewal Theorem). Let N_t be a renewal process with uncertain interarrival times $\xi_1, \xi_2, \cdots$ Then*

$$\lim_{t \to \infty} \frac{E[N_t]}{t} = E\left[\frac{1}{\xi_1}\right]. \tag{5.11}$$

If those interarrival times have a common uncertainty distribution Φ, then

$$\lim_{t \to \infty} \frac{E[N_t]}{t} = \int_0^{+\infty} \Phi\left(\frac{1}{x}\right) dx. \tag{5.12}$$

If the uncertainty distribution Φ is regular, then

$$\lim_{t \to \infty} \frac{E[N_t]}{t} = \int_0^1 \frac{1}{\Phi^{-1}(\alpha)} d\alpha. \tag{5.13}$$

Proof: The uncertainty distribution Υ_t of N_t has been given by Theorem 5.1. It follows from the operational law that the uncertainty distribution of N_t/t is

$$\Psi_t(x) = \Upsilon_t(tx) = 1 - \Phi\left(\frac{t}{\lfloor tx \rfloor + 1}\right)$$

where $\lfloor tx \rfloor$ represents the maximal integer less than or equal to tx. Thus

$$\frac{E[N_t]}{t} = \int_0^{+\infty} (1 - \Psi_t(x)) dx.$$

On the other hand, $1/\xi_1$ has an uncertainty distribution $1 - \Phi(1/x)$ whose expected value is

$$E\left[\frac{1}{\xi_1}\right] = \int_0^{+\infty} \Phi\left(\frac{1}{x}\right) dx.$$

Note that

$$(1 - \Psi_t(x)) \leq \Phi\left(\frac{1}{x}\right), \quad \forall t, x$$

and

$$\lim_{t\to\infty} (1 - \Psi_t(x)) = \Phi\left(\frac{1}{x}\right), \quad \forall x.$$

It follows from Lebesgue dominated convergence theorem that

$$\lim_{t\to\infty} \frac{E[N_t]}{t} = \lim_{t\to\infty} \int_0^{+\infty} (1 - \Psi_t(x))dx = \int_0^{+\infty} \Phi\left(\frac{1}{x}\right) dx = E\left[\frac{1}{\xi_1}\right].$$

Furthermore, since the inverse uncertainty distribution of $1/\xi$ is $1/\Phi^{-1}(1-\alpha)$, we get

$$E\left[\frac{1}{\xi}\right] = \int_0^1 \frac{1}{\Phi^{-1}(1-\alpha)}d\alpha = \int_0^1 \frac{1}{\Phi^{-1}(\alpha)}d\alpha.$$

The theorem is proved.

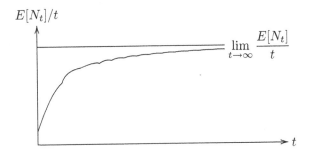

Figure 5.3: Average Renewal Number $E[N_t]/t$

Example 5.1: A renewal process N_t is called a *linear renewal process* if $\xi_1, \xi_2, \cdots$ are iid linear uncertain variables $\mathcal{L}(a, b)$ with $a > 0$. It follows from the renewal theorem that

$$\lim_{t\to\infty} \frac{E[N_t]}{t} = \frac{\ln b - \ln a}{b - a}. \tag{5.14}$$

Example 5.2: A renewal process N_t is called a *zigzag renewal process* if $\xi_1, \xi_2, \cdots$ are iid zigzag uncertain variables $\mathcal{Z}(a, b, c)$ with $a > 0$. It follows from the renewal theorem that

$$\lim_{t \to \infty} \frac{E[N_t]}{t} = \frac{1}{2} \left(\frac{\ln b - \ln a}{b - a} + \frac{\ln c - \ln b}{c - b} \right). \tag{5.15}$$

Example 5.3: A renewal process N_t is called a *lognormal renewal process* if $\xi_1, \xi_2, \cdots$ are iid lognormal uncertain variables $\mathcal{LOGN}(e, \sigma)$. If $\sigma < \pi/\sqrt{3}$, then

$$\lim_{t \to \infty} \frac{E[N_t]}{t} = \sqrt{3}\sigma \exp(-e) \csc(\sqrt{3}\sigma). \tag{5.16}$$

Otherwise, we have

$$\lim_{t \to \infty} \frac{E[N_t]}{t} = +\infty. \tag{5.17}$$

Renewal Reward Process

Let $(\xi_1, \eta_1), (\xi_2, \eta_2), \cdots$ be a sequence of pairs of uncertain variables. We shall interpret η_i as the rewards (or costs) associated with the i-th interarrival times ξ_i for $i = 1, 2, \cdots$, respectively.

Definition 5.8. *Let $\xi_1, \xi_2, \cdots$ be iid uncertain interarrival times, and let $\eta_1, \eta_2, \cdots$ be iid uncertain rewards. It is also assumed that $\xi_1, \eta_1, \xi_2, \eta_2, \cdots$ are independent. Then*

$$R_t = \sum_{i=1}^{N_t} \eta_i \tag{5.18}$$

is called a renewal reward process, where N_t is the renewal process.

A renewal reward process R_t denotes the total reward earned by time t. In addition, if $\eta_i \equiv 1$, then R_t degenerates to a renewal process.

Theorem 5.4. *Let R_t be a renewal reward process with uncertain interarrival times $\xi_1, \xi_2, \cdots$ and uncertain rewards $\eta_1, \eta_2, \cdots$ Assume those interarrival times and rewards have uncertainty distributions Φ and Ψ, respectively. Then R_t has an uncertainty distribution*

$$\Upsilon_t(x) = \max_{k \geq 0} \left(1 - \Phi \left(\frac{t}{k+1} \right) \right) \wedge \Psi \left(\frac{x}{k} \right). \tag{5.19}$$

Here we set $x/k = +\infty$ and $\Psi(x/k) = 1$ when $k = 0$.

Proof: It follows from the definition of renewal reward process that the renewal process N_t is independent of uncertain rewards $\eta_1, \eta_2, \cdots$, and

$$\Upsilon_t(x) = \mathcal{M}\{R_t \le x\} = \mathcal{M}\left\{\sum_{i=1}^{N_t} \eta_i \le x\right\}$$

$$= \mathcal{M}\left\{\bigcup_{k=0}^{\infty}(N_t = k) \cap \sum_{i=1}^{k} \eta_i \le x\right\}$$

$$= \mathcal{M}\left\{\bigcup_{k=0}^{\infty}(N_t = k) \cap \left(\eta_1 \le \frac{x}{k}\right)\right\}$$

$$= \max_{k \ge 0}\mathcal{M}\left\{(N_t \le k) \cap \left(\eta_1 \le \frac{x}{k}\right)\right\}$$

$$= \max_{k \ge 0}\mathcal{M}\{N_t \le k\} \wedge \mathcal{M}\left\{\eta_1 \le \frac{x}{k}\right\}$$

$$= \max_{k \ge 0}\left(1 - \Phi\left(\frac{t}{k+1}\right)\right) \wedge \Psi\left(\frac{x}{k}\right).$$

The theorem is proved.

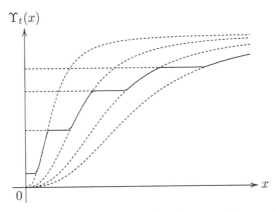

Figure 5.4: Uncertainty Distribution $\Upsilon_t(x)$ of Renewal Reward Process R_t in which the dashed horizontal lines are $1 - \Phi(t/(k+1))$ and the dashed curves are $\Psi(x/k)$.

Theorem 5.5 *(Renewal Reward Theorem). Assume that R_t is a renewal reward process with uncertain interarrival times $\xi_1, \xi_2, \cdots$ and uncertain rewards $\eta_1, \eta_2, \cdots$ If $E[\eta_1/\xi_1]$ exists, then*

$$\lim_{t \to \infty} \frac{E[R_t]}{t} = E\left[\frac{\eta_1}{\xi_1}\right]. \tag{5.20}$$

If those interarrival times and rewards have regular uncertainty distributions Φ and Ψ, respectively, then

$$\lim_{t \to \infty} \frac{E[R_t]}{t} = \int_0^1 \frac{\Psi^{-1}(\alpha)}{\Phi^{-1}(1-\alpha)} d\alpha. \tag{5.21}$$

Proof: It follows from Theorem 5.4 that the uncertainty distribution of R_t is

$$\Upsilon_t(x) = \max_{k \geq 0} \left(1 - \Phi\left(\frac{t}{k+1} \right) \right) \wedge \Psi\left(\frac{x}{k} \right).$$

Then R_t/t has an uncertainty distribution

$$\Upsilon_t(tx) = \max_{k \geq 0} \left(1 - \Phi\left(\frac{t}{k+1} \right) \right) \wedge \Psi\left(\frac{tx}{k} \right).$$

When $t \to \infty$, we have

$$\Upsilon_t(tx) \to \sup_{y \geq 0} (1 - \Phi(y)) \wedge \Psi(xy)$$

which is just the uncertainty distribution of η_1/ξ_1. Thus the equation (5.20) follows from the existence of $E[\eta_1/\xi_1]$. In addition, since the inverse uncertainty distribution of η_1/ξ_1 is just $\Psi^{-1}(\alpha)/\Phi^{-1}(1-\alpha)$, the equation (5.21) follows from Theorem 1.32 immediately.

5.3 Martingale

Definition 5.9. *An uncertain process X_t is called martingale if it has independent increments whose expected values are zero.*

5.4 Markov Process

Definition 5.10. *An uncertain process X_t is called Markov if, given the value of X_t, the uncertain variables X_s and X_u are independent for any $s > t > u$.*

5.5 Stationary Process

Definition 5.11. *An uncertain process X_t is called stationary if for any positive integer k and any times $t_1, t_2, \cdots, t_k$ and s, the uncertain vectors*

$$(X_{t_1}, X_{t_2}, \cdots, X_{t_k}) \quad \text{and} \quad (X_{t_1+s}, X_{t_2+s}, \cdots, X_{t_k+s}) \tag{5.22}$$

are identically distributed.

Chapter 6

Uncertain Calculus

Uncertain calculus, invented by Liu [123] in 2009, is a branch of mathematics that deals with differentiation and integration of function of uncertain processes. This chapter will introduce canonical process, uncertain integral, chain rule, and integration by parts.

6.1 Canonical Process

Definition 6.1 *(Liu [123]). An uncertain process C_t is said to be a canonical process if*
(i) $C_0 = 0$ and almost all sample paths are Lipschitz continuous,
(ii) C_t has stationary and independent increments,
(iii) every increment $C_{s+t} - C_s$ is a normal uncertain variable with expected value 0 and variance t^2, whose uncertainty distribution is

$$\Phi(x) = \left(1 + \exp\left(-\frac{\pi x}{\sqrt{3}t}\right)\right)^{-1}. \tag{6.1}$$

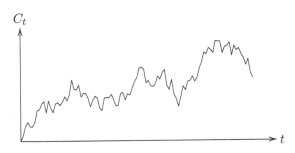

Figure 6.1: A Sample Path of Canonical Process

Note that almost all sample paths of canonical process are Lipschitz continuous functions, but almost all sample paths of Brownian motion are continuous but non-Lipschitz functions. If we say Brownian motion describes the irregular movement of pollen with infinite speed, then we may say the canonical process describes the irregular movement of pollen with finite speed.

Theorem 6.1 *(Existence Theorem). There is a canonical process.*

B. Liu: Uncertainty Theory: A Branch of Mathematics, SCI 300, pp. 139–145.
springerlink.com   © Springer-Verlag Berlin Heidelberg 2010

Proof: Without loss of generality, we only prove that there is a canonical process on the range of $t \in [0, 1]$. Let

$$\{\xi(r) \mid r \text{ represents rational numbers in } [0, 1]\}$$

be a countable sequence of independently normal uncertain variables with expected value zero and variance one. For each integer n, we define an uncertain process

$$X_n(t) = \begin{cases} \dfrac{1}{n} \sum_{i=1}^{k} \xi\left(\dfrac{i}{n}\right), & \text{if } t = \dfrac{k}{n} \quad (k = 0, 1, \cdots, n) \\ \text{linear}, & \text{otherwise.} \end{cases}$$

Since the limit

$$\lim_{n \to \infty} X_n(t)$$

exists almost surely, we may verify that the limit meets the conditions of canonical process. Hence there is a canonical process.

Theorem 6.2. *Let C_t be a canonical process. Then for each time $t > 0$, the ratio C_t/t is a normal uncertain variable with expected value 0 and variance 1. That is,*

$$\frac{C_t}{t} \sim \mathcal{N}(0, 1) \tag{6.2}$$

for any $t > 0$.

Proof: It follows from the definition of canonical process that at each time t, C_t is a normal uncertain variable with uncertainty distribution

$$\Phi(x) = \left(1 + \exp\left(-\frac{\pi x}{\sqrt{3}t}\right)\right)^{-1}.$$

Thus C_t/t has an uncertainty distribution

$$\Psi(x) = \Phi(tx) = \left(1 + \exp\left(-\frac{\pi x}{\sqrt{3}}\right)\right)^{-1}.$$

Hence C_t/t is a normal uncertain variable with expected value 0 and variance 1. The theorem is verified.

Theorem 6.3. *Let C_t be a canonical process. Then for any level $x \in \Re$ and any time $t > 0$, we have*

$$\mathcal{M}\{C_t \leq x\} = \left(1 + \exp\left(-\frac{\pi x}{\sqrt{3}t}\right)\right)^{-1}, \tag{6.3}$$

$$\mathcal{M}\{C_t \geq x\} = \left(1 + \exp\left(\frac{\pi x}{\sqrt{3}t}\right)\right)^{-1}. \tag{6.4}$$

Proof: Since C_t is a normal uncertain variable with expected value 0 and variance t^2, we get (6.3) immediately. The equation (6.4) may be derived from $\mathcal{M}\{C_t \geq x\} = 1 - \mathcal{M}\{C_t \leq x\}$.

Arithmetic Canonical Process

Definition 6.2. *Let C_t be a canonical process. Then for any real numbers e and σ,*

$$A_t = et + \sigma C_t \tag{6.5}$$

is called an arithmetic canonical process, where e is called the drift and σ is called the diffusion.

At each time t, the arithmetic canonical process A_t is a normal uncertain variable, i.e.,

$$A_t \sim \mathcal{N}(et, \sigma t). \tag{6.6}$$

That is, the expected value $E[A_t] = et$ and variance $V[A_t] = \sigma^2 t^2$ at any time t.

Geometric Canonical Process

Definition 6.3. *Let C_t be a canonical process. Then for any real numbers e and σ,*

$$G_t = \exp(et + \sigma C_t) \tag{6.7}$$

is called a geometric canonical process, where e is called the log-drift and σ is called the log-diffusion.

At each time t, the geometric canonical process G_t is a lognormal uncertain variable, i.e.,

$$G_t \sim \mathcal{LOGN}(et, \sigma t). \tag{6.8}$$

If $t < \pi/(\sigma\sqrt{3})$, then $E[G_t] = \sqrt{3}\sigma t \exp(et) \csc(\sqrt{3}\sigma t)$. However, when t arrives at $\pi/(\sigma\sqrt{3})$, we have $E[G_t] = +\infty$.

6.2   Uncertain Integral

Definition 6.4 *(Liu [123]). Let X_t be an uncertain process and let C_t be a canonical process. For any partition of closed interval $[a, b]$ with $a = t_1 < t_2 < \cdots < t_{k+1} = b$, the mesh is written as*

$$\Delta = \max_{1 \le i \le k} |t_{i+1} - t_i|. \tag{6.9}$$

Then the uncertain integral of X_t with respect to C_t is

$$\int_a^b X_t \mathrm{d}C_t = \lim_{\Delta \to 0} \sum_{i=1}^{k} X_{t_i} \cdot (C_{t_{i+1}} - C_{t_i}) \tag{6.10}$$

provided that the limit exists almost surely and is an uncertain variable.

Example 6.1: Let C_t be a canonical process. Then for any partition $0 = t_1 < t_2 < \cdots < t_{k+1} = s$, we have

$$\int_0^s dC_t = \lim_{\Delta \to 0} \sum_{i=1}^{k} (C_{t_{i+1}} - C_{t_i}) \equiv C_s - C_0 = C_s.$$

Example 6.2: Let C_t be a canonical process. Then for any partition $0 = t_1 < t_2 < \cdots < t_{k+1} = s$, we have

$$C_s^2 = \sum_{i=1}^{k} \left(C_{t_{i+1}}^2 - C_{t_i}^2 \right)$$

$$= \sum_{i=1}^{k} \left(C_{t_{i+1}} - C_{t_i} \right)^2 + 2 \sum_{i=1}^{k} C_{t_i} \left(C_{t_{i+1}} - C_{t_i} \right)$$

$$\to 0 + 2 \int_0^s C_t dC_t$$

as $\Delta \to 0$. That is,

$$\int_0^s C_t dC_t = \frac{1}{2} C_s^2.$$

Theorem 6.4. *Let C_t be a canonical process and let $f(t)$ be a determinstic and integrable function with respect to t. Then the uncertain integral*

$$\int_0^s f(t) dC_t \tag{6.11}$$

is a normal uncertain variable at each time s, i.e.,

$$\int_0^s f(t) dC_t \sim \mathcal{N} \left(0, \int_0^s |f(t)| dt \right). \tag{6.12}$$

Proof: Since the canonical process has stationary and independent increments and every increment is a normal uncertain variable, for any partition of closed interval $[0, s]$ with $0 = t_1 < t_2 < \cdots < t_{k+1} = s$, it follows from Theorem 1.23 that

$$\sum_{i=1}^{k} f(t_i)(C_{t_{i+1}} - C_{t_i}) \sim \mathcal{N} \left(0, \sum_{i=1}^{k} |f(t_i)|(t_{i+1} - t_i) \right).$$

That is, the sum is also a normal uncertain variable. Since f is an integrable function, we have

$$\sum_{i=1}^{k} |f(t_i)|(t_{i+1} - t_i) \to \int_0^s |f(t)| dt$$

as the mesh $\Delta \to 0$. Hence we obtain

$$\int_0^s f(t)\mathrm{d}C_t = \lim_{\Delta \to 0} \sum_{i=1}^k f(t_i)(C_{t_{i+1}} - C_{t_i}) \sim \mathcal{N}\left(0, \int_0^s |f(t)|\mathrm{d}t\right).$$

The theorem is proved.

Example 6.3: Let C_t be a canonical process. Then for any number α $(0 < \alpha < 1)$, the uncertain process

$$F_s = \int_0^s (s-t)^{-\alpha}\mathrm{d}C_t \tag{6.13}$$

is called a *fractional canonical process* with index α. At each time s, it follows from Theorem 6.4 that F_s is a normal uncertain variable, i.e.,

$$F_s \sim \mathcal{N}\left(0, \frac{s^{1-\alpha}}{1-\alpha}\right). \tag{6.14}$$

6.3 Chain Rule

Theorem 6.5 *(Liu [123]). Let C_t be a canonical process, and let $h(t, c)$ be a continuously differentiable function. Define $X_t = h(t, C_t)$. Then we have the following chain rule*

$$\mathrm{d}X_t = \frac{\partial h}{\partial t}(t, C_t)\mathrm{d}t + \frac{\partial h}{\partial c}(t, C_t)\mathrm{d}C_t. \tag{6.15}$$

Proof: Write $\Delta C_t = C_{t+\Delta t} - C_t = C_{\Delta t}$. Then Δt and ΔC_t are infinitesimals with the same order. Since the function h is continuously differentiable, by using Taylor series expansion, the infinitesimal increment of X_t has a first-order approximation

$$\Delta X_t = \frac{\partial h}{\partial t}(t, C_t)\Delta t + \frac{\partial h}{\partial c}(t, C_t)\Delta C_t.$$

Hence we obtain the chain rule because it makes

$$X_s = X_0 + \int_0^s \frac{\partial h}{\partial t}(t, C_t)\mathrm{d}t + \int_0^s \frac{\partial h}{\partial c}(t, C_t)\mathrm{d}C_t$$

for any $s \geq 0$.

Remark 6.1: The infinitesimal increment $\mathrm{d}C_t$ in (6.15) may be replaced with the derived canonical process

$$\mathrm{d}Y_t = u_t\mathrm{d}t + v_t\mathrm{d}C_t \tag{6.16}$$

where u_t and v_t are absolutely integrable uncertain processes, thus producing

$$\mathrm{d}h(t, Y_t) = \frac{\partial h}{\partial t}(t, Y_t)\mathrm{d}t + \frac{\partial h}{\partial c}(t, Y_t)\mathrm{d}Y_t. \tag{6.17}$$

Example 6.4: Applying the chain rule, we obtain the following formula

$$\mathrm{d}(tC_t) = C_t\mathrm{d}t + t\mathrm{d}C_t.$$

Hence we have

$$sC_s = \int_0^s \mathrm{d}(tC_t) = \int_0^s C_t\mathrm{d}t + \int_0^s t\mathrm{d}C_t.$$

That is,

$$\int_0^s t\mathrm{d}C_t = sC_s - \int_0^s C_t\mathrm{d}t.$$

Example 6.5: Applying the chain rule, we obtain the following formula

$$\mathrm{d}(C_t^2) = 2C_t\mathrm{d}C_t.$$

Then we have

$$C_s^2 = \int_0^s \mathrm{d}(C_t^2) = 2\int_0^s C_t\mathrm{d}C_t.$$

It follows that

$$\int_0^s C_t\mathrm{d}C_t = \frac{1}{2}C_s^2.$$

Example 6.6: Applying the chain rule, we obtain the following formula

$$\mathrm{d}(C_t^3) = 3C_t^2\mathrm{d}C_t.$$

Thus we get

$$C_s^3 = \int_0^s \mathrm{d}(C_t^3) = 3\int_0^s C_t^2\mathrm{d}C_t.$$

That is

$$\int_0^s C_t^2\mathrm{d}C_t = \frac{1}{3}C_s^3.$$

6.4 Integration by Parts

Theorem 6.6 *(Integration by Parts). Suppose that C_t is a canonical process and $F(t)$ is an absolutely continuous function. Then*

$$\int_0^s F(t)\mathrm{d}C_t = F(s)C_s - \int_0^s C_t\mathrm{d}F(t). \tag{6.18}$$

Proof: By defining $h(t, C_t) = F(t)C_t$ and using the chain rule, we get

$$\mathrm{d}(F(t)C_t) = C_t\mathrm{d}F(t) + F(t)\mathrm{d}C_t.$$

Thus

$$F(s)C_s = \int_0^s d(F(t)C_t) = \int_0^s C_t dF(t) + \int_0^s F(t)dC_t$$

which is just (6.18).

Example 6.7: Assume $F(t) \equiv 1$. Then by using the integration by parts, we immediately obtain

$$\int_0^s dC_t = C_s.$$

Example 6.8: Assume $F(t) = t$. Then by using the integration by parts, we immediately obtain

$$\int_0^s t dC_t = sC_s - \int_0^s C_t dt.$$

Example 6.9: Assume $F(t) = t^2$. Then by using the integration by parts, we obtain

$$\int_0^s t^2 dC_t = s^2 C_s - \int_0^s C_t dt^2 = s^2 C_s - 2\int_0^s tC_t dt$$

$$= (s^2 - 2s)C_s + 2\int_0^s C_t dt.$$

Example 6.10: Assume $F(t) = \sin t$. Then by using the integration by parts, we obtain

$$\int_0^s \sin t dC_t = C_s \sin s - \int_0^s C_t d\sin t = C_s \sin s - \int_0^s C_t \cos t dt.$$

Chapter 7

Uncertain Differential Equation

Uncertain differential equation, proposed by Liu [121] in 2008, is a type of differential equation driven by canonical process. Uncertain differential equation was then introduced into finance by Liu [123] in 2009. After that, an existence and uniqueness theorem of solution of uncertain differential equation was proved by Chen and Liu [17], and a stability theorem was showed by Chen [20].

This chapter will discuss the existence, uniqueness and stability of solutions of uncertain differential equations. This chapter will also provide a 99-method to solve uncertain differential equations numerically. Finally, some applications of uncertain differential equation in finance are documented.

7.1 Uncertain Differential Equation

Definition 7.1 *(Liu [121]). Suppose C_t is a canonical process, and f and g are some given functions. Then*

$$\mathrm{d}X_t = f(t, X_t)\mathrm{d}t + g(t, X_t)\mathrm{d}C_t \tag{7.1}$$

is called an uncertain differential equation. A solution is an uncertain process X_t that satisfies (7.1) identically in t.

Remark 7.1: Note that there is no precise definition for the terms $\mathrm{d}X_t$, $\mathrm{d}t$ and $\mathrm{d}C_t$ in the uncertain differential equation (7.1). The mathematically meaningful form is the uncertain integral equation

$$X_s = X_0 + \int_0^s f(t, X_t)\mathrm{d}t + \int_0^s g(t, X_t)\mathrm{d}C_t. \tag{7.2}$$

However, the differential form is more convenient for us. This is the main reason why we accept the differential form.

Example 7.1: Let C_t be a canonical process. Then the uncertain differential equation

$$\mathrm{d}X_t = a\mathrm{d}t + b\mathrm{d}C_t \tag{7.3}$$

B. Liu: Uncertainty Theory: A Branch of Mathematics, SCI 300, pp. 147–161.
springerlink.com © Springer-Verlag Berlin Heidelberg 2010

has a solution

$$X_t = at + bC_t \tag{7.4}$$

which is just an arithmetic canonical process.

Example 7.2: Let C_t be a canonical process. Then the uncertain differential equation

$$dX_t = aX_t dt + bX_t dC_t \tag{7.5}$$

has a solution

$$X_t = \exp(at + bC_t) \tag{7.6}$$

which is just a geometric canonical process.

Example 7.3: Let C_t be a canonical process. Then the uncertain differential equation

$$dX_t = (m - aX_t)dt + \sigma dC_t \tag{7.7}$$

has a solution

$$X_t = \frac{m}{a} + \exp(-at)\left(X_0 - \frac{m}{a}\right) + \sigma \exp(-at) \int_0^t \exp(as)dC_s \tag{7.8}$$

provided that $a \neq 0$. It follows from Theorem 6.4 that X_t is a normal uncertain variable, i.e.,

$$X_t \sim \mathcal{N}\left(\frac{m}{a} + \exp(-at)\left(X_0 - \frac{m}{a}\right), \frac{\sigma}{a} - \exp(-at)\frac{\sigma}{a}\right). \tag{7.9}$$

Example 7.4: Let u_t and v_t be some continuous functions with respect to t. Consider the *homogeneous linear uncertain differential equation*

$$dX_t = u_t X_t dt + v_t X_t dC_t. \tag{7.10}$$

It follows from the chain rule that

$$d \ln X_t = \frac{dX_t}{X_t} = u_t dt + v_t dC_t.$$

Integration of both sides yields

$$\ln X_t - \ln X_0 = \int_0^t u_s ds + \int_0^t v_s dC_s.$$

Therefore the solution of (7.10) is

$$X_t = X_0 \exp\left(\int_0^t u_s ds + \int_0^t v_s dC_s\right). \tag{7.11}$$

Example 7.5: Suppose $u_{1t}, u_{2t}, v_{1t}, v_{2t}$ are continuous functions with respect to t. Consider the *linear uncertain differential equation*

$$dX_t = (u_{1t}X_t + u_{2t})dt + (v_{1t}X_t + v_{2t})dC_t. \tag{7.12}$$

At first, we define two uncertain processes U_t and V_t via

$$dU_t = u_{1t}U_t dt + v_{1t}U_t dC_t, \quad dV_t = \frac{u_{2t}}{U_t}dt + \frac{v_{2t}}{U_t}dC_t.$$

Then we have $X_t = U_t V_t$ because

$$\begin{aligned}
dX_t &= V_t dU_t + U_t dV_t \\
&= (u_{1t}U_t V_t + u_{2t})dt + (v_{1t}U_t V_t + v_{2t})dC_t \\
&= (u_{1t}X_t + u_{2t})dt + (v_{1t}X_t + v_{2t})dC_t.
\end{aligned}$$

Note that

$$U_t = U_0 \exp\left(\int_0^t u_{1s}ds + \int_0^t v_{1s}dC_s\right),$$

$$V_t = V_0 + \int_0^t \frac{u_{2s}}{U_s}ds + \int_0^t \frac{v_{2s}}{U_s}dC_s.$$

Taking $U_0 = 1$ and $V_0 = X_0$, we get a solution of the linear uncertain differential equation as follows,

$$X_t = U_t\left(X_0 + \int_0^t \frac{u_{2s}}{U_s}ds + \int_0^t \frac{v_{2s}}{U_s}dC_s\right) \tag{7.13}$$

where

$$U_t = \exp\left(\int_0^t u_{1s}ds + \int_0^t v_{1s}dC_s\right). \tag{7.14}$$

7.2 Existence and Uniqueness Theorem

Theorem 7.1 *(Chen and Liu [17], Existence and Uniqueness Theorem). The uncertain differential equation*

$$dX_t = f(t, X_t)dt + g(t, X_t)dC_t \tag{7.15}$$

has a unique solution if the coefficients $f(x,t)$ and $g(x,t)$ satisfy the Lipschitz condition

$$|f(x,t) - f(y,t)| + |g(x,t) - g(y,t)| \le L|x-y|, \quad \forall x,y \in \Re, t \ge 0 \tag{7.16}$$

and linear growth condition

$$|f(x,t)| + |g(x,t)| \le L(1+|x|), \quad \forall x \in \Re, t \ge 0 \tag{7.17}$$

for some constant L. Moreover, the solution is sample-continuous.

Proof: We first prove the existence of solution by a successive approximation method. Define $X_t^{(0)} = X_0$, and

$$X_t^{(n)} = X_0 + \int_0^t f\left(X_s^{(n-1)}, s\right)ds + \int_0^t g\left(X_s^{(n-1)}, s\right)dC_s$$

for $n = 1, 2, \cdots$ and write

$$D_t^{(n)}(\gamma) = \max_{0 \le s \le t} \left| X_s^{(n+1)}(\gamma) - X_s^{(n)}(\gamma) \right|$$

for each $\gamma \in \Gamma$. It follows from the Lipschitz condition and linear growth condition that

$$D_t^{(0)}(\gamma) = \max_{0 \le s \le t} \left| \int_0^s f(X_0, v) dv + \int_0^s g(X_0, v) dC_v(\gamma) \right|$$

$$\le \int_0^t |f(X_0, v)|\, dv + K_\gamma \int_0^t |g(X_0, v)|\, dv$$

$$\le (1 + |X_0|) L (1 + K_\gamma) t$$

where K_γ is the Lipschitz constant to the sample path $C_t(\gamma)$. In fact, by using the induction method, we may verify

$$D_t^{(n)}(\gamma) \le (1 + |X_0|) \frac{L^{n+1}(1 + K_\gamma)^{n+1}}{(n+1)!} t^{n+1}$$

for each n. This means that, for each sample γ, the paths $X_t^{(k)}(\gamma)$ converges uniformly on any given interval $[0, T]$. Write the limit by $X_t(\gamma)$ that is just a solution of the uncertain differential equation because

$$X_t = X_0 + \int_0^t f(X_s, s) ds + \int_0^t g(X_s, s) ds.$$

Next we prove that the solution is unique. Assume that both X_t and X_t^* are solutions of the uncertain differential equation. Then for each $\gamma \in \Gamma$, it follows from the Lipschitz condition and linear growth condition that

$$|X_t(\gamma) - X_t^*(\gamma)| \le L(1 + K_\gamma) \int_0^t |X_v(\gamma) - X_v^*(\gamma)| dv.$$

By using Gronwall inequality, we obtain

$$|X_t(\gamma) - X_t^*(\gamma)| \le 0 \cdot \exp(L(1 + K_\gamma)t) = 0.$$

Hence $X_t = X_t^*$. The uniqueness is proved. Finally, let us prove the sample-continuity of X_t. The Lipschitz condition and linear growth condition may produce

$$|X_t(\gamma) - X_s(\gamma)| = \left| \int_s^t f(X_v(\gamma), v) dv + \int_s^t g(X_v(\gamma), v) dC_v(\gamma) \right|$$

$$\le (1 + K_\gamma)(1 + |X_0|) \exp(L(1 + K_\gamma)t)(t - s)$$

$$\to 0 \text{ as } s \to t.$$

Thus X_t is sample-continuous and the theorem is proved.

7.3   Stability Theorem

Definition 7.2 *(Liu [123]). An uncertain differential equation is said to be stable if for any given numbers $\kappa > 0$ and $\varepsilon > 0$, there exists a number $\delta > 0$ such that for any solutions X_t and Y_t, we have*

$$\mathcal{M}\{|X_t - Y_t| > \kappa\} < \varepsilon, \quad \forall t > 0 \tag{7.18}$$

whenever $|X_0 - Y_0| < \delta$.

In other words, an uncertain differential equation is stable if for any given number $\kappa > 0$, we have

$$\lim_{|X_0 - Y_0| \to 0} \mathcal{M}\{|X_t - Y_t| > \kappa\} = 0, \quad \forall t > 0. \tag{7.19}$$

Example 7.6: The uncertain differential equation $dX_t = adt + bdC_t$ is stable since for any given numbers $\kappa > 0$ and $\varepsilon > 0$, we may take $\delta = \kappa$ and have

$$\mathcal{M}\{|X_t - Y_t| > \kappa\} = \mathcal{M}\{|X_0 - Y_0| > \kappa\} = \mathcal{M}\{\emptyset\} = 0 < \varepsilon$$

for any time $t > 0$ whenever $|X_0 - Y_0| < \delta$.

Example 7.7: The uncertain differential equation $dX_t = X_t dt + bdC_t$ is unstable since for any given number $\kappa > 0$ and any different initial solutions X_0 and Y_0, we have

$$\mathcal{M}\{|X_t - Y_t| > \kappa\} = \mathcal{M}\{\exp(t)|X_0 - Y_0| > \kappa\} = 1$$

provided that t is sufficiently large.

Theorem 7.2 *(Chen [20], Stability Theorem). Suppose u_t and v_t are continuous functions such that*

$$\sup_{s \geq 0} \int_0^s u_t dt < +\infty, \quad \int_0^{+\infty} |v_t| dt < +\infty. \tag{7.20}$$

Then the uncertain differential equation

$$dX_t = u_t X_t dt + v_t X_t dC_t \tag{7.21}$$

is stable.

Proof: It has been proved that the unique solution of the uncertain differential equation $dX_t = u_t X_t dt + v_t X_t dC_t$ is

$$X_t = X_0 \exp\left(\int_0^t u_s ds + \int_0^t v_s dC_s\right).$$

Thus for any given number $\kappa > 0$, we have

$$\mathcal{M}\{|X_t - Y_t| > \kappa\} = \mathcal{M}\left\{|X_0 - Y_0| \exp\left(\int_0^t u_s ds + \int_0^t v_s dC_s\right) > \kappa\right\}$$

$$= \mathcal{M}\left\{\int_0^t v_s dC_s > \ln\frac{\kappa}{|X_0 - Y_0|} - \int_0^t u_s ds\right\} \to 0$$

as $|X_0 - Y_0| \to 0$ because

$$\int_0^t v_s dC_s \sim \mathcal{N}\left(0, \int_0^t |v_s| ds\right)$$

is a normal uncertain variable with expected value 0 and finite variance, and

$$\ln\frac{\kappa}{|X_0 - Y_0|} - \int_0^t u_s ds \to +\infty.$$

The theorem is proved.

7.4   Numerical Method

It is almost impossible to find analytic solutions for general uncertain differential equations. This fact provides a motivation to design numerical methods to solve uncertain differential equations.

Definition 7.3. *Let α be a number with $0 < \alpha < 1$. An uncertain differential equation*

$$dX_t = f(t, X_t)dt + g(t, X_t)dC_t \tag{7.22}$$

is said to have an α-path X_t^α if it solves the corresponding ordinary differential equation

$$dX_t^\alpha = f(t, X_t^\alpha)dt + g(t, X_t^\alpha)\Phi^{-1}(\alpha)dt \tag{7.23}$$

where $\Phi^{-1}(\alpha)$ is the inverse uncertainty distribution of standard normal uncertain variable, i.e.,

$$\Phi^{-1}(\alpha) = \frac{\sqrt{3}}{\pi} \ln\frac{\alpha}{1 - \alpha}. \tag{7.24}$$

Example 7.8: The uncertain differential equation $dX_t = a dt + b dC_t$ with $X_0 = 0$ has an α-path

$$X_t^\alpha = at + b\Phi^{-1}(\alpha)t. \tag{7.25}$$

Example 7.9: The uncertain differential equation $dX_t = aX_t dt + bX_t dC_t$ with $X_0 = 1$ has an α-path

$$X_t^\alpha = \exp\left(at + b\Phi^{-1}(\alpha)t\right). \tag{7.26}$$

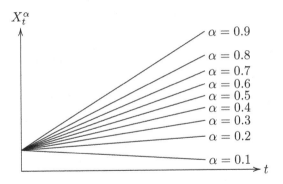

Figure 7.1: A Spectrum of α-Paths of $dX_t = adt + bdC_t$

Definition 7.4. *The uncertain differential equation (7.22) is said to be monotone increasing if for any $\alpha \in (0,1)$ and any $t \geq 0$, we have*

$$\mathcal{M}\{X_t \leq X_t^\alpha\} = \alpha \qquad (7.27)$$

where X_t and X_t^α are the solution and α-path of (7.22), respectively.

Example 7.10: The homogeneous linear uncertain differential equation

$$dX_t = aX_t dt + bX_t dC_t \qquad (7.28)$$

is monotone increasing whenever $b > 0$.

Example 7.11: The special linear uncertain differential equation

$$dX_t = (m - aX_t)dt + \sigma dC_t \qquad (7.29)$$

is monotone increasing whenever $\sigma > 0$.

Theorem 7.3. *If an uncertain differential equation is monotone increasing, then its α-path X_t^α is increasing with respect to α at each time t. That is,*

$$X_t^\alpha \leq X_t^\beta \qquad (7.30)$$

at each time t whenever $\alpha < \beta$.

Proof: Since the uncertain differential equation is monotone increasing, we immediately have

$$\mathcal{M}\{X_t \leq X_t^\alpha\} = \alpha < \beta = \mathcal{M}\{X_t \leq X_t^\beta\}.$$

It follows from the monotonicity of uncertain measure that $X_t^\alpha \leq X_t^\beta$.

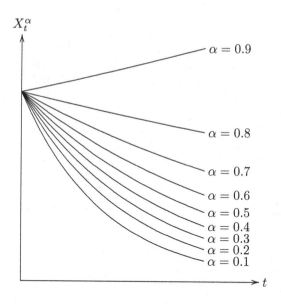

Figure 7.2: A Spectrum of α-Paths of $dX_t = aX_t dt + bX_t dC_t$

Definition 7.5. *The uncertain differential equation (7.22) is said to be monotone decreasing if for any $\alpha \in (0,1)$ and any $t \geq 0$, we have*

$$\mathcal{M}\{X_t \leq X_t^\alpha\} = 1 - \alpha \tag{7.31}$$

where X_t and X_t^α are the solution and α-path of (7.22), respectively.

Theorem 7.4. *If an uncertain differential equation is monotone decreasing, then its α-path X_t^α is decreasing with respect to α at each time t. That is,*

$$X_t^\alpha \geq X_t^\beta \tag{7.32}$$

at each time t whenever $\alpha < \beta$.

Proof: Since the uncertain differential equation is monotone decreasing, we immediately have

$$\mathcal{M}\{X_t \leq X_t^\alpha\} = 1 - \alpha > 1 - \beta = \mathcal{M}\{X_t \leq X_t^\beta\}.$$

It follows from the monotonicity of uncertain measure that $X_t^\alpha \geq X_t^\beta$.

99-Method for Solving $dX_t = f(t, X_t)dt + g(t, X_t)dC_t$

For solving a monotone uncertain differential equation, a key point is to obtain a 99-table of its solution X_s. In order to do so, a 99-method is designed as follows:

Step 1. Fix a time s and set $\alpha = 0$.

Step 2. Set $\alpha \leftarrow \alpha + 0.01$.

Step 3. Employ a classical numerical method to solve the corresponding ordinary differential equation $dX_t^\alpha = f(t, X_t^\alpha)dt + g(t, X_t^\alpha)\Phi^{-1}(\alpha)dt$ and obtain X_s^α.

Step 4. Repeat the second and third steps until $\alpha = 0.99$.

Step 5. For a monotone increasing equation, the solution X_s has a 99-table,

0.01	0.02	$\cdots$	0.99
$X_s^{0.01}$	$X_s^{0.02}$	$\cdots$	$X_s^{0.99}$

$$(7.33)$$

Step 6. For a monotone decreasing equation, the solution X_s has a 99-table,

0.01	0.02	$\cdots$	0.99
$X_s^{0.99}$	$X_s^{0.98}$	$\cdots$	$X_s^{0.01}$

$$(7.34)$$

Note that the 99-method works only when the uncertain differential equation is almost monotone. In addition, the 99-method may be extended to the 999-method if a more precise result is needed. It is suggested that the ordinary differential equations in Step 3 are approximated by the recursion formula

$$X_{i+1}^\alpha = X_i^\alpha + f(t_i, X_i^\alpha)\Delta + g(t_i, X_i^\alpha)\Phi^{-1}(\alpha)\Delta \qquad (7.35)$$

where Δ is the step length.

Example 7.12: Consider a monotone increasing uncertain differential equation

$$dX_t = X_t dt + X_t dC_t, \quad X_0 = 1 \qquad (7.36)$$

whose solution is $X_t = \exp(t + C_t)$. The 99-method may solve this equation successfully and obtain a 99-table of X_t at time $t = 1$ shown in Figure 7.3. The computer program is available at http://orsc.edu.cn/liu/resources.htm.

Example 7.13: Consider a monotone increasing uncertain differential equation

$$dX_t = (1 - X_t)dt + dC_t, \quad X_0 = 1 \qquad (7.37)$$

whose solution is

$$X_t = 1 + \int_0^t \exp(s - t)dC_s. \qquad (7.38)$$

The 99-method obtains a 99-table of X_t at time $t = 1$ shown in Figure 7.4.

Example 7.14: Consider a nonlinear uncertain differential equation

$$dX_t = (t + X_t)dt + \sqrt{1 + X_t}dC_t, \quad X_0 = 2. \qquad (7.39)$$

This equation is not completely monotone, even is not well defined because $1 + X_t$ may take negative values on some extreme sample paths. However,

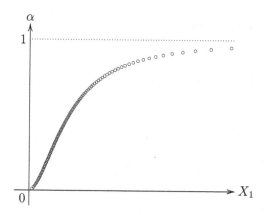

Figure 7.3: The 99-Table of $dX_t = X_t dt + X_t dC_t$ with $X_0 = 1$

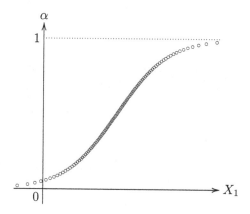

Figure 7.4: The 99-Table of $dX_t = (1 - X_t)dt + dC_t$ with $X_0 = 1$

this blemish may be ignored and the 99-method is still valid. The 99-method obtains a 99-table of X_t at time $t = 1$ shown in Figure 7.5.

Open Problem: A necessary condition of monotone uncertain differential equation is that its α-path X_t^α is monotone with respect to α. What is a sufficient condition?

7.5 Uncertain Differential Equation with Jumps

In many cases the stock price is not continuous because of economic crisis or war. In order to incorporate those into stock model, we should develop an uncertain calculus with jump process. For many applications, a renewal

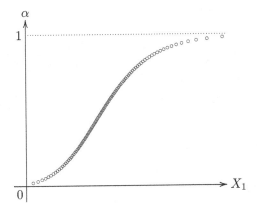

Figure 7.5: The 99-Table of $dX_t = (t + X_t)dt + \sqrt{1 + X_t}dC_t$ with $X_0 = 2$

process N_t is sufficient. The uncertain integral of uncertain process X_t with respect to N_t is

$$\int_a^b X_t dN_t = \lim_{\Delta \to 0} \sum_{i=1}^k X_{t_i} \cdot (N_{t_{i+1}} - N_{t_i}) = \sum_{a \le t \le b} X_t \cdot (N_t - N_{t-}). \quad (7.40)$$

Definition 7.6. *Suppose C_t is a canonical process, N_t is a renewal process, and f, g, h are some given functions. Then*

$$dX_t = f(t, X_t)dt + g(t, X_t)dC_t + h(t, X_t)dN_t \quad (7.41)$$

is called an uncertain differential equation with jumps. A solution is an uncertain process X_t that satisfies (7.41) identically in t.

Example 7.15: Let C_t be a canonical process and N_t a renewal process. Then the uncertain differential equation with jumps

$$dX_t = adt + bdC_t + cdN_t$$

has a solution $X_t = at + bC_t + cN_t$ which is just a jump process.

Example 7.16: Let C_t be a canonical process and N_t a renewal process. Then the uncertain differential equation with jumps

$$dX_t = aX_tdt + bX_tdC_t + cX_tdN_t$$

has a solution $X_t = \exp(at + bC_t + cN_t)$ which may be employed to model stock price with jumps.

7.6   Uncertain Finance

If we assume that the stock price follows some uncertain differential equation, then we may produce a new topic of uncertain finance. As an example, Liu [123] supposed that the stock price follows geometric canonical process and presented a *stock model* in which the bond price X_t and the stock price Y_t are determined by

$$\begin{cases} \mathrm{d}X_t = rX_t\mathrm{d}t \\ \mathrm{d}Y_t = eY_t\mathrm{d}t + \sigma Y_t\mathrm{d}C_t \end{cases} \tag{7.42}$$

where r is the riskless interest rate, e is the stock drift, σ is the stock diffusion, and C_t is a canonical process.

European Call Option Price

A European call option gives the holder the right to buy a stock at a specified time for specified price. Assume that the option has strike price K and expiration time s. Then the payoff from such an option is $(Y_s - K)^+$. Considering the time value of money resulted from the bond, the present value of this payoff is $\exp(-rs)(Y_s - K)^+$. Hence the European call option price should be the expected present value of the payoff,

$$f_c = \exp(-rs)E[(Y_s - K)^+]. \tag{7.43}$$

It is clear that the option price is a decreasing function of interest rate r. That is, the European call option will devaluate if the interest rate is raised; and the European call option will appreciate in value if the interest rate is reduced. In addition, the option price is also a decreasing function of strike price K.

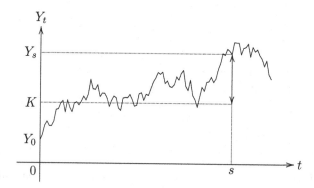

Figure 7.6: Payoff $(Y_s - K)^+$ from European Call Option

Let us consider the financial market described by the stock model (7.42). The European call option price is

$$
\begin{aligned}
f_c &= \exp(-rs)E[(Y_0 \exp(es + \sigma C_s) - K)^+] \\
&= \exp(-rs) \int_0^{+\infty} \mathcal{M}\{Y_0 \exp(es + \sigma C_s) - K \geq x\} \mathrm{d}x \\
&= \exp(-rs)Y_0 \int_{K/Y_0}^{+\infty} \mathcal{M}\{\exp(es + \sigma C_s) \geq y\} \mathrm{d}y \\
&= \exp(-rs)Y_0 \int_{K/Y_0}^{+\infty} \mathcal{M}\{es + \sigma C_s \geq \ln y\} \mathrm{d}y \\
&= \exp(-rs)Y_0 \int_{K/Y_0}^{+\infty} \left(1 + \exp\left(\frac{\pi(\ln y - es)}{\sqrt{3}\sigma s}\right)\right)^{-1} \mathrm{d}y.
\end{aligned}
$$

Thus Liu [123] derived the following European call option price formula,

$$
f_c = \exp(-rs)Y_0 \int_{K/Y_0}^{+\infty} \left(1 + \exp\left(\frac{\pi(\ln y - es)}{\sqrt{3}\sigma s}\right)\right)^{-1} \mathrm{d}y. \tag{7.44}
$$

European Put Option Price

A European put option gives the holder the right to sell a stock at a specified time for specified price. Assume that the option has strike price K and expiration time s. Then the payoff from such an option is $(K - Y_s)^+$. Considering the time value of money resulted from the bond, the present value of this payoff is $\exp(-rs)(K - Y_s)^+$. Hence the European put option price should be the expected present value of the payoff,

$$
f_p = \exp(-rs)E[(K - Y_s)^+]. \tag{7.45}
$$

It is easy to verify that the option price is a decreasing function of interest rate r, and is an increasing function of strike price K.

Let us consider the financial market described by the stock model (7.42). The European put option price is

$$
\begin{aligned}
f_p &= \exp(-rs)E[(K - Y_0 \exp(es + \sigma C_s))^+] \\
&= \exp(-rs) \int_0^{+\infty} \mathcal{M}\{K - Y_0 \exp(es + \sigma C_s) \geq x\} \mathrm{d}x \\
&= \exp(-rs)Y_0 \int_{K/Y_0}^{+\infty} \mathcal{M}\{\exp(es + \sigma C_s) \leq y\} \mathrm{d}y \\
&= \exp(-rs)Y_0 \int_0^{K/Y_0} \mathcal{M}\{es + \sigma C_s \leq \ln y\} \mathrm{d}y \\
&= \exp(-rs)Y_0 \int_0^{K/Y_0} \left(1 + \exp\left(\frac{\pi(es - \ln y)}{\sqrt{3}\sigma s}\right)\right)^{-1} \mathrm{d}y.
\end{aligned}
$$

Thus Liu [123] derived the following European put option price formula,

$$f_p = \exp(-rs)Y_0 \int_0^{K/Y_0} \left(1 + \exp\left(\frac{\pi(es - \ln y)}{\sqrt{3}\sigma s}\right)\right)^{-1} dy. \qquad (7.46)$$

Multi-factor Stock Model

Now we assume that there are multiple stocks whose prices are determined by multiple canonical processes. For this case, we have a *multi-factor stock model* in which the bond price X_t and the stock prices Y_{it} are determined by

$$\begin{cases} dX_t = rX_t dt \\ dY_{it} = e_i Y_{it} dt + \sum_{j=1}^{n} \sigma_{ij} Y_{it} dC_{jt}, \ i = 1, 2, \cdots, m \end{cases} \qquad (7.47)$$

where r is the riskless interest rate, e_i are the stock drift coefficients, σ_{ij} are the stock diffusion coefficients, C_{it} are independent canonical processes, $i = 1, 2, \cdots, m$, $j = 1, 2, \cdots, n$.

Portfolio Selection

For the stock model (7.47), we have the choice of $m+1$ different investments. At each instant t we may choose a portfolio $(\beta_t, \beta_{1t}, \cdots, \beta_{mt})$ (i.e., the investment fractions meeting $\beta_t + \beta_{1t} + \cdots + \beta_{mt} = 1$). Then the wealth Z_t at time t should follow the uncertain differential equation

$$dZ_t = r\beta_t Z_t dt + \sum_{i=1}^{m} e_i \beta_{it} Z_t dt + \sum_{i=1}^{m} \sum_{j=1}^{n} \sigma_{ij} \beta_{it} Z_t dC_{jt}. \qquad (7.48)$$

Portfolio selection problem is to find an optimal portfolio $(\beta_t, \beta_{1t}, \cdots, \beta_{mt})$ such that the expected wealth $E[Z_s]$ is maximized.

No-Arbitrage

The stock model (7.47) is said to be *no-arbitrage* if there is no portfolio $(\beta_t, \beta_{1t}, \cdots, \beta_{mt})$ such that for some time $s > 0$, we have

$$\mathcal{M}\{\exp(-rs)Z_s \geq Z_0\} = 1 \qquad (7.49)$$

and

$$\mathcal{M}\{\exp(-rs)Z_s > Z_0\} > 0 \qquad (7.50)$$

where Z_t is determined by (7.48) and represents the wealth at time t. We may prove that the stock model (7.47) is no-arbitrage if and only if its diffusion matrix

$$\begin{pmatrix} \sigma_{11} & \sigma_{12} & \cdots & \sigma_{1n} \\ \sigma_{21} & \sigma_{22} & \cdots & \sigma_{2n} \\ \vdots & \vdots & \ddots & \vdots \\ \sigma_{m1} & \sigma_{m2} & \cdots & \sigma_{mn} \end{pmatrix}$$

has rank m, i.e., the row vectors are linearly independent.

Stock Model with Mean-Reverting Process

Peng [175] assumed that the stock price follows a type of mean-reverting uncertain process and proposed the following stock model,

$$\begin{cases} \mathrm{d}X_t = rX_t\mathrm{d}t \\ \mathrm{d}Y_t = (m - aY_t)\mathrm{d}t + \sigma\mathrm{d}C_t \end{cases} \tag{7.51}$$

where r, m, a, σ are given constants.

Stock Model with Periodic Dividends

Liu [134] assumed that a dividend of a fraction δ of the stock price is paid at deterministic times $T_1, T_2, \cdots$ and presented a stock model with periodic dividends,

$$\begin{cases} X_t = X_0 \exp(rt) \\ Y_t = Y_0(1 - \delta)^{n_t} \exp(et + \sigma C_t) \end{cases} \tag{7.52}$$

where $n_t = \max\{i : T_i \leq t\}$ is the number of dividend payments made by time t.

Currency Models

Liu [133] assumed that the exchange rate follows a geometric canonical process and proposed a currency model with uncertain exchange rate,

$$\begin{cases} \mathrm{d}X_t = eX_t\mathrm{d}t + \sigma X_t\mathrm{d}C_t & \text{(Exchange rate)} \\ \mathrm{d}Y_t = uY_t\mathrm{d}t & \text{(Yuan Bond)} \\ \mathrm{d}Z_t = vZ_t\mathrm{d}t & \text{(Dollar Bond)} \end{cases} \tag{7.53}$$

where e, σ, u, v are constants. If the exchange rate follows a mean-reverting uncertain process, then the currency model with uncertain exchange rate is

$$\begin{cases} \mathrm{d}X_t = (m - \alpha X_t)\mathrm{d}t + \sigma\mathrm{d}C_t & \text{(Exchange rate)} \\ \mathrm{d}Y_t = uY_t\mathrm{d}t & \text{(Yuan Bond)} \\ \mathrm{d}Z_t = vZ_t\mathrm{d}t & \text{(Dollar Bond)} \end{cases} \tag{7.54}$$

where m, α, σ, u, v are constants.

Chapter 8

Uncertain Logic

Uncertain logic is a generalization of mathematical logic for dealing with uncertain knowledge via uncertainty theory. The first model is uncertain propositional logic designed by Li and Liu [96] in which the truth value of an uncertain proposition is defined as the uncertain measure that the proposition is true. An important contribution is the truth value theorem by Chen and Ralescu [18] that provides a numerical method for calculating the truth value of uncertain formulas. The second model is uncertain predicate logic proposed by Zhang and Peng [227] in which an uncertain predicate proposition is defined as a sequence of uncertain propositions indexed by one or more parameters.

One advantage of uncertain logic is the well consistency with classical logic. For example, uncertain logic obeys the law of truth conservation and is consistent with the law of excluded middle and the law of contradiction. This chapter will introduce uncertain propositional logic and uncertain predicate logic.

8.1   Uncertain Proposition

Definition 8.1 *(Li and Liu [96]). An uncertain proposition is a statement whose truth value is quantified by an uncertain measure.*

That is, if we use ξ to express an uncertain proposition and use c to express its truth value in uncertain measure, then the uncertain proposition ξ is essentially an uncertain variable

$$\xi = \begin{cases} 1 \text{ with uncertain measure } c \\ 0 \text{ with uncertain measure } 1 - c \end{cases} \tag{8.1}$$

where $\xi = 1$ means ξ is true and $\xi = 0$ means ξ is false.

Example 8.1: "Tom is tall with truth value 0.7" is an uncertain proposition, where "Tom is tall" is a statement, and its truth value is 0.7 in uncertain measure.

Example 8.2: "Beijing is a big city with truth value 0.9" is an uncertain proposition, where "Beijing is a big city" is a statement, and its truth value is 0.9 in uncertain measure.

B. Liu: Uncertainty Theory: A Branch of Mathematics, SCI 300, pp. 163–175.
springerlink.com   © Springer-Verlag Berlin Heidelberg 2010

Definition 8.2. *Uncertain propositions are called independent if they are independent uncertain variables.*

Example 8.3: If ξ and η are independent uncertain propositions, then for any x and y taking values either 0 or 1, we have

$$\mathcal{M}\{(\xi = x) \cap (\eta = y)\} = \mathcal{M}\{\xi = x\} \wedge \mathcal{M}\{\eta = y\},$$
$$\mathcal{M}\{(\xi = x) \cup (\eta = y)\} = \mathcal{M}\{\xi = x\} \vee \mathcal{M}\{\eta = y\}.$$

8.2 Connective Symbols

In addition to the proposition symbols ξ and η, we also need the negation symbol $\neg$, conjunction symbol $\wedge$, disjunction symbol $\vee$, conditional symbol $\rightarrow$, and biconditional symbol $\leftrightarrow$. Note that

$$\neg\xi \text{ means "not } \xi\text{"}; \tag{8.2}$$

$$\xi \vee \eta \text{ means "}\xi \text{ or } \eta\text{"}; \tag{8.3}$$

$$\xi \wedge \eta = \neg(\neg\xi \vee \neg\eta) \text{ means "}\xi \text{ and } \eta\text{"}; \tag{8.4}$$

$$\xi \rightarrow \eta = (\neg\xi) \vee \eta \text{ means "if } \xi \text{ then } \eta\text{"}; \tag{8.5}$$

$$\xi \leftrightarrow \eta = (\xi \rightarrow \eta) \wedge (\eta \rightarrow \xi) \text{ means "}\xi \text{ if and only if } \eta\text{"}. \tag{8.6}$$

8.3 Uncertain Formula

An *uncertain formula* is a finite sequence of uncertain propositions and connective symbols that must make sense. For example, let ξ, η, τ be uncertain propositions. Then

$$X = \neg\xi, \quad X = \xi \wedge \eta, \quad X = (\xi \vee \eta) \rightarrow \tau$$

are all uncertain formulas. However, $\neg \vee \xi$, $\xi \rightarrow \vee$ and $\xi\eta \rightarrow \tau$ are not formulas. Note that an uncertain formula X is essentially an uncertain variable taking values 0 or 1. If $X = 1$, then X is true; if $X = 0$, then X is false.

Definition 8.3. *Uncertain formulas are called independent if they are independent uncertain variables.*

8.4 Truth Function

Assume X is a formula containing propositions $\xi_1, \xi_2, \cdots, \xi_n$. It is well-known that there is a truth function $f : \{0,1\}^n \rightarrow \{0,1\}$ such that $X = 1$ if and only if $f(\xi_1, \xi_2, \cdots, \xi_n) = 1$.

Example 8.4: The truth function of formula $\xi_1 \vee \xi_2$ (ξ_1 or ξ_2) is

$$f(1,1) = 1, \quad f(1,0) = 1, \quad f(0,1) = 1, \quad f(0,0) = 0.$$

Example 8.5: The truth function of formula $\xi_1 \wedge \xi_2$ (ξ_1 and ξ_2) is
$$f(1,1) = 1, \quad f(1,0) = 0, \quad f(0,1) = 0, \quad f(0,0) = 0.$$

Example 8.6: The truth function of formula $\xi_1 \to \xi_2$ (if ξ_1 then ξ_2) is
$$f(1,1) = 1, \quad f(1,0) = 0, \quad f(0,1) = 1, \quad f(0,0) = 1.$$

Example 8.7: The truth function of $\xi_1 \leftrightarrow \xi_2$ (ξ_1 if and only if ξ_2) is
$$f(1,1) = 1, \quad f(1,0) = 0, \quad f(0,1) = 0, \quad f(0,0) = 1.$$

Example 8.8: The truth function of $\xi_1 \vee \xi_2 \to \xi_3$ is given by
$$f(1,1,1) = 1, \quad f(1,0,1) = 1, \quad f(0,1,1) = 1, \quad f(0,0,1) = 1,$$
$$f(1,1,0) = 0, \quad f(1,0,0) = 0, \quad f(0,1,0) = 0, \quad f(0,0,0) = 1.$$

8.5   Truth Value

Truth value is a key concept in uncertain logic, and is defined as the uncertain measure that the uncertain formula is true.

Definition 8.4 *(Li and Liu [96]). Let X be an uncertain formula. Then the truth value of X is defined as the uncertain measure that the uncertain formula X is true, i.e.,*
$$T(X) = \mathcal{M}\{X = 1\}. \tag{8.7}$$

The truth value is nothing but an uncertain measure. The higher the truth value is, the more true the uncertain formula is.

An uncertain formula X is called a *tautology* if $T(X) = 1$. For this case, X is certainly true. In other words, it is always true for all possible combinations of the values assigned to its propositions.

An uncertain formula X is called a *contradiction* if $T(X) = 0$. For this case, X is certainly false. In other words, it is always false for all possible combinations of the values assigned to its propositions.

An uncertain formula X is called a *contingency* if $0 < T(X) < 1$. For this case, X can be made either true or false based on the values assigned to its propositions.

Example 8.9: Let ξ and η be two independent uncertain propositions with truth values a and b, respectively. Then
$$T(\xi) = \mathcal{M}\{\xi = 1\} = a, \tag{8.8}$$
$$T(\neg\xi) = \mathcal{M}\{\xi = 0\} = 1 - a, \tag{8.9}$$
$$T(\xi \vee \eta) = \mathcal{M}\{\xi \vee \eta = 1\} = \mathcal{M}\{(\xi = 1) \cup (\eta = 1)\} = a \vee b, \tag{8.10}$$
$$T(\xi \wedge \eta) = \mathcal{M}\{\xi \wedge \eta = 1\} = \mathcal{M}\{(\xi = 1) \cap (\eta = 1)\} = a \wedge b, \tag{8.11}$$
$$T(\xi \to \eta) = T(\neg\xi \vee \eta) = (1 - a) \vee b. \tag{8.12}$$

8.6   Truth Value Theorem

Theorem 8.1 *(Law of Excluded Middle). Let ξ be an uncertain proposition. Then $\xi \vee \neg\xi$ is a tautology, i.e.,*

$$T(\xi \vee \neg\xi) = 1. \tag{8.13}$$

Proof: It follows from the definition of truth value and property of uncertain measure that

$$T(\xi \vee \neg\xi) = \mathcal{M}\{\xi \vee \neg\xi = 1\} = \mathcal{M}\{(\xi = 1) \cup (\xi = 0)\} = 1.$$

The theorem is proved.

Theorem 8.2 *(Law of Contradiction). Let ξ be an uncertain proposition. Then $\xi \wedge \neg\xi$ is a contradiction, i.e.,*

$$T(\xi \wedge \neg\xi) = 0. \tag{8.14}$$

Proof: It follows from the definition of truth value and property of uncertain measure that

$$T(\xi \wedge \neg\xi) = \mathcal{M}\{\xi \wedge \neg\xi = 1\} = \mathcal{M}\{(\xi = 1) \cap (\xi = 0)\} = \mathcal{M}\{\emptyset\} = 0.$$

The theorem is proved.

Theorem 8.3 *(Law of Truth Conservation). Let ξ be an uncertain proposition. Then we have*

$$T(\xi) + T(\neg\xi) = 1. \tag{8.15}$$

Proof: It follows from the self-duality of uncertain measure that

$$T(\neg\xi) = \mathcal{M}\{\neg\xi = 1\} = \mathcal{M}\{\xi = 0\} = 1 - \mathcal{M}\{\xi = 1\} = 1 - T(\xi).$$

The theorem is proved.

Theorem 8.4 *(De Morgan's Law). For any uncertain propositions ξ and η, we have*

$$T(\neg(\xi \wedge \eta)) = T((\neg\xi) \vee (\neg\eta)), \tag{8.16}$$

$$T(\neg(\xi \vee \eta)) = T((\neg\xi) \wedge (\neg\eta)). \tag{8.17}$$

Proof: It follows from the basic properties of uncertain measure that

$$T(\neg(\xi \wedge \eta)) = \mathcal{M}\{\xi \wedge \eta = 0\} = \mathcal{M}\{(\xi = 0) \cup (\eta = 0)\}$$
$$= \mathcal{M}\{(\neg\xi) \vee (\neg\eta) = 1\} = T((\neg\xi) \vee (\neg\eta))$$

which proves the first equality. A similar way may verify the second equality. The theorem is proved.

Theorem 8.5 *(Law of Contraposition). For any uncertain propositions ξ and η, we have*

$$T(\xi \to \eta) = T(\neg\eta \to \neg\xi). \tag{8.18}$$

Proof: It follows from the definition of conditional symbol and basic properties of uncertain measure that

$$T(\xi \to \eta) = \mathcal{M}\{(\neg\xi) \vee \eta = 1\} = \mathcal{M}\{(\xi = 0) \cup (\eta = 1)\}$$
$$= \mathcal{M}\{\eta \vee (\neg\xi) = 1\} = T(\neg\eta \to \neg\xi).$$

The theorem is proved.

Theorem 8.6 *(Monotonicity and Subadditivity). For any uncertain propositions ξ and η, we have*

$$T(\xi) \vee T(\eta) \leq T(\xi \vee \eta) \leq T(\xi) + T(\eta). \tag{8.19}$$

Proof: It follows from the monotonicity of uncertain measure that

$$T(\xi \vee \eta) = \mathcal{M}\{\xi \vee \eta = 1\} = \mathcal{M}\{(\xi = 1) \cup (\eta = 1)\}$$
$$\geq \mathcal{M}\{\xi = 1\} \vee \mathcal{M}\{\eta = 1\} = T(\xi) \vee T(\eta).$$

It follows from the subadditivity of uncertain measure that

$$T(\xi \vee \eta) = \mathcal{M}\{\xi \vee \eta = 1\} = \mathcal{M}\{(\xi = 1) \cup (\eta = 1)\}$$
$$\leq \mathcal{M}\{\xi = 1\} + \mathcal{M}\{\eta = 1\} = T(\xi) + T(\eta).$$

The theorem is verified.

Theorem 8.7. *For any uncertain propositions ξ and η, we have*

$$T(\xi) + T(\eta) - 1 \leq T(\xi \wedge \eta) \leq T(\xi) \wedge T(\eta). \tag{8.20}$$

Proof: It follows from the monotonicity of truth value that

$$T(\xi \wedge \eta) = 1 - T(\neg\xi \vee \neg\eta) \leq 1 - T(\neg\xi) \vee T(\neg\eta)$$
$$= (1 - T(\neg\xi)) \wedge (1 - T(\neg\eta)) = T(\xi) \wedge T(\eta).$$

It follows from the subadditivity of truth value that

$$T(\xi \wedge \eta) = 1 - T(\neg\xi \vee \neg\eta) \geq 1 - (T(\neg\xi) + T(\neg\eta))$$
$$= 1 - (1 - T(\xi)) - (1 - T(\eta)) = T(\xi) + T(\eta) - 1.$$

The theorem is proved.

Theorem 8.8. *Let ξ be an uncertain proposition. Then $\xi \to \xi$ is a tautology, i.e.,*

$$T(\xi \to \xi) = 1. \tag{8.21}$$

Proof: It follows from the definition of conditional symbol and the law of excluded middle that

$$T(\xi \to \xi) = T(\neg \xi \vee \xi) = 1.$$

The theorem is proved.

Theorem 8.9. *Let ξ be an uncertain proposition. Then $\xi \leftrightarrow \xi$ is a tautology, i.e.,*

$$T(\xi \leftrightarrow \xi) = 1. \tag{8.22}$$

Proof: It follows from the definition of biconditional symbol and Theorem 8.8 that

$$T(\xi \leftrightarrow \xi) = T((\xi \to \xi) \wedge (\xi \to \xi)) = T(\xi \to \xi) = 1.$$

The theorem is proved.

Theorem 8.10. *Let ξ be an uncertain proposition. Then we have*

$$T(\xi \to \neg\xi) = 1 - T(\xi). \tag{8.23}$$

Proof: It follows from the definition of conditional symbol and the law of truth conservation that

$$T(\xi \to \neg\xi) = T(\neg\xi \vee \neg\xi) = T(\neg\xi) = 1 - T(\xi).$$

The theorem is proved.

Theorem 8.11. *If two uncertain propositions ξ and η are independent, then we have*

$$T(\xi \vee \eta) = T(\xi) \vee T(\eta), \quad T(\xi \wedge \eta) = T(\xi) \wedge T(\eta). \tag{8.24}$$

Proof: Since ξ and η are independent uncertain propositions, they are independent uncertain variables. Hence

$$T(\xi \vee \eta) = \mathcal{M}\{\xi \vee \eta = 1\} = \mathcal{M}\{\xi = 1\} \vee \mathcal{M}\{\eta = 1\} = T(\xi) \vee T(\eta),$$
$$T(\xi \wedge \eta) = \mathcal{M}\{\xi \wedge \eta = 1\} = \mathcal{M}\{\xi = 1\} \wedge \mathcal{M}\{\eta = 1\} = T(\xi) \wedge T(\eta).$$

The theorem is proved.

Theorem 8.12. *If two uncertain propositions ξ and η are independent, then we have*

$$T(\xi \to \eta) = (1 - T(\xi)) \vee T(\eta). \tag{8.25}$$

Proof: Since ξ and η are independent, the uncertain propositions $\neg\xi$ and η are also independent. It follows that

$$T(\xi \to \eta) = T(\neg\xi \vee \eta) = T(\neg\xi) \vee T(\eta) = (1 - T(\xi)) \vee T(\eta)$$

which proves the theorem.

Theorem 8.13 *(Chen and Ralescu [18], Truth Value Theorem). Assume that $\xi_1, \xi_2, \cdots, \xi_n$ are independent uncertain propositions with truth values $a_1, a_2, \cdots, a_n$, respectively. If X is an uncertain formula containing $\xi_1, \xi_2, \cdots, \xi_n$ with truth function f, then the truth value of X is*

$$T(X) = \begin{cases} \sup\limits_{f(x_1, x_2, \cdots, x_n) = 1} \min\limits_{1 \le i \le n} \nu_i(x_i), \\ \qquad if \quad \sup\limits_{f(x_1, x_2, \cdots, x_n) = 1} \min\limits_{1 \le i \le n} \nu_i(x_i) < 0.5 \\ 1 - \sup\limits_{f(x_1, x_2, \cdots, x_n) = 0} \min\limits_{1 \le i \le n} \nu_i(x_i), \\ \qquad if \quad \sup\limits_{f(x_1, x_2, \cdots, x_n) = 1} \min\limits_{1 \le i \le n} \nu_i(x_i) \ge 0.5 \end{cases} \qquad (8.26)$$

where x_i take values either 0 or 1, and ν_i are defined by

$$\nu_i(x_i) = \begin{cases} a_i, & if \ x_i = 1 \\ 1 - a_i, & if \ x_i = 0 \end{cases} \qquad (8.27)$$

for $i = 1, 2, \cdots, n$, respectively.

Proof: Since $X = 1$ if and only if $f(\xi_1, \xi_2, \cdots, \xi_n) = 1$, we immediately have

$$T(X) = \mathcal{M}\{f(\xi_1, \xi_2, \cdots, \xi_n) = 1\}.$$

Thus the equation (8.26) follows from Theorem 1.27 immediately.

Example 8.10: Let $\xi_1, \xi_2, \cdots, \xi_n$ be independent uncertain propositions with truth values $a_1, a_2, \cdots, a_n$, respectively. Then

$$X = \xi_1 \wedge \xi_2 \wedge \cdots \wedge \xi_n \qquad (8.28)$$

is an uncertain formula whose truth function is

$$f(x_1, x_2, \cdots, x_n) = x_1 \wedge x_2 \wedge \cdots \wedge x_n$$

It follows from the truth value theorem or Theorem 1.28 that the truth value is

$$T(\xi_1 \wedge \xi_2 \wedge \cdots \wedge \xi_n) = a_1 \wedge a_2 \wedge \cdots \wedge a_n. \qquad (8.29)$$

Example 8.11: Let $\xi_1, \xi_2, \cdots, \xi_n$ be independent uncertain propositions with truth values $a_1, a_2, \cdots, a_n$, respectively. Then

$$X = \xi_1 \vee \xi_2 \vee \cdots \vee \xi_n \qquad (8.30)$$

is an uncertain formula whose truth function is

$$f(x_1, x_2, \cdots, x_n) = x_1 \vee x_2 \vee \cdots \vee x_n$$

It follows from the truth value theorem or Theorem 1.29 that the truth value is

$$T(\xi_1 \vee \xi_2 \vee \cdots \vee \xi_n) = a_1 \vee a_2 \vee \cdots \vee a_n. \tag{8.31}$$

Example 8.12: Let $\xi_1, \xi_2, \cdots, \xi_n$ be independent uncertain propositions with truth values $a_1, a_2, \cdots, a_n$, respectively. For any integer k with $1 \le k \le n$,

$$X = \text{"at least } k \text{ propositions of } \xi_1, \xi_2, \cdots, \xi_n \text{ are true"} \tag{8.32}$$

is an uncertain formula whose truth function is

$$f(x_1, x_2, \cdots, x_n) = \begin{cases} 1, & \text{if } x_1 + x_2 + \cdots + x_n \ge k \\ 0, & \text{if } x_1 + x_2 + \cdots + x_n < k. \end{cases}$$

It follows from the truth value theorem or Theorem 1.30 that the truth value is

$$T(X) = \text{the } k\text{th largest value of } a_1, a_2, \cdots, a_n. \tag{8.33}$$

Example 8.13: Let ξ_1 and ξ_2 be independent uncertain propositions with truth values a_1 and a_2, respectively. Then

$$X = \xi_1 \leftrightarrow \xi_2 \tag{8.34}$$

is an uncertain formula whose truth function is

$$f(1,1) = 1, \quad f(1,0) = 0, \quad f(0,1) = 0, \quad f(0,0) = 1.$$

Then we have

$$\sup_{f(x_1, x_2)=1} \min_{1 \le i \le 2} \nu_i(x_i) = \max\{a_1 \wedge a_2, (1 - a_1) \wedge (1 - a_2)\},$$

$$\sup_{f(x_1, x_2)=0} \min_{1 \le i \le 2} \nu_i(x_i) = \max\{(1 - a_1) \wedge a_2, a_1 \wedge (1 - a_2)\}.$$

When $a_1 \ge 0.5$ and $a_2 \ge 0.5$, we have

$$\sup_{f(x_1, x_2)=1} \min_{1 \le i \le 2} \nu_i(x_i) = a_1 \wedge a_2 \ge 0.5.$$

It follows from the truth value theorem that

$$T(X) = 1 - \sup_{f(x_1, x_2)=0} \min_{1 \le i \le 2} \nu_i(x_i) = 1 - (1 - a_1) \vee (1 - a_2) = a_1 \wedge a_2.$$

When $a_1 \ge 0.5$ and $a_2 < 0.5$, we have

$$\sup_{f(x_1, x_2)=1} \min_{1 \le i \le 2} \nu_i(x_i) = (1 - a_1) \vee a_2 \le 0.5.$$

It follows from the truth value theorem that

$$T(X) = \sup_{f(x_1,x_2)=1} \min_{1 \le i \le 2} \nu_i(x_i) = (1 - a_1) \vee a_2.$$

When $a_1 < 0.5$ and $a_2 \ge 0.5$, we have

$$\sup_{f(x_1,x_2)=1} \min_{1 \le i \le 2} \nu_i(x_i) = a_1 \vee (1 - a_2) \le 0.5.$$

It follows from the truth value theorem that

$$T(X) = \sup_{f(x_1,x_2)=1} \min_{1 \le i \le 2} \nu_i(x_i) = a_1 \vee (1 - a_2).$$

When $a_1 < 0.5$ and $a_2 < 0.5$, we have

$$\sup_{f(x_1,x_2)=1} \min_{1 \le i \le 2} \nu_i(x_i) = (1 - a_1) \wedge (1 - a_2) > 0.5.$$

It follows from the truth value theorem that

$$T(X) = 1 - \sup_{f(x_1,x_2)=0} \min_{1 \le i \le 2} \nu_i(x_i) = 1 - a_1 \vee a_2 = (1 - a_1) \wedge (1 - a_2).$$

Thus we have

$$T(X) = \begin{cases} a_1 \wedge a_2, & \text{if } a_1 \ge 0.5 \text{ and } a_2 \ge 0.5 \\ (1 - a_1) \vee a_2, & \text{if } a_1 \ge 0.5 \text{ and } a_2 < 0.5 \\ a_1 \vee (1 - a_2), & \text{if } a_1 < 0.5 \text{ and } a_2 \ge 0.5 \\ (1 - a_1) \wedge (1 - a_2), & \text{if } a_1 < 0.5 \text{ and } a_2 < 0.5. \end{cases} \tag{8.35}$$

Example 8.14: Let ξ_1 and ξ_2 be independent uncertain propositions with truth values a_1 and a_2, respectively. Then

$$X = \text{``}\xi_1 \text{ or } \xi_2 \text{ and not both''} \tag{8.36}$$

is an uncertain formula whose truth function is

$$f(1,1) = 0, \quad f(1,0) = 1, \quad f(0,1) = 1, \quad f(0,0) = 0.$$

Then we have

$$\sup_{f(x_1,x_2)=1} \min_{1 \le i \le 2} \nu_i(x_i) = \max\{a_1 \wedge (1 - a_2), (1 - a_1) \wedge a_2\},$$

$$\sup_{f(x_1,x_2)=0} \min_{1 \le i \le 2} \nu_i(x_i) = \max\{a_1 \wedge a_2, (1 - a_1) \wedge (1 - a_2)\}.$$

When $a_1 \ge 0.5$ and $a_2 \ge 0.5$, we have

$$\sup_{f(x_1,x_2)=1} \min_{1 \le i \le 2} \nu_i(x_i) = (1 - a_1) \vee (1 - a_2) \le 0.5.$$

It follows from the truth value theorem that

$$T(X) = \sup_{f(x_1,x_2)=1} \min_{1 \le i \le 2} \nu_i(x_i) = (1 - a_1) \vee (1 - a_2).$$

When $a_1 \ge 0.5$ and $a_2 < 0.5$, we have

$$\sup_{f(x_1,x_2)=1} \min_{1 \le i \le 2} \nu_i(x_i) = a_1 \wedge (1 - a_2) \ge 0.5.$$

It follows from the truth value theorem that

$$T(X) = 1 - \sup_{f(x_1,x_2)=0} \min_{1 \le i \le 2} \nu_i(x_i) = 1 - (1 - a_1) \vee a_2 = a_1 \wedge (1 - a_2).$$

When $a_1 < 0.5$ and $a_2 \ge 0.5$, we have

$$\sup_{f(x_1,x_2)=1} \min_{1 \le i \le 2} \nu_i(x_i) = (1 - a_1) \wedge a_2 \ge 0.5.$$

It follows from the truth value theorem that

$$T(X) = 1 - \sup_{f(x_1,x_2)=0} \min_{1 \le i \le 2} \nu_i(x_i) = 1 - a_1 \vee (1 - a_2) = (1 - a_1) \wedge a_2.$$

When $a_1 < 0.5$ and $a_2 < 0.5$, we have

$$\sup_{f(x_1,x_2)=1} \min_{1 \le i \le 2} \nu_i(x_i) = a_1 \vee a_2 < 0.5.$$

It follows from the truth value theorem that

$$T(X) = \sup_{f(x_1,x_2)=1} \min_{1 \le i \le 2} \nu_i(x_i) = a_1 \vee a_2.$$

Thus we have

$$T(X) = \begin{cases} (1 - a_1) \vee (1 - a_2), & \text{if } a_1 \ge 0.5 \text{ and } a_2 \ge 0.5 \\ a_1 \wedge (1 - a_2), & \text{if } a_1 \ge 0.5 \text{ and } a_2 < 0.5 \\ (1 - a_1) \wedge a_2, & \text{if } a_1 < 0.5 \text{ and } a_2 \ge 0.5 \\ a_1 \vee a_2, & \text{if } a_1 < 0.5 \text{ and } a_2 < 0.5. \end{cases} \tag{8.37}$$

Exercise 8.1: Let ξ, η, τ be independent uncertain propositions with truth values a, b, c, respectively. What is $T(\xi \vee \eta \to \tau)$?

Exercise 8.2: Let ξ, η, τ be independent uncertain propositions with truth values a, b, c, respectively. What is $T(\xi \to \eta \wedge \tau)$?

8.7   Truth Value Solver

Truth Value Solver is a software for computing the truth values of uncertain formula based on the truth value theorem. This software may be downloaded from http://orsc.edu.cn/liu/resources.htm. Now let us perform it via some numerical examples.

Example 8.15:   Assume that $\xi_1, \xi_2, \xi_3, \xi_4, \xi_5$ are independent uncertain propositions with truth values $0.1, 0.3, 0.5, 0.7, 0.9$, respectively. Let

$$X = (\xi_1 \wedge \xi_2) \vee (\xi_2 \wedge \xi_3) \vee (\xi_3 \wedge \xi_4) \vee (\xi_4 \wedge \xi_5). \qquad (8.38)$$

It is clear that the truth function is

$$f(x_1, x_2, x_3, x_4, x_5) = \begin{cases} 1, & \text{if } x_1 + x_2 = 2 \\ 1, & \text{if } x_2 + x_3 = 2 \\ 1, & \text{if } x_3 + x_4 = 2 \\ 1, & \text{if } x_4 + x_5 = 2 \\ 0, & \text{otherwise.} \end{cases}$$

A run of the truth value solver shows that the truth value of X is 0.7 in uncertain measure.

Example 8.16:   Assume that $\xi_1, \xi_2, \xi_3, \xi_4, \xi_5$ are independent uncertain propositions with truth values $0.1, 0.3, 0.5, 0.7, 0.9$, respectively. Let

$$X = \text{"only 4 propositions of } \xi_1, \xi_2, \xi_3, \xi_4, \xi_5 \text{ are true"}. \qquad (8.39)$$

It is clear that the truth function is

$$f(x_1, x_2, x_3, x_4, x_5) = \begin{cases} 1, & \text{if } x_1 + x_2 + x_3 + x_4 + x_5 = 4 \\ 0, & \text{if } x_1 + x_2 + x_3 + x_4 + x_5 \neq 4. \end{cases}$$

A run of the truth value solver shows that the truth value of X is 0.3 in uncertain measure.

Example 8.17:   Assume that $\xi_1, \xi_2, \xi_3, \xi_4, \xi_5$ are independent uncertain propositions with truth values $0.1, 0.3, 0.5, 0.7, 0.9$, respectively. Let

$$X = \text{"only odd number of propositions of } \xi_1, \xi_2, \xi_3, \xi_4, \xi_5 \text{ are true"}. \quad (8.40)$$

It is clear that the truth function is

$$f(x_1, x_2, x_3, x_4, x_5) = \begin{cases} 1, & \text{if } x_1 + x_2 + x_3 + x_4 + x_5 \in \{1, 3, 5\} \\ 0, & \text{if } x_1 + x_2 + x_3 + x_4 + x_5 \in \{0, 2, 4\}. \end{cases}$$

A run of the truth value solver shows that the truth value is 0.5 in uncertain measure.

8.8   Uncertain Predicate Logic

Uncertain predicate logic, proposed by Zhang and Peng [227], is a generalization of classical predicate logic for dealing with uncertain knowledge via uncertainty theory.

Consider the following propositions: "Beijing is a big city", and "Tianjin is a big city". Uncertain propositional logic treats them as unrelated propositions. However, uncertain predicate logic represents them by a predicate proposition $\xi(a)$. If a represents Beijing, then

$$\xi(a) = \text{"Beijing is a big city".} \tag{8.41}$$

If a represents Tianjin, then

$$\xi(a) = \text{"Tianjin is a big city".} \tag{8.42}$$

Definition 8.5 *(Zhang and Peng [227]). Uncertain predicate proposition is a sequence of uncertain propositions indexed by one or more parameters.*

That is, if we use $\xi(a)$ to express an uncertain predicate proposition where the parameter a is called a variable, then for each fixed a^*, we obtain an uncertain proposition $\xi(a^*)$ in the sense of uncertain propositional logic. In other words, an uncertain predicate proposition may be regarded as a function on $a \in A$ that takes values of uncertain propositions in the sense of Definition 8.1.

In order to deal with uncertain predicate propositions, we need a universal quantifier $\forall$ and an existential quantifier $\exists$. If $\xi(a)$ is a predicate proposition defined by (8.41) and (8.42), then

$$(\forall a)\xi(a) = \text{"Both Beijing and Tianjin are big cities",} \tag{8.43}$$

$$(\exists a)\xi(a) = \text{"At least one of Beijing and Tianjin is a big city".} \tag{8.44}$$

Note that $(\forall a)\xi(a)$ and $(\exists a)\neg\xi(a)$ are essentially uncertain propositions in the sense of uncertain propositional logic. In addition, it is easy to verify that

$$\neg(\forall a)\xi(a) = (\exists a)\neg\xi(a). \tag{8.45}$$

Thus we have

$$T((\forall a)\xi(a) \vee (\exists a)\neg\xi(a)) = 1, \tag{8.46}$$

$$T((\forall a)\xi(a) \wedge (\exists a)\neg\xi(a)) = 0, \tag{8.47}$$

$$T((\forall a)\xi(a)) + T((\exists a)\neg\xi(a)) = 1. \tag{8.48}$$

Theorem 8.14 *(Zhang and Peng [227]). Let $\xi(a)$ be an uncertain predicate proposition such that $\{\xi(a)|a \in A\}$ is a class of independent uncertain propositions. Then we have*

$$T((\forall a)\xi(a)) = \inf_{a \in A} T(\xi(a)), \tag{8.49}$$

$$T((\exists a)\xi(a)) = \sup_{a \in A} T(\xi(a)). \tag{8.50}$$

Proof: For each uncertain predicate proposition $\xi(a)$, by the meaning of universal quantifier, we obtain

$$T((\forall a)\xi(a)) = \mathcal{M}\{(\forall a)\xi(a) = 1\} = \mathcal{M}\left\{\bigcap_{a \in A}(\xi(a) = 1)\right\}.$$

Since $\{\xi(a)|a \in A\}$ is a class of independent uncertain propositions, we get

$$T((\forall a)\xi(a)) = \inf_{a \in A}\mathcal{M}\{\xi(a) = 1\} = \inf_{a \in A} T(\xi(a)).$$

The first equation is verified. Similarly, by the meaning of existential quantifier, we obtain

$$T((\exists a)\xi(a)) = \mathcal{M}\{(\exists a)\xi(a) = 1\} = \mathcal{M}\left\{\bigcup_{a \in A}(\xi(a) = 1)\right\}.$$

Since $\{\xi(a)|a \in A\}$ is a class of independent uncertain propositions, we get

$$T((\exists a)\xi(a)) = \sup_{a \in A}\mathcal{M}\{\xi(a) = 1\} = \sup_{a \in A} T(\xi(a)).$$

The second equation is proved.

Theorem 8.15 *(Zhang and Peng [227]). Let $\xi(a, b)$ be an uncertain predicate proposition such that $\{\xi(a, b)|a \in A, b \in B\}$ is a class of independent uncertain propositions. Then we have*

$$T((\forall a)(\exists b)\xi(a, b)) = \inf_{a \in A}\sup_{b \in B} T(\xi(a, b)), \tag{8.51}$$

$$T((\exists a)(\forall b)\xi(a, b)) = \sup_{a \in A}\inf_{b \in B} T(\xi(a, b)). \tag{8.52}$$

Proof: Since $\{\xi(a, b)|a \in A, b \in B\}$ is a class of independent uncertain propositions, both $\{(\exists b)\xi(a, b)|a \in A\}$ and $\{(\forall b)\xi(a, b)|a \in A\}$ are two classes of independent uncertain propositions. It follows from Theorem 8.14 that

$$T((\forall a)(\exists b)\xi(a, b)) = \inf_{a \in A} T((\exists b)\xi(a, b)) = \inf_{a \in A}\sup_{b \in B} T(\xi(a, b)),$$

$$T((\exists a)(\forall b)\xi(a, b)) = \sup_{a \in A} T((\forall b)\xi(a, b)) = \sup_{a \in A}\inf_{b \in B} T(\xi(a, b)).$$

The theorem is proved.

Chapter 9

Uncertain Entailment

Uncertain entailment, developed by Liu [124] in 2009, is a methodology for calculating the truth value of an uncertain formula via the maximum uncertainty principle when the truth values of other uncertain formulas are given. In order to solve this problem, this chapter will introduce an entailment model. As applications of uncertain entailment, this chapter will also discuss modus ponens, modus tollens, and hypothetical syllogism.

9.1 Entailment Model

Assume $\xi_1, \xi_2, \cdots, \xi_n$ are independent uncertain propositions with *unknown* truth values $\alpha_1, \alpha_2, \cdots, \alpha_n$, respectively. Also assume that $X_1, X_2, \cdots, X_m$ are uncertain formulas containing $\xi_1, \xi_2, \cdots, \xi_n$ with *known* truth values β_1, $\beta_2, \cdots, \beta_m$, respectively. Now let X be an additional uncertain formula containing $\xi_1, \xi_2, \cdots, \xi_n$. What is the truth value of X?

This is just the uncertain entailment problem. In order to solve it, let us consider what values $\alpha_1, \alpha_2, \cdots, \alpha_n$ may take. The first constraint is

$$0 \leq \alpha_j \leq 1, \quad j = 1, 2, \cdots, n. \tag{9.1}$$

We also hope

$$T(X_i) = \beta_i, \quad i = 1, 2, \cdots, m \tag{9.2}$$

where each $T(X_i)$ $(1 \leq i \leq m)$ is determined by the truth function f_i as follows,

$$T(X_i) = \begin{cases} \sup\limits_{f_i(x_1, x_2, \cdots, x_n)=1} \min\limits_{1 \leq j \leq n} \nu_j(x_j), \\ \quad \text{if } \sup\limits_{f_i(x_1, x_2, \cdots, x_n)=1} \min\limits_{1 \leq j \leq n} \nu_j(x_j) < 0.5 \\ 1 - \sup\limits_{f_i(x_1, x_2, \cdots, x_n)=0} \min\limits_{1 \leq j \leq n} \nu_j(x_j), \\ \quad \text{if } \sup\limits_{f_i(x_1, x_2, \cdots, x_n)=1} \min\limits_{1 \leq j \leq n} \nu_j(x_j) \geq 0.5 \end{cases} \tag{9.3}$$

and

$$\nu_j(x_j) = \begin{cases} \alpha_j, & \text{if } x_j = 1 \\ 1 - \alpha_j, & \text{if } x_j = 0 \end{cases} \tag{9.4}$$

for $j = 1, 2, \cdots, n$.

B. Liu: Uncertainty Theory: A Branch of Mathematics, SCI 300, pp. 177–186.
springerlink.com

Based on the truth values $\alpha_1, \alpha_2, \cdots, \alpha_n$ and truth function f, the truth value of X is

$$
T(X) = \begin{cases}
\sup\limits_{f(x_1,x_2,\cdots,x_n)=1} \min\limits_{1\leq j\leq n} \nu_j(x_j), \\
\qquad \text{if} \quad \sup\limits_{f(x_1,x_2,\cdots,x_n)=1} \min\limits_{1\leq j\leq n} \nu_j(x_j) < 0.5 \\
1 - \sup\limits_{f(x_1,x_2,\cdots,x_n)=0} \min\limits_{1\leq j\leq n} \nu_j(x_j), \\
\qquad \text{if} \quad \sup\limits_{f(x_1,x_2,\cdots,x_n)=1} \min\limits_{1\leq j\leq n} \nu_j(x_j) \geq 0.5.
\end{cases} \tag{9.5}
$$

Since the truth values $\alpha_1, \alpha_2, \cdots, \alpha_n$ are not uniquely determined, the truth value $T(X)$ is not unique too. For this case, we have to use the maximum uncertainty principle to determine the truth value $T(X)$. That is, $T(X)$ should be assigned the value as close to 0.5 as possible. In other words, we should minimize the value $|T(X) - 0.5|$ via choosing appreciate values of $\alpha_1, \alpha_2, \cdots, \alpha_n$.

Entailment Model (Liu [124]). *Let $\xi_1, \xi_2, \cdots, \xi_n$ be independent uncertain propositions with unknown truth values $\alpha_1, \alpha_2, \cdots, \alpha_n$, respectively. Assume $X_1, X_2, \cdots, X_m$ are uncertain formulas containing $\xi_1, \xi_2, \cdots, \xi_n$ with known truth values $\beta_1, \beta_2, \cdots, \beta_m$, respectively. Then the truth value $T(X)$ of an additional uncertain formula X containing $\xi_1, \xi_2, \cdots, \xi_n$ solves*

$$
\begin{cases}
\min |T(X) - 0.5| \\
subject\ to: \\
\quad T(X_i) = \beta_i, \quad i = 1, 2, \cdots, m \\
\quad 0 \leq \alpha_j \leq 1, \quad j = 1, 2, \cdots, n
\end{cases} \tag{9.6}
$$

where $T(X_1), T(X_2), \cdots, T(X_m), T(X)$ are functions of $\alpha_1, \alpha_2, \cdots, \alpha_n$ via (9.3) and (9.5).

If the entailment model (9.6) has no feasible solution, then the truth values $\beta_1, \beta_2, \cdots, \beta_m$ are inconsistent with each other. For this case, we cannot entail anything on the uncertain formula X.

If the entailment model (9.6) has an optimal solution $(\alpha_1^*, \alpha_2^*, \cdots, \alpha_n^*)$, then the truth value of X is just (9.5) with

$$
\nu_j(x_j) = \begin{cases}
\alpha_j^*, & \text{if } x_j = 1 \\
1 - \alpha_j^*, & \text{if } x_j = 0
\end{cases} \tag{9.7}
$$

for $j = 1, 2, \cdots, n$.

Example 9.1: Let ξ_1 and ξ_2 be independent uncertain propositions with unknown truth values α_1 and α_2, respectively. It is known that

$$
T(\xi_1 \vee \xi_2) = \beta_1, \quad T(\xi_1 \wedge \xi_2) = \beta_2. \tag{9.8}
$$

What is the truth value of $\xi_1 \rightarrow \xi_2$? In order to answer this question, we write

$$X_1 = \xi_1 \vee \xi_2, \quad X_2 = \xi_1 \wedge \xi_2, \quad X = \xi_1 \rightarrow \xi_2.$$

Then we have

$$T(X_1) = \alpha_1 \vee \alpha_2 = \beta_1,$$
$$T(X_2) = \alpha_1 \wedge \alpha_2 = \beta_2,$$
$$T(X) = (1 - \alpha_1) \vee \alpha_2.$$

For this case, the entailment model (9.6) becomes

$$\begin{cases} \min |(1 - \alpha_1) \vee \alpha_2 - 0.5| \\ \text{subject to:} \\ \quad \alpha_1 \vee \alpha_2 = \beta_1 \\ \quad \alpha_1 \wedge \alpha_2 = \beta_2 \\ \quad 0 \leq \alpha_1 \leq 1 \\ \quad 0 \leq \alpha_2 \leq 1. \end{cases} \quad (9.9)$$

When $\beta_1 \geq \beta_2$, there are only two feasible solutions $(\alpha_1, \alpha_2) = (\beta_1, \beta_2)$ and $(\alpha_1, \alpha_2) = (\beta_2, \beta_1)$. If $\beta_1 + \beta_2 < 1$, the optimal solution produces

$$T(X) = (1 - \alpha_1^*) \vee \alpha_2^* = 1 - \beta_1;$$

if $\beta_1 + \beta_2 = 1$, the optimal solution produces

$$T(X) = (1 - \alpha_1^*) \vee \alpha_2^* = \beta_1 \text{ or } \beta_2;$$

if $\beta_1 + \beta_2 > 1$, the optimal solution produces

$$T(X) = (1 - \alpha_1^*) \vee \alpha_2^* = \beta_2.$$

When $\beta_1 < \beta_2$, there is no feasible solution and the truth values are ill-assigned. As a summary, we have

$$T(\xi_1 \rightarrow \xi_2) = \begin{cases} 1 - \beta_1, & \text{if } \beta_1 \geq \beta_2 \text{ and } \beta_1 + \beta_2 < 1 \\ \beta_1 \text{ or } \beta_2, & \text{if } \beta_1 \geq \beta_2 \text{ and } \beta_1 + \beta_2 = 1 \\ \beta_2, & \text{if } \beta_1 \geq \beta_2 \text{ and } \beta_1 + \beta_2 > 1 \\ \text{illness}, & \text{if } \beta_1 < \beta_2. \end{cases} \quad (9.10)$$

Example 9.2: Let ξ_1, ξ_2, ξ_3 be independent uncertain propositions with unknown truth values $\alpha_1, \alpha_2, \alpha_3$, respectively. It is known that

$$T(\xi_1 \rightarrow \xi_2) = \beta_1, \quad T(\xi_2 \rightarrow \xi_3) = \beta_2. \quad (9.11)$$

What is the truth value of ξ_2? In order to answer this question, we write

$$X_1 = \xi_1 \rightarrow \xi_2, \quad X_2 = \xi_2 \rightarrow \xi_3, \quad X = \xi_2.$$

Then we have

$$T(X_1) = (1 - \alpha_1) \vee \alpha_2 = \beta_1,$$
$$T(X_2) = (1 - \alpha_2) \vee \alpha_3 = \beta_2,$$
$$T(X) = \alpha_2.$$

For this case, the entailment model (9.6) becomes

$$\begin{cases} \min |\alpha_2 - 0.5| \\ \text{subject to:} \\ \quad (1 - \alpha_1) \vee \alpha_2 = \beta_1 \\ \quad (1 - \alpha_2) \vee \alpha_3 = \beta_2 \\ \quad 0 \le \alpha_1 \le 1 \\ \quad 0 \le \alpha_2 \le 1 \\ \quad 0 \le \alpha_3 \le 1. \end{cases} \tag{9.12}$$

The optimal solution $(\alpha_1^*, \alpha_2^*, \alpha_3^*)$ produces

$$T(\xi_2) = \begin{cases} \beta_1, & \text{if } \beta_1 + \beta_2 \ge 1 \text{ and } \beta_1 < 0.5 \\ 1 - \beta_2, & \text{if } \beta_1 + \beta_2 \ge 1 \text{ and } \beta_2 < 0.5 \\ 0.5, & \text{if } \beta_1 \ge 0.5 \text{ and } \beta_2 \ge 0.5 \\ \text{illness}, & \text{if } \beta_1 + \beta_2 < 1. \end{cases} \tag{9.13}$$

Example 9.3: Let ξ_1, ξ_2, ξ_3 be independent uncertain propositions with unknown truth values $\alpha_1, \alpha_2, \alpha_3$, respectively. It is known that

$$T(\xi_1 \to \xi_2) = \beta_1, \quad T(\xi_1 \to \xi_3) = \beta_2. \tag{9.14}$$

What is the truth value of $\xi_1 \to \xi_2 \wedge \xi_3$? In order to answer this question, we write

$$X_1 = \xi_1 \to \xi_2, \quad X_2 = \xi_1 \to \xi_3, \quad X = \xi_1 \to \xi_2 \wedge \xi_3.$$

Then we have

$$T(X_1) = (1 - \alpha_1) \vee \alpha_2 = \beta_1,$$
$$T(X_2) = (1 - \alpha_1) \vee \alpha_3 = \beta_2,$$
$$T(X) = (1 - \alpha_1) \vee (\alpha_2 \wedge \alpha_3).$$

For this case, the entailment model (9.6) becomes

$$\begin{cases} \min |(1 - \alpha_1) \vee (\alpha_2 \wedge \alpha_3) - 0.5| \\ \text{subject to:} \\ \quad (1 - \alpha_1) \vee \alpha_2 = \beta_1 \\ \quad (1 - \alpha_1) \vee \alpha_3 = \beta_2 \\ \quad 0 \le \alpha_1 \le 1 \\ \quad 0 \le \alpha_2 \le 1 \\ \quad 0 \le \alpha_3 \le 1. \end{cases} \tag{9.15}$$

The optimal solution $(\alpha_1^*, \alpha_2^*, \alpha_3^*)$ produces $T(\xi_1 \to \xi_2 \wedge \xi_3) = \beta_1 \wedge \beta_2$.

Example 9.4: Let ξ_1, ξ_2, ξ_3 be independent uncertain propositions with unknown truth values $\alpha_1, \alpha_2, \alpha_3$, respectively. It is known that

$$T(\xi_1 \to \xi_2) = \beta_1, \quad T(\xi_1 \to \xi_3) = \beta_2. \qquad (9.16)$$

What is the truth value of $\xi_1 \to \xi_2 \vee \xi_3$? In order to answer this question, we write

$$X_1 = \xi_1 \to \xi_2, \quad X_2 = \xi_1 \to \xi_3, \quad X = \xi_1 \to \xi_2 \vee \xi_3.$$

Then we have

$$T(X_1) = (1 - \alpha_1) \vee \alpha_2 = \beta_1,$$

$$T(X_2) = (1 - \alpha_1) \vee \alpha_3 = \beta_2,$$

$$T(X) = (1 - \alpha_1) \vee \alpha_2 \vee \alpha_3.$$

For this case, the entailment model (9.6) becomes

$$\begin{cases} \min |(1 - \alpha_1) \vee \alpha_2 \vee \alpha_3 - 0.5| \\ \text{subject to:} \\ \quad (1 - \alpha_1) \vee \alpha_2 = \beta_1 \\ \quad (1 - \alpha_1) \vee \alpha_3 = \beta_2 \\ \quad 0 \leq \alpha_1 \leq 1 \\ \quad 0 \leq \alpha_2 \leq 1 \\ \quad 0 \leq \alpha_3 \leq 1. \end{cases} \qquad (9.17)$$

The optimal solution $(\alpha_1^*, \alpha_2^*, \alpha_3^*)$ produces $T(\xi_1 \to \xi_2 \vee \xi_3) = \beta_1 \vee \beta_2$.

Example 9.5: Let ξ_1, ξ_2, ξ_3 be independent uncertain propositions with unknown truth values $\alpha_1, \alpha_2, \alpha_3$, respectively. It is known that

$$T(\xi_1 \to \xi_3) = \beta_1, \quad T(\xi_2 \to \xi_3) = \beta_2. \qquad (9.18)$$

What is the truth value of $\xi_1 \vee \xi_2 \to \xi_3$? In order to answer this question, we write

$$X_1 = \xi_1 \to \xi_3, \quad X_2 = \xi_2 \to \xi_3, \quad X = \xi_1 \vee \xi_2 \to \xi_3.$$

Then we have

$$T(X_1) = (1 - \alpha_1) \vee \alpha_3 = \beta_1,$$

$$T(X_2) = (1 - \alpha_2) \vee \alpha_3 = \beta_2,$$

$$T(X) = (1 - \alpha_1 \vee \alpha_2) \vee \alpha_3.$$

For this case, the entailment model (9.6) becomes

$$\begin{cases} \min |(1 - \alpha_1 \vee \alpha_2) \vee \alpha_3 - 0.5| \\ \text{subject to:} \\ \quad (1 - \alpha_1) \vee \alpha_3 = \beta_1 \\ \quad (1 - \alpha_2) \vee \alpha_3 = \beta_2 \\ \quad 0 \le \alpha_1 \le 1 \\ \quad 0 \le \alpha_2 \le 1 \\ \quad 0 \le \alpha_3 \le 1. \end{cases} \qquad (9.19)$$

The optimal solution $(\alpha_1^*, \alpha_2^*, \alpha_3^*)$ produces $T(\xi_1 \vee \xi_2 \to \xi_3) = \beta_1 \wedge \beta_2$.

Example 9.6: Let ξ_1, ξ_2, ξ_3 be independent uncertain propositions with unknown truth values $\alpha_1, \alpha_2, \alpha_3$, respectively. It is known that

$$T(\xi_1 \to \xi_3) = \beta_1, \quad T(\xi_2 \to \xi_3) = \beta_2. \qquad (9.20)$$

What is the truth value of $\xi_1 \wedge \xi_2 \to \xi_3$? In order to answer this question, we write

$$X_1 = \xi_1 \to \xi_3, \quad X_2 = \xi_2 \to \xi_3, \quad X = \xi_1 \wedge \xi_2 \to \xi_3.$$

Then we have

$$T(X_1) = (1 - \alpha_1) \vee \alpha_3 = \beta_1,$$
$$T(X_2) = (1 - \alpha_2) \vee \alpha_3 = \beta_2,$$
$$T(X) = (1 - \alpha_1 \wedge \alpha_2) \vee \alpha_3.$$

For this case, the entailment model (9.6) becomes

$$\begin{cases} \min |(1 - \alpha_1 \wedge \alpha_2) \vee \alpha_3 - 0.5| \\ \text{subject to:} \\ \quad (1 - \alpha_1) \vee \alpha_3 = \beta_1 \\ \quad (1 - \alpha_2) \vee \alpha_3 = \beta_2 \\ \quad 0 \le \alpha_1 \le 1 \\ \quad 0 \le \alpha_2 \le 1 \\ \quad 0 \le \alpha_3 \le 1. \end{cases} \qquad (9.21)$$

The optimal solution $(\alpha_1^*, \alpha_2^*, \alpha_3^*)$ produces $T(\xi_1 \wedge \xi_2 \to \xi_3) = \beta_1 \vee \beta_2$.

9.2 Modus Ponens

Classical modus ponens tells us that if both ξ and $\xi \to \eta$ are true, then η is true. This section provides a version of modus ponens in the framework of uncertain logic.

Theorem 9.1 *(Liu [124], Modus Ponens). Let ξ and η be independent uncertain propositions. Suppose ξ and $\xi \to \eta$ are two uncertain formulas with truth values β_1 and β_2, respectively. Then the truth value of η is*

$$T(\eta) = \begin{cases} \beta_2, & \text{if } \beta_1 + \beta_2 > 1 \\ 0.5 \wedge \beta_2, & \text{if } \beta_1 + \beta_2 = 1 \\ \text{illness}, & \text{if } \beta_1 + \beta_2 < 1. \end{cases} \qquad (9.22)$$

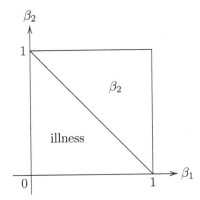

Figure 9.1: Modus Ponens

Proof: Denote the truth values of ξ and η by α_1 and α_2, respectively, and write

$$X_1 = \xi, \quad X_2 = \xi \to \eta, \quad X = \eta.$$

It is clear that

$$T(X_1) = \alpha_1 = \beta_1,$$
$$T(X_2) = (1 - \alpha_1) \vee \alpha_2 = \beta_2,$$
$$T(X) = \alpha_2.$$

For this case, the entailment model (9.6) becomes

$$\begin{cases} \min |\alpha_2 - 0.5| \\ \text{subject to:} \\ \quad \alpha_1 = \beta_1 \\ \quad (1 - \alpha_1) \vee \alpha_2 = \beta_2 \\ \quad 0 \le \alpha_1 \le 1 \\ \quad 0 \le \alpha_2 \le 1. \end{cases} \qquad (9.23)$$

When $\beta_1 + \beta_2 > 1$, there is only one feasible solution and then the optimal solution is

$$\alpha_1^* = \beta_1, \quad \alpha_2^* = \beta_2.$$

Thus $T(\eta) = \alpha_2^* = \beta_2$. When $\beta_1 + \beta_2 = 1$, the feasible set is $\{\beta_1\} \times [0, \beta_2]$ and the optimal solution is

$$\alpha_1^* = \beta_1, \quad \alpha_2^* = 0.5 \wedge \beta_2.$$

Thus $T(\eta) = \alpha_2^* = 0.5 \wedge \beta_2$. When $\beta_1 + \beta_2 < 1$, there is no feasible solution and the truth values are ill-assigned. The theorem is proved.

Remark 9.1: Different from the classical logic, the uncertain propositions ξ and η in $\xi \to \eta$ are statements with some truth values rather than pure statements. Thus the truth value of $\xi \to \eta$ is understood as

$$T(\xi \to \eta) = (1 - T(\xi)) \vee T(\eta). \tag{9.24}$$

Remark 9.2: Note that $T(\eta)$ in (9.22) does not necessarily represent the objective truth degree of η. For example, if $T(\xi)$ is small, then $T(\eta)$ is the truth value that η might (not must) be true.

9.3   Modus Tollens

Classical modus tollens tells us that if $\xi \to \eta$ is true and η is false, then ξ is false. This section provides a version of modus tollens in the framework of uncertain logic.

Theorem 9.2 (Liu [124], Modus Tollens). *Let ξ and η be independent uncertain propositions. Suppose $\xi \to \eta$ and η are two uncertain formulas with truth values β_1 and β_2, respectively. Then the truth value of ξ is*

$$T(\xi) = \begin{cases} 1 - \beta_1, & \text{if } \beta_1 > \beta_2 \\ (1 - \beta_1) \vee 0.5, & \text{if } \beta_1 = \beta_2 \\ \text{illness}, & \text{if } \beta_1 < \beta_2. \end{cases} \tag{9.25}$$

Proof: Denote the truth values of ξ and η by α_1 and α_2, respectively, and write

$$X_1 = \xi \to \eta, \quad X_2 = \eta, \quad X = \xi.$$

It is clear that

$$T(X_1) = (1 - \alpha_1) \vee \alpha_2 = \beta_1,$$
$$T(X_2) = \alpha_2 = \beta_2,$$
$$T(X) = \alpha_1.$$

For this case, the entailment model (9.6) becomes

$$\begin{cases} \min |\alpha_1 - 0.5| \\ \text{subject to:} \\ \quad (1 - \alpha_1) \vee \alpha_2 = \beta_1 \\ \quad \alpha_2 = \beta_2 \\ \quad 0 \leq \alpha_1 \leq 1 \\ \quad 0 \leq \alpha_2 \leq 1. \end{cases} \tag{9.26}$$

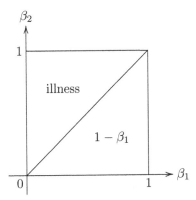

Figure 9.2: Modus Tollens

When $\beta_1 > \beta_2$, there is only one feasible solution and then the optimal solution is

$$\alpha_1^* = 1 - \beta_1, \quad \alpha_2^* = \beta_2.$$

Thus $T(\xi) = \alpha_1^* = 1 - \beta_1$. When $\beta_1 = \beta_2$, the feasible set is $[1 - \beta_1, 1] \times \{\beta_2\}$ and the optimal solution is

$$\alpha_1^* = (1 - \beta_1) \vee 0.5, \quad \alpha_2^* = \beta_2.$$

Thus $T(\xi) = \alpha_1^* = (1 - \beta_1) \vee 0.5$. When $\beta_1 < \beta_2$, there is no feasible solution and the truth values are ill-assigned. The theorem is proved.

9.4   Hypothetical Syllogism

Classical hypothetical syllogism tells us that if both $\xi \to \eta$ and $\eta \to \tau$ are true, then $\xi \to \tau$ is true. This section provides a version of hypothetical syllogism in the framework of uncertain logic.

Theorem 9.3 (Liu [124], Hypothetical Syllogism). *Let ξ, η, τ be independent uncertain propositions. Suppose $\xi \to \eta$ and $\eta \to \tau$ are two uncertain formulas with truth values β_1 and β_2, respectively. Then the truth value of $\xi \to \tau$ is*

$$T(\xi \to \tau) = \begin{cases} \beta_1 \wedge \beta_2, & \text{if } \beta_1 \wedge \beta_2 \geq 0.5 \\ 0.5, & \text{if } \beta_1 + \beta_2 \geq 1 \text{ and } \beta_1 \wedge \beta_2 < 0.5 \\ \text{illness}, & \text{if } \beta_1 + \beta_2 < 1. \end{cases} \qquad (9.27)$$

Proof: Denote the truth values of ξ, η, τ by $\alpha_1, \alpha_2, \alpha_3$, respectively, and write

$$X_1 = \xi \to \eta, \quad X_2 = \eta \to \tau, \quad X = \xi \to \tau.$$

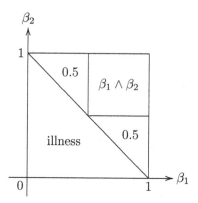

Figure 9.3: Hypothetical Syllogism

It is clear that

$$T(X_1) = (1 - \alpha_1) \vee \alpha_2 = \beta_1,$$
$$T(X_2) = (1 - \alpha_2) \vee \alpha_3 = \beta_2,$$
$$T(X) = (1 - \alpha_1) \vee \alpha_3.$$

For this case, the entailment model (9.6) becomes

$$\begin{cases} \min |(1 - \alpha_1) \vee \alpha_3 - 0.5| \\ \text{subject to:} \\ \quad (1 - \alpha_1) \vee \alpha_2 = \beta_1 \\ \quad (1 - \alpha_2) \vee \alpha_3 = \beta_2 \\ \quad 0 \leq \alpha_1 \leq 1 \\ \quad 0 \leq \alpha_2 \leq 1 \\ \quad 0 \leq \alpha_3 \leq 1. \end{cases} \tag{9.28}$$

When $\beta_1 \wedge \beta_2 \geq 0.5$, we have

$$T(\xi \to \tau) = (1 - \alpha_1^*) \vee \alpha_3^* = \beta_1 \wedge \beta_2.$$

When $\beta_1 + \beta_2 \geq 1$ and $\beta_1 \wedge \beta_2 < 0.5$, we have

$$T(\xi \to \tau) = (1 - \alpha_1^*) \vee \alpha_3^* = 0.5.$$

When $\beta_1 + \beta_2 < 1$, there is no feasible solution and the truth values are ill-assigned. The theorem is proved.

9.5   Automatic Entailment Machine

Automatic Entailment Machine is a software for solving the entailment model. This software may be downloaded from http://orsc.edu.cn/liu/resources.htm.

Chapter 10

Uncertain Set Theory

Uncertain set theory was proposed by Liu [125] in 2010 as a generalization of uncertainty theory to the domain of uncertain sets. This chapter will introduce the concepts of uncertain set, membership degree, membership function, uncertainty distribution, independence, operational law, expected value, critical values, Hausdorff distance, and conditional uncertain set.

10.1 Uncertain Set

Roughly speaking, an uncertain set is a set-valued function on an uncertainty space. Thus uncertain set is neither a random set nor a fuzzy set. A formal definition is given as follows.

Definition 10.1 *(Liu [125]). An uncertain set is a measurable function ξ from an uncertainty space $(\Gamma, \mathcal{L}, \mathcal{M})$ to a collection of sets of real numbers, i.e., for any Borel set B of real numbers, the set*

$$\{\xi \subset B\} = \{\gamma \in \Gamma \mid \xi(\gamma) \subset B\} \tag{10.1}$$

is an event.

Example 10.1: Take an uncertainty space $(\Gamma, \mathcal{L}, \mathcal{M})$ to be $\{\gamma_1, \gamma_2, \gamma_3\}$ with power set $\mathcal{L}$. Then the set-valued function

$$\xi(\gamma) = \begin{cases} [1,3], & \text{if } \gamma = \gamma_1 \\ [2,4], & \text{if } \gamma = \gamma_2 \\ [3,5], & \text{if } \gamma = \gamma_3 \end{cases} \tag{10.2}$$

is an uncertain set on $(\Gamma, \mathcal{L}, \mathcal{M})$.

Example 10.2: Take an uncertainty space $(\Gamma, \mathcal{L}, \mathcal{M})$ to be $\Re$ with Borel algebra $\mathcal{L}$. Then the set-valued function

$$\xi(\gamma) = [\gamma, \gamma + 1], \quad \forall \gamma \in \Gamma \tag{10.3}$$

is an uncertain set on $(\Gamma, \mathcal{L}, \mathcal{M})$.

B. Liu: Uncertainty Theory: A Branch of Mathematics, SCI 300, pp. 187–213.
springerlink.com   © Springer-Verlag Berlin Heidelberg 2010

Example 10.3: Take an uncertain space $(\Gamma, \mathcal{L}, \mathcal{M})$ to be $[0, +\infty)$ with Borel algebra $\mathcal{L}$. Then the set-valued function

$$\xi(\gamma) = \left[-\frac{1}{1+\gamma^2}, \frac{1}{1+\gamma^2} \right], \quad \forall \gamma \in \Gamma \tag{10.4}$$

is an uncertain set on $(\Gamma, \mathcal{L}, \mathcal{M})$.

Example 10.4: Any uncertain variable in the sense of Definition 1.5 is a special uncertain set in the sense of Definition 10.1.

Theorem 10.1. *Let ξ be an uncertain set and let B be a Borel set of real numbers. Then*

$$\{\xi \not\subset B\} = \{\gamma \in \Gamma \mid \xi(\gamma) \not\subset B\} \tag{10.5}$$

is an event.

Proof: Since ξ is an uncertain set and B is a Borel set, the set $\{\xi \subset B\}$ is an event. Thus $\{\xi \not\subset B\}$ is an event by using the relation $\{\xi \not\subset B\} = \{\xi \subset B\}^c$.

Theorem 10.2. *Let ξ be an uncertain set and let B be a Borel set. Then*

$$\{\xi \cap B = \emptyset\} = \{\gamma \in \Gamma \mid \xi(\gamma) \cap B = \emptyset\} \tag{10.6}$$

is an event.

Proof: Since ξ is an uncertain set and B is a Borel set, the set $\{\xi \subset B^c\}$ is an event. Thus $\{\xi \cap B = \emptyset\}$ is an event by using the relation $\{\xi \cap B = \emptyset\} = \{\xi \subset B^c\}$.

Theorem 10.3. *Let ξ be an uncertain set and let B be a Borel set. Then*

$$\{\xi \cap B \neq \emptyset\} = \{\gamma \in \Gamma \mid \xi(\gamma) \cap B \neq \emptyset\} \tag{10.7}$$

is an event.

Proof: Since ξ is an uncertain set and B is a Borel set, the set $\{\xi \cap B = \emptyset\}$ is an event. Thus $\{\xi \cap B \neq \emptyset\}$ is an event by using the relation $\{\xi \cap B \neq \emptyset\} = \{\xi \cap B = \emptyset\}^c$.

Theorem 10.4. *Let ξ be an uncertain set and let a be a real number. Then*

$$\{a \in \xi\} = \{\gamma \in \Gamma \mid a \in \xi(\gamma)\} \tag{10.8}$$

is an event.

Proof: Since ξ is an uncertain set and a is a real number, the set $\{\xi \not\subset \{a\}^c\}$ is an event. Thus $\{a \in \xi\}$ is an event by using the relation $\{a \in \xi\} = \{\xi \not\subset \{a\}^c\}$.

Theorem 10.5. *Let ξ be an uncertain set and let a be a real number. Then*

$$\{a \notin \xi\} = \{\gamma \in \Gamma \mid a \notin \xi(\gamma)\} \tag{10.9}$$

is an event.

Proof: Since ξ is an uncertain set and a is a real number, the set $\{a \in \xi\}$ is an event. Thus $\{a \notin \xi\}$ is an event by using the relation $\{a \notin \xi\} = \{a \in \xi\}^c$.

Definition 10.2. *Let ξ and η be two uncertain sets on the uncertainty space $(\Gamma, \mathcal{L}, \mathcal{M})$. Then the complement ξ^c of uncertain set ξ is*

$$\xi^c(\gamma) = \xi(\gamma)^c, \quad \forall \gamma \in \Gamma. \tag{10.10}$$

The union $\xi \cup \eta$ of uncertain sets ξ and η is

$$(\xi \cup \eta)(\gamma) = \xi(\gamma) \cup \eta(\gamma), \quad \forall \gamma \in \Gamma. \tag{10.11}$$

The intersection $\xi \cap \eta$ of uncertain sets ξ and η is

$$(\xi \cap \eta)(\gamma) = \xi(\gamma) \cap \eta(\gamma), \quad \forall \gamma \in \Gamma. \tag{10.12}$$

Theorem 10.6 *(Law of Excluded Middle). Let ξ be an uncertain set and let ξ^c be its complement. Then*

$$\xi \cup \xi^c \equiv \Re. \tag{10.13}$$

Proof: For each $\gamma \in \Gamma$, it follows from the definition of ξ and ξ^c that the union is

$$(\xi \cup \xi^c)(\gamma) = \xi(\gamma) \cup \xi^c(\gamma) = \xi(\gamma) \cup \xi(\gamma)^c = \Re.$$

Thus we have $\xi \cup \xi^c \equiv \Re$.

Theorem 10.7 *(Law of Contradiction). Let ξ be an uncertain set and let ξ^c be its complement. Then*

$$\xi \cap \xi^c \equiv \emptyset. \tag{10.14}$$

Proof: For each $\gamma \in \Gamma$, it follows from the definition of ξ and ξ^c that the intersection is

$$(\xi \cap \xi^c)(\gamma) = \xi(\gamma) \cap \xi^c(\gamma) = \xi(\gamma) \cap \xi(\gamma)^c = \emptyset.$$

Thus we have $\xi \cap \xi^c \equiv \emptyset$.

Theorem 10.8 *(Double-Negation Law). Let ξ be an uncertain set. Then we have*

$$(\xi^c)^c = \xi. \tag{10.15}$$

Proof: For each $\gamma \in \Gamma$, it follows from the definition of complement that

$$(\xi^c)^c(\gamma) = (\xi^c(\gamma))^c = (\xi(\gamma)^c)^c = \xi(\gamma).$$

Thus we have $(\xi^c)^c = \xi$.

Theorem 10.9 *(De Morgan's Law). Let ξ and η be uncertain sets. Then*

$$(\xi \cup \eta)^c = \xi^c \cap \eta^c, \quad (\xi \cap \eta)^c = \xi^c \cup \eta^c. \tag{10.16}$$

Proof: For each $\gamma \in \Gamma$, it follows from the definition of complement that

$$(\xi \cup \eta)^c(\gamma) = ((\xi(\gamma) \cup \eta(\gamma))^c = \xi(\gamma)^c \cap \eta(\gamma)^c = (\xi^c \cap \eta^c)(\gamma).$$

Thus we have $(\xi \cup \eta)^c = \xi^c \cap \eta^c$. In addition, since

$$(\xi \cap \eta)^c(\gamma) = ((\xi(\gamma) \cap \eta(\gamma))^c = \xi(\gamma)^c \cup \eta(\gamma)^c = (\xi^c \cup \eta^c)(\gamma),$$

we get $(\xi \cap \eta)^c = \xi^c \cup \eta^c$.

Definition 10.3. *Let $\xi_1, \xi_2, \cdots, \xi_n$ be uncertain sets on the uncertainty space $(\Gamma, \mathcal{L}, \mathcal{M})$, and f a measurable function. Then $\xi = f(\xi_1, \xi_2, \cdots, \xi_n)$ is an uncertain set defined by*

$$\xi(\gamma) = f(\xi_1(\gamma), \xi_2(\gamma), \cdots, \xi_n(\gamma)), \quad \forall \gamma \in \Gamma. \tag{10.17}$$

Example 10.5: Let ξ and η be two uncertain sets on the uncertainty space $(\Gamma, \mathcal{L}, \mathcal{M})$. Then

$$(\xi + \eta)(\gamma) = \xi(\gamma) + \eta(\gamma), \quad \forall \gamma \in \Gamma, \tag{10.18}$$

$$(\xi - \eta)(\gamma) = \xi(\gamma) - \eta(\gamma), \quad \forall \gamma \in \Gamma, \tag{10.19}$$

$$(\xi \times \eta)(\gamma) = \xi(\gamma) \times \eta(\gamma), \quad \forall \gamma \in \Gamma, \tag{10.20}$$

$$(\xi \div \eta)(\gamma) = \xi(\gamma) \div \eta(\gamma), \quad \forall \gamma \in \Gamma. \tag{10.21}$$

Definition 10.4. *Let ξ and η be two uncertain sets. We say ξ is included in η (i.e., $\xi \subset \eta$) if $\xi(\gamma) \subset \eta(\gamma)$ for almost all $\gamma \in \Gamma$ in the sense of classical set theory.*

Definition 10.5. *Let ξ and η be two uncertain sets. We say ξ is equal to η (i.e., $\xi = \eta$) if $\xi(\gamma) = \eta(\gamma)$ for almost all $\gamma \in \Gamma$ in the sense of classical set theory.*

Definition 10.6. *An uncertain set ξ is said to be nonempty if $\xi(\gamma) \neq \emptyset$ for almost all $\gamma \in \Gamma$ in the sense of classical set theory.*

10.2 Membership Degree

Let ξ and η be two nonempty uncertain sets. What is the degree that η is included in ξ? In other words, what is the degree that η is a subset of ξ? Unfortunately, this problem is not as simple as you think. In order to discuss this issue, we introduce some symbols. At first, the set

$$\{\eta \subset \xi\} = \{\gamma \in \Gamma \mid \eta(\gamma) \subset \xi(\gamma)\} \qquad (10.22)$$

is an event that η is strongly included in ξ; and the set

$$\{\eta \not\subset \xi^c\} = \{\gamma \in \Gamma \mid \eta(\gamma) \not\subset \xi(\gamma)^c\} = \{\gamma \in \Gamma \mid \eta(\gamma) \cap \xi(\gamma) \neq \emptyset\} \qquad (10.23)$$

is an event that η is weakly included in ξ. It is easy to verify that

$$\{\eta \subset \xi\} \subset \{\eta \not\subset \xi^c\}. \qquad (10.24)$$

That is, "strong inclusion" is a subset of "weak inclusion".

Definition 10.7. *Let ξ and η be two nonempty uncertain sets. Then the strong membership degree of η to ξ is defined as the uncertain measure that η is strongly included in ξ, i.e., $\mathcal{M}\{\eta \subset \xi\}$.*

Definition 10.8. *Let ξ and η be two nonempty uncertain sets. Then the weak membership degree of η to ξ is defined as the uncertain measure that η is weakly included in ξ, i.e., $\mathcal{M}\{\eta \not\subset \xi^c\}$.*

What is the appropriate event that η is included in ξ? Intuitively, it is too conservative if we take the strong inclusion $\{\eta \subset \xi\}$, and it is too adventurous if we take the weak inclusion $\{\eta \not\subset \xi^c\}$. Thus we have to introduce a new symbol $\rhd$ to represent this inclusion relationship called *imaginary inclusion*. That is, $\eta \rhd \xi$ represents the event that η is imaginarily included in ξ.

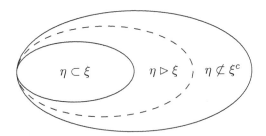

Figure 10.1: Strong Inclusion, Weak Inclusion and Imaginary Inclusion

How do we determine $\mathcal{M}\{\eta \rhd \xi\}$? It is too conservative if we take the strong membership degree $\mathcal{M}\{\eta \subset \xi\}$, and it is too adventurous if we take the weak membership degree $\mathcal{M}\{\eta \not\subset \xi^c\}$. In fact, it is reasonable to take the middle value between $\mathcal{M}\{\eta \subset \xi\}$ and $\mathcal{M}\{\eta \not\subset \xi^c\}$.

Definition 10.9 *(Liu [125]). Let ξ and η be two nonempty uncertain sets. Then the membership degree of η to ξ is defined as the average of strong and weak membership degrees, i.e.,*

$$\mathcal{M}\{\eta \rhd \xi\} = \frac{1}{2}\left(\mathcal{M}\{\eta \subset \xi\} + \mathcal{M}\{\eta \not\subset \xi^c\}\right). \qquad (10.25)$$

The membership degree is understood as the uncertain measure that η is imaginarily included in ξ.

For any uncertain sets ξ and η, the membership degree $\mathcal{M}\{\eta \rhd \xi\}$ reflects the truth degree that η is a subset of ξ. If $\mathcal{M}\{\eta \rhd \xi\} = 1$, then η is completely included in ξ. If $\mathcal{M}\{\eta \rhd \xi\} = 0$, then η and ξ have no intersection at all. It is always true that

$$\mathcal{M}\{\eta \subset \xi\} \leq \mathcal{M}\{\eta \rhd \xi\} \leq \mathcal{M}\{\eta \not\subset \xi^c\}. \tag{10.26}$$

In addition, the membership degree is asymmetrical, i.e., generally speaking,

$$\mathcal{M}\{\eta \rhd \xi\} \neq \mathcal{M}\{\xi \rhd \eta\}. \tag{10.27}$$

Furthermore, any uncertain set is included in itself completely, i.e.,

$$\mathcal{M}\{\xi \rhd \xi\} \equiv 1. \tag{10.28}$$

Theorem 10.10. *Let ξ be a nonempty uncertain set, and let A be a Borel set of real numbers. Then*

$$\mathcal{M}\{\xi \rhd A\} + \mathcal{M}\{\xi \rhd A^c\} = 1. \tag{10.29}$$

Proof: Since A is a special uncertain set, it follows from Definition 10.9 that

$$\mathcal{M}\{\xi \rhd A\} = \frac{1}{2}\left(\mathcal{M}\{\xi \subset A\} + \mathcal{M}\{\xi \not\subset A^c\}\right),$$

$$\mathcal{M}\{\xi \rhd A^c\} = \frac{1}{2}\left(\mathcal{M}\{\xi \subset A^c\} + \mathcal{M}\{\xi \not\subset A\}\right).$$

By using the self-duality of uncertain measure, we get

$$\mathcal{M}\{\xi \rhd A\} + \mathcal{M}\{\xi \rhd A^c\}$$

$$= \frac{1}{2}\left(\mathcal{M}\{\xi \subset A\} + \mathcal{M}\{\xi \not\subset A^c\}\right) + \frac{1}{2}\left(\mathcal{M}\{\xi \subset A^c\} + \mathcal{M}\{\xi \not\subset A\}\right)$$

$$= \frac{1}{2}\left(\mathcal{M}\{\xi \subset A\} + \mathcal{M}\{\xi \not\subset A\}\right) + \frac{1}{2}\left(\mathcal{M}\{\xi \subset A^c\} + \mathcal{M}\{\xi \not\subset A^c\}\right)$$

$$= \frac{1}{2} + \frac{1}{2} = 1.$$

The theorem is verified.

10.3   Membership Function

This section will introduce a concept of membership function for a special type of uncertain set that takes values in a nested class of sets. Keep in mind that only some special uncertain sets have their own membership functions.

Definition 10.10. *A real-valued function μ is called a membership function if*

$$0 \leq \mu(x) \leq 1, \quad x \in \mathfrak{R}. \tag{10.30}$$

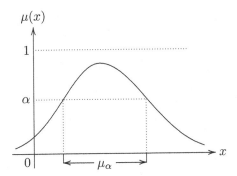

Figure 10.2: The α-Cut μ_α of Membership Function $\mu(x)$

The link between membership function and uncertain set will be discussed later.

Definition 10.11. *Let μ be a membership function. Then for any number $\alpha \in [0,1]$, the set*

$$\mu_\alpha = \big\{ x \in \Re \mid \mu(x) \geq \alpha \big\} \tag{10.31}$$

is called the α-cut of μ.

Theorem 10.11. *The α-cut μ_α is a monotone decreasing set with respect to α. That is, for any real numbers α and β in $[0,1]$ with $\alpha > \beta$, we have $\mu_\alpha \subset \mu_\beta$.*

Proof: For any $x \in \mu_\alpha$, we have $\mu(x) \geq \alpha$. Since $\alpha > \beta$, we have $\mu(x) > \beta$ and $x \in \mu_\beta$. Hence $\mu_\alpha \subset \mu_\beta$.

Definition 10.12. *Let μ be a membership function. Then for any number $\alpha \in [0,1]$, the set*

$$W_\alpha = \big\{ \mu_\beta \mid \beta \leq \alpha \big\} \tag{10.32}$$

is called the α-class of μ. Especially, the 1-class is called the total class of μ.

Note that each element in W_α is a β-cut of μ where β is a number less than or equal to α. In addition, μ_{β_1} and μ_{β_2} are regarded as distinct elements in W_α whenever $\beta_1 \neq \beta_2$. Each α-class (including total class) forms a family of nested sets. In the sense that the universe is assumed to be the total class, the complement W_α^c is the class of β-cuts with $\beta > \alpha$, i.e.,

$$W_\alpha^c = \big\{ \mu_\beta \mid \beta > \alpha \big\}. \tag{10.33}$$

Thus $W_\alpha \cup W_\alpha^c$ is just the total class.

Now it is ready to assign a membership function to an uncertain set. Roughly speaking, an uncertain set ξ is said to have a membership function μ if ξ takes values in the total class of μ and contains each α-cut with uncertain measure α. Precisely, we have the following definition.

Definition 10.13 *(Liu [125]). An uncertain set ξ is said to have a membership function μ if the range of ξ is just the total class of μ, and*

$$\mathcal{M}\{\xi \in W_\alpha\} = \alpha, \quad \forall \alpha \in [0,1] \tag{10.34}$$

where W_α is the α-class of μ.

Since W_α^c is the complement of W_α, it follows from the self-duality of uncertain measure that

$$\mathcal{M}\{\xi \in W_\alpha^c\} = 1 - \alpha, \quad \forall \alpha \in [0,1]. \tag{10.35}$$

In addition, it is easy to verify that $\{\xi \notin W_\alpha\} = \{\xi \in W_\alpha^c\}$. Hence

$$\mathcal{M}\{\xi \notin W_\alpha\} = 1 - \alpha, \quad \forall \alpha \in [0,1]. \tag{10.36}$$

If you think that Definition 10.13 is hard-to-understand, you may accept the following representation theorem.

Theorem 10.12 *((Liu [125]), Representation Theorem). Let ξ be an uncertain set with membership function μ. Then ξ may be represented by*

$$\xi = \bigcup_{0 \leq \alpha \leq 1} \alpha \cdot \mu_\alpha \tag{10.37}$$

where μ_α is the α-cut of membership function μ.

Proof: The representation theorem is essentially nothing but an alternative explanation of membership function. The equation (10.37) tells us that the range of ξ is just the total class of μ, and $\mathcal{M}\{\xi \in W_\alpha\} = \alpha$ for any $\alpha \in [0,1]$.

Remark 10.1: What uncertain set does the representation theorem stand for? Take an uncertainty space $(\Gamma, \mathcal{L}, \mathcal{M})$ to be $[0,1]$ with $\mathcal{M}\{[0,\gamma]\} = \gamma$ for each $\gamma \in [0,1]$. Then the set-valued function

$$\xi(\gamma) = \mu_\gamma \tag{10.38}$$

on the uncertainty space $(\Gamma, \mathcal{L}, \mathcal{M})$ is just the uncertain set.

Remark 10.2: It is not true that any uncertain set has its own membership function. For example, the uncertain set

$$\xi = \begin{cases} [1,2] \text{ with uncertain measure } 0.5 \\ [2,3] \text{ with uncertain measure } 0.5 \end{cases}$$

has no membership function.

Remark 10.3: Although an uncertain variable is a special uncertain set, it has no membership function.

Example 10.6: The set $\Re$ of real numbers is a special uncertain set $\xi(\gamma) \equiv \Re$. Such an uncertain set ξ has a membership function

$$\mu(x) \equiv 1, \quad \forall x \in \Re. \tag{10.39}$$

For this case, the membership function μ is identical with the characteristic function of $\Re$.

Example 10.7: The empty set $\emptyset$ is a special uncertain set $\xi(\gamma) \equiv \emptyset$. Such an uncertain set ξ has a membership function

$$\mu(x) \equiv 0, \quad \forall x \in \Re. \tag{10.40}$$

For this case, the membership function μ is identical with the characteristic function of $\emptyset$.

Example 10.8: Let a be a number in $\Re$ and let α be a number in $(0, 1)$. Then the membership function

$$\mu(x) = \begin{cases} \alpha, & \text{if } x = a \\ 0, & \text{if } x \neq a \end{cases} \tag{10.41}$$

represents the uncertain set

$$\xi = \begin{cases} \{a\} & \text{with uncertain measure } \alpha \\ \emptyset & \text{with uncertain measure } 1 - \alpha \end{cases} \tag{10.42}$$

that takes values either the singleton $\{a\}$ or the empty set $\emptyset$. This means that uncertainty exists even when there is a unique element in the universal set.

Example 10.9: By a *rectangular uncertain set* we mean the uncertain set fully determined by the pair (a, b) of crisp numbers with $a < b$, whose membership function is
$$\mu(x) = 1, \quad a \leq x \leq b.$$

Example 10.10: By a *triangular uncertain set* we mean the uncertain set fully determined by the triplet (a, b, c) of crisp numbers with $a < b < c$, whose membership function is

$$\mu(x) = \begin{cases} \dfrac{x - a}{b - a}, & \text{if } a \leq x \leq b \\ \dfrac{x - c}{b - c}, & \text{if } b \leq x \leq c. \end{cases}$$

Example 10.11: By a *trapezoidal uncertain set* we mean the uncertain set fully determined by the quadruplet (a, b, c, d) of crisp numbers with $a < b < c < d$, whose membership function is

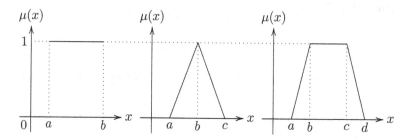

Figure 10.3: Rectangular, Triangular and Trapezoidal Membership Functions

$$\mu(x) = \begin{cases} \dfrac{x-a}{b-a}, & \text{if } a \le x \le b \\[2mm] 1, & \text{if } b \le x \le c \\[2mm] \dfrac{x-d}{c-d}, & \text{if } c \le x \le d. \end{cases}$$

Theorem 10.13. *Let ξ be a nonempty uncertain set with membership function μ. Then for any number $x \in \Re$, we have*

$$\mathcal{M}\{x \in \xi\} = \mu(x), \quad \mathcal{M}\{x \not\in \xi\} = 1 - \mu(x), \\ \mathcal{M}\{x \not\in \xi^c\} = \mu(x), \quad \mathcal{M}\{x \in \xi^c\} = 1 - \mu(x). \tag{10.43}$$

Proof: Since μ is the membership function of ξ, we have $\{x \in \xi\} = \{\xi \in W_\alpha\}$ where $\alpha = \mu(x)$. Thus

$$\mathcal{M}\{x \in \xi\} = \mathcal{M}\{\xi \in W_\alpha\} = \alpha = \mu(x).$$

In addition, it follows from the self-duality of uncertain measure that

$$\mathcal{M}\{x \not\in \xi\} = 1 - \mathcal{M}\{x \in \xi\} = 1 - \mu(x).$$

Finally, it is easy to verify that $\{x \in \xi^c\} = \{x \not\in \xi\}$. Hence $\mathcal{M}\{x \in \xi^c\} = 1 - \mu(x)$ and $\mathcal{M}\{x \not\in \xi^c\} = \mu(x)$.

Theorem 10.14. *Let ξ be a nonempty uncertain set with membership function μ, and let x be a constant. Then*

$$\mathcal{M}\{x \rhd \xi\} = \mu(x). \tag{10.44}$$

Proof: Note that $\mathcal{M}\{x \in \xi\} = \mu(x)$ and $\mathcal{M}\{x \not\in \xi^c\} = \mu(x)$. It follows that

$$\mathcal{M}\{x \rhd \xi\} = \frac{1}{2} \left(\mathcal{M}\{x \in \xi\} + \mathcal{M}\{x \not\in \xi^c\} \right) = \mu(x).$$

Theorem 10.15. *Let ξ be a nonempty uncertain set with membership function μ. Then for any number α, we have*

$$\mathcal{M}\{\mu_\alpha \subset \xi\} = \alpha, \quad \mathcal{M}\{\mu_\alpha \not\subset \xi\} = 1 - \alpha. \tag{10.45}$$

Proof: Since $\{\mu_\alpha \subset \xi\}$ is just the α-class of μ, we immediately have $\mathcal{M}\{\mu_\alpha \subset \xi\} = \alpha$. In addition, by the self-duality of uncertain measure, we obtain $\mathcal{M}\{\mu_\alpha \not\subset \xi\} = 1 - \mathcal{M}\{\mu_\alpha \subset \xi\} = 1 - \alpha$.

Theorem 10.16. *Let ξ be a nonempty uncertain set with membership function μ, and let A be a set of real numbers. Then*

$$\begin{aligned}
\mathcal{M}\{A \subset \xi\} &= \inf_{x \in A} \mu(x), \quad \mathcal{M}\{A \not\subset \xi\} = 1 - \inf_{x \in A} \mu(x), \\
\mathcal{M}\{A \not\subset \xi^c\} &= \sup_{x \in A} \mu(x), \quad \mathcal{M}\{A \subset \xi^c\} = 1 - \sup_{x \in A} \mu(x).
\end{aligned} \tag{10.46}$$

Proof: Since μ is the membership function of ξ, we immediately have

$$\{A \subset \xi\} = \{\xi \in W_\alpha\}, \text{ with } \alpha = \inf_{x \in A} \mu(x).$$

Thus we get

$$\mathcal{M}\{A \subset \xi\} = \mathcal{M}\{\xi \in W_\alpha\} = \alpha = \inf_{x \in A} \mu(x).$$

Since $\{A \not\subset \xi\} = \{A \subset \xi\}^c$, it follows from the self-duality of uncertain measure that

$$\mathcal{M}\{A \not\subset \xi\} = 1 - \mathcal{M}\{A \subset \xi\} = 1 - \inf_{x \in A} \mu(x).$$

In addition, we have

$$\{A \not\subset \xi^c\} = \{\xi \in W_\alpha\}, \text{ with } \alpha = \sup_{x \in A} \mu(x).$$

Thus

$$\mathcal{M}\{A \not\subset \xi^c\} = \mathcal{M}\{\xi \in W_\alpha\} = \alpha = \sup_{x \in A} \mu(x).$$

Since $\{A \subset \xi^c\} = \{A \not\subset \xi^c\}^c$, it follows from the self-duality of uncertain measure that

$$\mathcal{M}\{A \subset \xi^c\} = 1 - \mathcal{M}\{A \not\subset \xi^c\} = 1 - \sup_{x \in A} \mu(x).$$

The theorem is verified.

Theorem 10.17. *Let ξ be an uncertain set with membership function μ, and let A be a set of real numbers. Then*

$$\mathcal{M}\{A \rhd \xi\} = \frac{1}{2}\left(\inf_{x \in A} \mu(x) + \sup_{x \in A} \mu(x)\right). \tag{10.47}$$

Proof: Since μ is the membership function of ξ, we immediately have

$$\mathcal{M}\{A \subset \xi\} = \inf_{x \in A} \mu(x), \quad \mathcal{M}\{A \not\subset \xi^c\} = \sup_{x \in A} \mu(x).$$

The theorem follows from the definition of membership degree directly.

Theorem 10.18. *Let ξ be an uncertain set with membership function μ, and let f be a strictly monotone function. Then $f(\xi)$ is an uncertain set with membership function*

$$\nu(x) = \mu(f^{-1}(x)). \tag{10.48}$$

Proof: At first, for each $\alpha \in [0,1]$, let x be a point in the α-cut of ν, i.e., $x \in \mu_\alpha$. Then

$$\nu(x) \geq \alpha \Rightarrow \mu(f^{-1}(x)) \geq \alpha \Rightarrow f^{-1}(x) \in \mu_\alpha \Rightarrow x \in f(\mu_\alpha) \Rightarrow \nu_\alpha \subset f(\mu_\alpha).$$

If x is a point in $f(\mu_\alpha)$, i.e., $x \in f(\mu_\alpha)$, then

$$f^{-1}(x) \in \mu_\alpha \Rightarrow \mu(f^{-1}(x)) \geq \alpha \Rightarrow \nu(x) \geq \alpha \Rightarrow x \in \nu_\alpha \Rightarrow f(\mu_\alpha) \subset \nu_\alpha.$$

Thus $\nu_\alpha = f(\mu_\alpha)$. In addition, since the range of ξ is the total class of μ, the range of $f(\xi)$ is the total class of ν. Finally, since the α-classes of μ and ν have the same preimagine, i.e.,

$$\mathcal{M}\{f(\xi) \in W_\alpha^\nu\} = \mathcal{M}\{\xi \in W_\alpha^\mu\} = \alpha$$

for each $\alpha \in [0,1]$, the membership function of $f(\xi)$ is just ν.

Example 10.12: Let ξ be an uncertain set with membership function μ. Then $-\xi$ is an uncertain set with membership function $\mu(-x)$.

Example 10.13: Let ξ be an uncertain set with membership function μ and let k be a real number. Then $\xi + k$ is an uncertain set with membership function $\mu(x - k)$.

Example 10.14: Let ξ be an uncertain set with membership function μ and let a be a positive number. Then $a\xi$ is an uncertain set with membership function $\mu(x/a)$.

Example 10.15: Let ξ be a positive uncertain set with membership function μ. Then $1/\xi$ is an uncertain set with membership function $\mu(1/x)$.

Membership Function is Frangible for Arithmetic Operations

Generally speaking, the complement ξ^c, union $\xi \cup \eta$, intersection $\xi \cap \eta$, sum $\xi + \eta$ and product $\xi \times \eta$ of uncertain sets have no membership functions even though the original uncertain sets have their own membership functions.

10.4   Uncertainty Distribution

This section introduces the concept of uncertainty distribution for nonempty uncertain sets, and gives a sufficient and necessary condition for uncertainty distribution.

Definition 10.14 *(Liu [125]). Let ξ be a nonempty uncertain set. Then the function*

$$\Phi(x) = \mathcal{M}\{\xi \triangleright (-\infty, x]\}, \quad \forall x \in \Re \tag{10.49}$$

is called the uncertainty distribution of ξ.

Example 10.16: The uncertainty distribution of the uncertain set $\xi \equiv \Re$ is

$$\Phi(x) \equiv 0.5. \tag{10.50}$$

Example 10.17: The uncertain set $\xi \equiv \emptyset$ has no uncertainty distribution because it is not a nonempty set.

Example 10.18: Let ξ be an uncertain set taking value $[1, 2]$ with uncertain measure 0.5 and value $[3, 4]$ with uncertain measure 0.5. That is,

$$\xi = \begin{cases} [1, 2] & \text{with uncertain measure } 0.5 \\ [3, 4] & \text{with uncertain measure } 0.5. \end{cases}$$

Then its uncertainty distribution is

$$\Phi(x) = \begin{cases} 0, & \text{if } x < 1 \\ 0.25, & \text{if } 1 \le x < 2 \\ 0.5, & \text{if } 2 \le x < 3 \\ 0.75, & \text{if } 3 \le x < 4 \\ 1, & \text{if } x \ge 4. \end{cases}$$

Theorem 10.19 *(Measure Inversion Theorem). Let ξ be a nonempty uncertain set with continuous uncertainty distribution Φ. Then*

$$\mathcal{M}\{\xi \triangleright (-\infty, x]\} = \Phi(x), \quad \mathcal{M}\{\xi \triangleright [x, +\infty)\} = 1 - \Phi(x) \tag{10.51}$$

for any $x \in \Re$.

Proof: The first equation follows from the definition of uncertainty distribution, and the second equation follows from the self-duality of uncertain measure.

Theorem 10.20 *(Sufficient and Necessary Condition for Uncertainty Distribution) A function $\Phi : \Re \to [0, 1]$ is an uncertainty distribution of uncertain set if and only if it is an increasing function except $\Phi(x) \equiv 0$ and $\Phi(x) \equiv 1$.*

Proof: Suppose Φ is an uncertainty distribution. Since an uncertain variable is a special uncertain set, it follows from Theorem 1.11 that Φ is an increasing function except $\Phi(x) \equiv 0$ and $\Phi(x) \equiv 1$. Conversely, suppose Φ is an increasing function but $\Phi(x) \not\equiv 0$ and $\Phi(x) \not\equiv 1$. Theorem 1.11 tells us that there is an uncertain variable (a degenerate uncertain set) whose uncertainty distribution is just Φ.

Theorem 10.21. *Let ξ be a nonempty uncertain set with continuous membership function μ. If x_0 is a point with $\mu(x_0) = 1$, then the uncertainty distribution of ξ is*

$$\Phi(x) = \begin{cases} \sup_{y \le x} \mu(y)/2, & \text{if } x \le x_0 \\ 1 - \sup_{y \ge x} \mu(y)/2, & \text{if } x \ge x_0. \end{cases} \tag{10.52}$$

Especially, if μ is unimodal, then

$$\Phi(x) = \begin{cases} \mu(x)/2, & \text{if } x \le x_0 \\ 1 - \mu(x)/2, & \text{if } x \ge x_0. \end{cases} \tag{10.53}$$

Proof: When $x \le x_0$, it follows from the continuity of membership function that

$$\mathcal{M}\{\xi \subset (-\infty, x]\} = 0, \quad \mathcal{M}\{\xi \not\subset (x, +\infty)\} = \sup_{y \le x} \mu(y).$$

Thus we have

$$\Phi(x) = \mathcal{M}\{\xi \rhd (-\infty, x]\} = \frac{1}{2}\left(0 + \sup_{y \le x} \mu(y)\right) = \sup_{y \le x} \mu(y)/2.$$

When $x \ge x_0$, we get

$$\mathcal{M}\{\xi \subset (-\infty, x]\} = 1 - \sup_{y \ge x} \mu(y), \quad \mathcal{M}\{\xi \not\subset (x, +\infty)\} = 1.$$

Thus we have

$$\Phi(x) = \mathcal{M}\{\xi \rhd (-\infty, x]\} = \frac{1}{2}\left(1 - \sup_{y \ge x} \mu(y) + 1\right) = 1 - \sup_{y \ge x} \mu(y)/2.$$

The theorem is proved.

Example 10.19: The rectangular uncertain set (a, b) has an uncertainty distribution

$$\Phi(x) = \begin{cases} 0, & \text{if } x < a \\ 0.5, & \text{if } a \le x < b \\ 1, & \text{if } x \ge b. \end{cases} \tag{10.54}$$

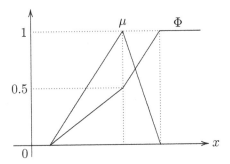

Figure 10.4: Membership Function and Uncertainty Distribution

Example 10.20: The triangular uncertain set (a, b, c) has an uncertainty distribution

$$\Phi(x) = \begin{cases} \dfrac{x - a}{2(b - a)}, & \text{if } a \le x \le b \\[2mm] \dfrac{c - 2b + x}{2(c - b)}, & \text{if } b \le x \le c. \end{cases} \tag{10.55}$$

Example 10.21: The trapezoidal uncertain set (a, b, c, d) has an uncertainty distribution

$$\Phi(x) = \begin{cases} \dfrac{x - a}{2(b - a)}, & \text{if } a \le x \le b \\[2mm] 0.5, & \text{if } b \le x \le c \\[2mm] \dfrac{d - 2c + x}{2(d - c)}, & \text{if } c \le x \le d. \end{cases} \tag{10.56}$$

10.5 Independence

Definition 10.15 *(Liu [125]). The uncertain sets $\xi_1, \xi_2, \cdots, \xi_m$ are said to be independent if*

$$\mathcal{M} \left\{ \bigcap_{i=1}^{m} (\xi_i \subset B_i) \right\} = \min_{1 \le i \le m} \mathcal{M} \{\xi_i \subset B_i\} \tag{10.57}$$

and

$$\mathcal{M} \left\{ \bigcup_{i=1}^{m} (\xi_i \subset B_i) \right\} = \max_{1 \le i \le m} \mathcal{M} \{\xi_i \subset B_i\} \tag{10.58}$$

for any Borel sets $B_1, B_2, \cdots, B_m$ of real numbers.

Theorem 10.22. *The uncertain sets $\xi_1, \xi_2, \cdots, \xi_m$ are independent if and only if*

$$\mathcal{M} \left\{ \bigcap_{i=1}^{m} (\xi_i \not\subset B_i) \right\} = \min_{1 \le i \le m} \mathcal{M} \{\xi_i \not\subset B_i\} \tag{10.59}$$

and

$$\mathcal{M}\left\{\bigcup_{i=1}^{m}(\xi_i \not\subset B_i)\right\} = \max_{1 \leq i \leq m} \mathcal{M}\{\xi_i \not\subset B_i\} \tag{10.60}$$

for any Borel sets $B_1, B_2, \cdots, B_m$ of real numbers.

Proof: Since $\xi_1, \xi_2, \cdots, \xi_m$ are independent uncertain sets, we immediately have (10.57) and (10.58). It follows from the self-duality of uncertain measure that

$$\mathcal{M}\left\{\bigcap_{i=1}^{m}(\xi_i \not\subset B_i)\right\} = 1 - \mathcal{M}\left\{\bigcup_{i=1}^{m}(\xi_i \subset B_i)\right\}$$

$$= 1 - \max_{1 \leq i \leq m} \mathcal{M}\{\xi_i \subset B_i\} = \min_{1 \leq i \leq m} \mathcal{M}\{\xi_i \not\subset B_i\}$$

and

$$\mathcal{M}\left\{\bigcup_{i=1}^{m}(\xi_i \not\subset B_i)\right\} = 1 - \mathcal{M}\left\{\bigcap_{i=1}^{m}(\xi_i \subset B_i)\right\}$$

$$= 1 - \min_{1 \leq i \leq m} \mathcal{M}\{\xi_i \subset B_i\} = \max_{1 \leq i \leq m} \mathcal{M}\{\xi_i \not\subset B_i\}.$$

Thus (10.59) and (10.60) are proved. Conversely, assume (10.59) and (10.60). Then

$$\mathcal{M}\left\{\bigcap_{i=1}^{m}(\xi_i \subset B_i)\right\} = 1 - \mathcal{M}\left\{\bigcup_{i=1}^{m}(\xi_i \not\subset B_i)\right\}$$

$$= 1 - \max_{1 \leq i \leq m} \mathcal{M}\{\xi_i \not\subset B_i\} = \min_{1 \leq i \leq m} \mathcal{M}\{\xi_i \subset B_i\}$$

and

$$\mathcal{M}\left\{\bigcup_{i=1}^{m}(\xi_i \subset B_i)\right\} = 1 - \mathcal{M}\left\{\bigcap_{i=1}^{m}(\xi_i \not\subset B_i)\right\}$$

$$= 1 - \min_{1 \leq i \leq m} \mathcal{M}\{\xi_i \not\subset B_i\} = \max_{1 \leq i \leq m} \mathcal{M}\{\xi_i \subset B_i\}.$$

Thus (10.57) and (10.58) are verified. The proof is complete.

Theorem 10.23. *The uncertain sets $\xi_1, \xi_2, \cdots, \xi_m$ are independent if and only if*

$$\mathcal{M}\left\{\bigcap_{i=1}^{m}(\xi_i \cap B_i = \emptyset)\right\} = \min_{1 \leq i \leq m} \mathcal{M}\{\xi_i \cap B_i = \emptyset\} \tag{10.61}$$

and

$$\mathcal{M}\left\{\bigcup_{i=1}^{m}(\xi_i \cap B_i = \emptyset)\right\} = \max_{1 \leq i \leq m} \mathcal{M}\{\xi_i \cap B_i = \emptyset\} \tag{10.62}$$

for any Borel sets $B_1, B_2, \cdots, B_m$ of real numbers.

Proof: The theorem follows from the fact that $\xi_i \cap B_i = \emptyset$ if and only if $\xi_i \subset B_i^c$ for each i.

Theorem 10.24. *The uncertain sets $\xi_1, \xi_2, \cdots, \xi_m$ are independent if and only if*

$$\mathcal{M}\left\{\bigcap_{i=1}^{m}(\xi_i \cap B_i \neq \emptyset)\right\} = \min_{1 \leq i \leq m} \mathcal{M}\{\xi_i \cap B_i \neq \emptyset\} \qquad (10.63)$$

and

$$\mathcal{M}\left\{\bigcup_{i=1}^{m}(\xi_i \cap B_i \neq \emptyset)\right\} = \max_{1 \leq i \leq m} \mathcal{M}\{\xi_i \cap B_i \neq \emptyset\} \qquad (10.64)$$

for any Borel sets $B_1, B_2, \cdots, B_m$ of real numbers.

Proof: The theorem follows from the fact that $\xi_i \cap B_i \neq \emptyset$ if and only if $\xi_i \not\subset B_i^c$ for each i.

10.6   Operational Law

This section will discuss the operational law on independent uncertain sets via uncertainty distributions.

Theorem 10.25 *(Liu [125], Operational Law). Let $\xi_1, \xi_2, \cdots, \xi_n$ be independent uncertain sets, and $f : \Re^n \to \Re$ a measurable function. Then $\xi = f(\xi_1, \xi_2, \cdots, \xi_n)$ is an uncertain set such that*

$$\mathcal{M}\{\xi \subset B\} = \begin{cases} \displaystyle\sup_{f(B_1, B_2, \cdots, B_n) \subset B} \min_{1 \leq k \leq n} \mathcal{M}_k\{\xi_k \subset B_k\}, \\ \qquad if \displaystyle\sup_{f(B_1, B_2, \cdots, B_n) \subset B} \min_{1 \leq k \leq n} \mathcal{M}_k\{\xi_k \subset B_k\} > 0.5 \\ 1 - \displaystyle\sup_{f(B_1, B_2, \cdots, B_n) \subset B^c} \min_{1 \leq k \leq n} \mathcal{M}_k\{\xi_k \subset B_k\}, \\ \qquad if \displaystyle\sup_{f(B_1, B_2, \cdots, B_n) \subset B^c} \min_{1 \leq k \leq n} \mathcal{M}_k\{\xi_k \subset B_k\} > 0.5 \\ 0.5, \qquad otherwise \end{cases}$$

for Borel sets $B, B_1, B_2, \cdots, B_n$ of real numbers.

Proof: Write $\Lambda = \{\xi \subset B\}$ and $\Lambda_k = \{\xi_k \subset B_k\}$ for $k = 1, 2, \cdots, n$. It is easy to verify that

$$\Lambda_1 \times \Lambda_2 \times \cdots \times \Lambda_n \subset \Lambda \text{ if and only if } f(B_1, B_2, \cdots, B_n) \subset B,$$

$$\Lambda_1 \times \Lambda_2 \times \cdots \times \Lambda_n \subset \Lambda^c \text{ if and only if } f(B_1, B_2, \cdots, B_n) \subset B^c.$$

Thus the operational law follows from the product measure axiom immediately.

Increasing Function of Uncertain Sets

Theorem 10.26 *(Liu [125]). Let $\xi_1, \xi_2, \cdots, \xi_n$ be independent uncertain sets with uncertainty distributions $\Phi_1, \Phi_2, \cdots, \Phi_n$, respectively. If $f : \Re^n \to \Re$ is a strictly increasing function, then*

$$\xi = f(\xi_1, \xi_2, \cdots, \xi_n) \tag{10.65}$$

is an uncertain set whose inverse uncertainty distribution is

$$\Phi^{-1}(\alpha) = f(\Phi_1^{-1}(\alpha), \Phi_2^{-1}(\alpha), \cdots, \Phi_n^{-1}(\alpha)) \tag{10.66}$$

for any α with $0 < \alpha < 1$.

Proof: For simplicity, we only prove the case $n = 2$. Since ξ_1 and ξ_2 are independent uncertain sets and f is a strictly increasing function, we have

$$\mathcal{M}\{\xi \triangleright (-\infty, \Phi^{-1}(\alpha)]\}$$

$$= \mathcal{M}\{f(\xi_1, \xi_2) \triangleright (-\infty, f(\Phi_1^{-1}(\alpha), \Phi_2^{-1}(\alpha))]\}$$

$$\geq \mathcal{M}\{(\xi_1 \triangleright (-\infty, \Phi_1^{-1}(\alpha)]) \cap (\xi_2 \triangleright (-\infty, \Phi_2^{-1}(\alpha)])\}$$

$$= \mathcal{M}\{\xi_1 \triangleright (-\infty, \Phi_1^{-1}(\alpha)]\} \wedge \mathcal{M}\{\xi_2 \triangleright (-\infty, \Phi_2^{-1}(\alpha)]\}$$

$$= \alpha \wedge \alpha = \alpha.$$

On the other hand, there exists some index i such that

$$\{f(\xi_1, \xi_2) \triangleright (-\infty, f(\Phi_1^{-1}(\alpha), \Phi_2^{-1}(\alpha))]\} \subset \{\xi_i \triangleright (-\infty, \Phi_i^{-1}(\alpha)]\}.$$

Thus

$$\mathcal{M}\{\xi \triangleright (-\infty, \Phi^{-1}(\alpha)]\} \leq \mathcal{M}\{\xi_i \triangleright (-\infty, \Phi_i^{-1}(\alpha)]\} = \alpha.$$

It follows that $\mathcal{M}\{\xi \triangleright (-\infty, \Phi^{-1}(\alpha)]\} = \alpha$. In other words, Φ is just the uncertainty distribution of ξ. The theorem is proved.

Example 10.22: Let ξ_1 and ξ_2 be independent uncertain sets with uncertainty distributions $\Phi_1(x)$ and $\Phi_2(x)$, respectively, and let a_1 and a_2 be nonnegative numbers. Then the inverse uncertainty distribution of the weighted sum $a_1\xi_1 + a_2\xi_2$ is

$$\Phi^{-1}(\alpha) = a_1\Phi_1^{-1}(\alpha) + a_2\Phi_2^{-1}(\alpha), \quad \forall \alpha \in (0, 1). \tag{10.67}$$

Example 10.23: Let ξ_1 and ξ_2 be independent and nonnegative uncertain sets with uncertainty distributions Φ_1 and Φ_2, respectively. Then the inverse uncertainty distribution of the product $\xi_1 \times \xi_2$ is

$$\Phi^{-1}(\alpha) = \Phi_1^{-1}(\alpha) \times \Phi_2^{-1}(\alpha), \quad 0 < \alpha < 1. \tag{10.68}$$

Example 10.24: Assume ξ_1, ξ_2, ξ_3 are independent and nonnegative uncertain sets with uncertainty distributions Φ_1, Φ_2, Φ_3, respectively. Then the inverse uncertainty distribution of $(\xi_1 + \xi_2)\xi_3$ is

$$\Phi^{-1}(\alpha) = \left(\Phi_1^{-1}(\alpha) + \Phi_2^{-1}(\alpha)\right)\Phi_3^{-1}(\alpha), \quad 0 < \alpha < 1. \tag{10.69}$$

Decreasing Function of Uncertain Sets

Theorem 10.27 *(Liu [125]). Let $\xi_1, \xi_2, \cdots, \xi_n$ be independent uncertain sets with uncertainty distributions $\Phi_1, \Phi_2, \cdots, \Phi_n$, respectively. If $f : \Re^n \to \Re$ is a strictly decreasing function, then*

$$\xi = f(\xi_1, \xi_2, \cdots, \xi_n) \tag{10.70}$$

is an uncertain set whose inverse uncertainty distribution is

$$\Phi^{-1}(\alpha) = f(\Phi_1^{-1}(1-\alpha), \Phi_2^{-1}(1-\alpha), \cdots, \Phi_n^{-1}(1-\alpha)) \tag{10.71}$$

for any α with $0 < \alpha < 1$.

Proof: For simplicity, we only prove the case $n = 2$. Since ξ_1 and ξ_2 are independent uncertain sets and f is a strictly decreasing function, we have

$$\mathcal{M}\{\xi \rhd (-\infty, \Phi^{-1}(\alpha)]\}$$
$$= \mathcal{M}\{f(\xi_1, \xi_2) \rhd (-\infty, f(\Phi_1^{-1}(1-\alpha), \Phi_2^{-1}(1-\alpha))]\}$$
$$\geq \mathcal{M}\{(\xi_1 \rhd [\Phi_1^{-1}(1-\alpha), +\infty)) \cap (\xi_2 \rhd [\Phi_2^{-1}(1-\alpha), +\infty))\}$$
$$= \mathcal{M}\{\xi_1 \rhd [\Phi_1^{-1}(1-\alpha), +\infty)\} \wedge \mathcal{M}\{\xi_2 \rhd [\Phi_2^{-1}(1-\alpha), +\infty)\}$$
$$= \alpha \wedge \alpha = \alpha.$$

On the other hand, there exists some index i such that

$$\{f(\xi_1, \xi_2) \rhd (-\infty, f(\Phi_1^{-1}(1-\alpha), \Phi_2^{-1}(1-\alpha))]\} \subset \{\xi_i \rhd [\Phi_i^{-1}(1-\alpha), +\infty)\}.$$

Thus

$$\mathcal{M}\{\xi \rhd (-\infty, \Phi^{-1}(\alpha)]\} \leq \mathcal{M}\{\xi_i \rhd [\Phi_i^{-1}(1-\alpha), +\infty)\} = \alpha.$$

It follows that $\mathcal{M}\{\xi \rhd (-\infty, \Phi^{-1}(\alpha)]\} = \alpha$. In other words, Φ is just the uncertainty distribution of ξ. The theorem is proved.

Alternating Monotone Function of Uncertain Sets

Theorem 10.28 *(Liu [125]). Let $\xi_1, \xi_2, \cdots, \xi_n$ be independent uncertain sets with uncertainty distributions $\Phi_1, \Phi_2, \cdots, \Phi_n$, respectively. If $f(x_1, x_2, \cdots, x_n)$ is strictly increasing with respect to $x_1, x_2, \cdots, x_m$ and strictly decreasing with respect to $x_{m+1}, x_{m+2}, \cdots, x_n$, then*

$$\xi = f(\xi_1, \xi_2, \cdots, \xi_n) \tag{10.72}$$

is an uncertain set whose inverse uncertainty distribution is

$$\Phi^{-1}(\alpha) = f(\Phi_1^{-1}(\alpha), \cdots, \Phi_m^{-1}(\alpha), \Phi_{m+1}^{-1}(1-\alpha), \cdots, \Phi_n^{-1}(1-\alpha)) \tag{10.73}$$

for any α with $0 < \alpha < 1$.

Proof: For simplicity, we only prove the case $n = 2$. Since ξ_1 and ξ_2 are independent uncertain sets and and the function $f(x_1, x_2)$ is strictly increasing with respect to x_1 and strictly decreasing with x_2, we have

$$\mathcal{M}\{\xi \rhd (-\infty, \Phi^{-1}(\alpha)]\}$$
$$= \mathcal{M}\{f(\xi_1, \xi_2) \rhd (-\infty, f(\Phi_1^{-1}(\alpha), \Phi_2^{-1}(1-\alpha))]\}$$
$$\geq \mathcal{M}\{(\xi_1 \rhd (-\infty, \Phi_1^{-1}(\alpha)]) \cap (\xi_2 \rhd [\Phi_2^{-1}(1-\alpha), +\infty))\}$$
$$= \mathcal{M}\{\xi_1 \rhd (-\infty, \Phi_1^{-1}(\alpha)]\} \wedge \mathcal{M}\{\xi_2 \rhd [\Phi_2^{-1}(1-\alpha), +\infty)\}$$
$$= \alpha \wedge \alpha = \alpha.$$

On the other hand, the event $\{\xi \rhd (-\infty, \Phi^{-1}(\alpha)]$ is a subset of either $\{\xi_1 \rhd (-\infty, \Phi_1^{-1}(\alpha)]\}$ or $\{\xi_2 \rhd [\Phi_2^{-1}(1-\alpha), +\infty)\}$. Thus

$$\mathcal{M}\{\xi \rhd (-\infty, \Phi^{-1}(\alpha)]\} \leq \alpha.$$

It follows that $\mathcal{M}\{\xi \rhd (-\infty, \Phi^{-1}(\alpha)]\} = \alpha$. In other words, Φ is just the uncertainty distribution of ξ. The theorem is proved.

Example 10.25: Let ξ_1 and ξ_2 be independent uncertain sets with uncertainty distributions Φ_1 and Φ_2, respectively. Then the inverse uncertainty distribution of the difference $\xi_1 - \xi_2$ is

$$\Phi^{-1}(\alpha) = \Phi_1^{-1}(\alpha) - \Phi_2^{-1}(1-\alpha), \quad 0 < \alpha < 1. \tag{10.74}$$

Example 10.26: Let ξ_1 and ξ_2 be independent and positive uncertain sets with uncertainty distributions Φ_1 and Φ_2, respectively. Then the inverse uncertainty distribution of the quotient ξ_1/ξ_2 is

$$\Phi^{-1}(\alpha) = \Phi_1^{-1}(\alpha)/\Phi_2^{-1}(1-\alpha), \quad 0 < \alpha < 1. \tag{10.75}$$

10.7 Expected Value

Definition 10.16 *(Liu [125]). Let ξ be a nonempty uncertain set. Then the expected value of ξ is defined by*

$$E[\xi] = \int_0^{+\infty} \mathcal{M}\{\xi \rhd [r, +\infty)\} dr - \int_{-\infty}^0 \mathcal{M}\{\xi \rhd (-\infty, r]\} dr \tag{10.76}$$

provided that at least one of the two integrals is finite.

Example 10.27: Consider an uncertain set ξ that has no membership function but may be represented by

$$\xi = \begin{cases} [1,2] \text{ with uncertain measure } 0.5 \\ [2,3] \text{ with uncertain measure } 0.5. \end{cases}$$

Intuitively, the expected value of ξ should be 2. Let us verify it by Definition 10.16. At first, we have

$$\mathcal{M}\{\xi \triangleright [r, +\infty)\} = \begin{cases} 1, & \text{if } 0 \leq r \leq 1 \\ 0.75, & \text{if } 1 < r \leq 2 \\ 0.25, & \text{if } 2 < r \leq 3 \\ 0, & \text{if } r > 3, \end{cases}$$

$$\mathcal{M}\{\xi \triangleright (-\infty, r]\} \equiv 0, \quad \forall r \leq 0.$$

Thus

$$E[\xi] = \int_0^1 1 \mathrm{d}r + \int_1^2 0.75 \mathrm{d}r + \int_2^3 0.25 \mathrm{d}r = 2.$$

Example 10.28: Let $(\Gamma, \mathcal{L}, \mathcal{M})$ be an uncertainty space with $\Gamma = [0, 1]$ and $\mathcal{M}\{[0, \gamma]\} = \gamma$ for each $\gamma \in [0, 1]$. Define an uncertain set

$$\xi(\gamma) = [\gamma, \gamma + 1], \quad \forall \gamma \in [0, 1].$$

At first, we have

$$\mathcal{M}\{\xi \triangleright [r, +\infty)\} = \begin{cases} 1 - r/2, & \text{if } 0 \leq r \leq 2 \\ 0, & \text{if } r > 2, \end{cases}$$

$$\mathcal{M}\{\xi \triangleright (-\infty, r]\} \equiv 0, \quad \forall r \leq 0.$$

Thus

$$E[\xi] = \int_0^2 (1 - r/2) \mathrm{d}r = 1.$$

Example 10.29: Let ξ be an uncertain variable (a degenerate uncertain set). Then we have

$$\mathcal{M}\{\xi \triangleright [r, +\infty)\} = \mathcal{M}\{\xi \geq r\}, \quad \mathcal{M}\{\xi \triangleright (-\infty, r]\} = \mathcal{M}\{\xi \leq r\}.$$

Thus

$$E[\xi] = \int_0^{+\infty} \mathcal{M}\{\xi \geq r\} \mathrm{d}r - \int_{-\infty}^0 \mathcal{M}\{\xi \leq r\} \mathrm{d}r.$$

That is, the expected value of uncertain set does coincide with that of uncertain variable.

Theorem 10.29. *Let ξ be a nonempty uncertain set with uncertainty distribution Φ. If ξ has a finite expected value, then*

$$E[\xi] = \int_0^{+\infty} (1 - \Phi(x)) \mathrm{d}x - \int_{-\infty}^0 \Phi(x) \mathrm{d}x. \tag{10.77}$$

Proof: The theorem follows immediately from $\Phi(x) = \mathcal{M}\{\xi \rhd (-\infty, x]\}$ and $1 - \Phi(x) = \mathcal{M}\{\xi \rhd (x, +\infty)\}$ for any $x \in \Re$.

Theorem 10.30. *Let ξ be a nonempty uncertain set with uncertainty distribution Φ. If ξ has a finite expected value, then*

$$E[\xi] = \int_0^1 \Phi^{-1}(\alpha) \mathrm{d}\alpha. \tag{10.78}$$

Proof: It follows from the definitions of expected value operator and uncertainty distribution that

$$
\begin{aligned}
E[\xi] &= \int_0^{+\infty} \mathcal{M}\{\xi \rhd [r, +\infty)\} \mathrm{d}r - \int_{-\infty}^0 \mathcal{M}\{\xi \rhd (-\infty, r]\} \mathrm{d}r \\
&= \int_{\Phi(0)}^1 \Phi^{-1}(\alpha) \mathrm{d}\alpha + \int_0^{\Phi(0)} \Phi^{-1}(\alpha) \mathrm{d}\alpha = \int_0^1 \Phi^{-1}(\alpha) \mathrm{d}\alpha.
\end{aligned}
$$

The theorem is proved.

Theorem 10.31. *Let ξ be a nonempty uncertain set with membership function μ. If ξ has a finite expected value and μ is a unimodal function about x_0 (i.e., increasing on $(-\infty, x_0)$ and decreasing on $(x_0, +\infty)$), then the expected value of ξ is*

$$E[\xi] = x_0 + \frac{1}{2} \int_{x_0}^{+\infty} \mu(x) \mathrm{d}x - \frac{1}{2} \int_{-\infty}^{x_0} \mu(x) \mathrm{d}x. \tag{10.79}$$

Proof: Since μ is increasing on $(-\infty, x_0)$ and decreasing on $(x_0, +\infty)$, it follows from the definition of membership degree that

$$
\mathcal{M}\{\xi \rhd (-\infty, x]\} =
\begin{cases}
\mu(x)/2, & \text{if } x \le x_0 \\
1 - \mu(x)/2, & \text{if } x \ge x_0
\end{cases}
\tag{10.80}
$$

and

$$
\mathcal{M}\{\xi \rhd [x, +\infty)\} =
\begin{cases}
1 - \mu(x)/2, & \text{if } x \le x_0 \\
\mu(x)/2, & \text{if } x \ge x_0
\end{cases}
\tag{10.81}
$$

for almost all $x \in \Re$. If $x_0 \ge 0$, we have

$$
\begin{aligned}
E[\xi] &= \int_0^{+\infty} \mathcal{M}\{\xi \rhd [x, +\infty)\} \mathrm{d}x - \int_{-\infty}^0 \mathcal{M}\{\xi \rhd (-\infty, x]\} \mathrm{d}x \\
&= \int_0^{x_0} (1 - \mu(x)/2) \, \mathrm{d}x + \int_{x_0}^{+\infty} \mu(x)/2 \mathrm{d}x - \int_{-\infty}^0 \mu(x)/2 \mathrm{d}x \\
&= x_0 - \frac{1}{2} \int_0^{x_0} \mu(x) \mathrm{d}x + \frac{1}{2} \int_{x_0}^{+\infty} \mu(x) \mathrm{d}x - \frac{1}{2} \int_{-\infty}^0 \mu(x) \mathrm{d}x \\
&= x_0 + \frac{1}{2} \int_{x_0}^{+\infty} \mu(x) \mathrm{d}x - \frac{1}{2} \int_{-\infty}^{x_0} \mu(x) \mathrm{d}x.
\end{aligned}
$$

If $x_0 < 0$, we have

$$
\begin{aligned}
E[\xi] &= \int_0^{+\infty} \mathcal{M}\{\xi \rhd [x, +\infty)\}\mathrm{d}x - \int_{-\infty}^0 \mathcal{M}\{\xi \rhd (-\infty, x]\}\mathrm{d}x \\
&= \int_0^{+\infty} \mu(x)/2\mathrm{d}x - \int_{-\infty}^{x_0} \mu(x)/2\mathrm{d}x - \int_{x_0}^0 (1 - \mu(x)/2)\mathrm{d}x \\
&= \frac{1}{2}\int_0^{+\infty} \mu(x)\mathrm{d}x - \frac{1}{2}\int_{-\infty}^{x_0} \mu(x)\mathrm{d}x + x_0 + \frac{1}{2}\int_{x_0}^0 \mu(x)\mathrm{d}x \\
&= x_0 + \frac{1}{2}\int_{x_0}^{+\infty} \mu(x)\mathrm{d}x - \frac{1}{2}\int_{-\infty}^{x_0} \mu(x)\mathrm{d}x.
\end{aligned}
$$

The theorem is thus proved.

Example 10.30: The rectangular uncertain set $\xi = (a, b)$ has an expected value

$$
E[\xi] = \frac{a+b}{2}. \tag{10.82}
$$

Example 10.31: The triangular uncertain set $\xi = (a, b, c)$ has an expected value

$$
E[\xi] = \frac{a + 2b + c}{4}. \tag{10.83}
$$

Example 10.32: The trapezoidal uncertain set $\xi = (a, b, c, d)$ has an expected value

$$
E[\xi] = \frac{a + b + c + d}{4}. \tag{10.84}
$$

Theorem 10.32 (Liu [125]). *Let ξ_1 and ξ_2 be independent nonempty uncertain sets with finite expected values. Then for any real numbers a_1 and a_2, we have*

$$
E[a_1\xi_1 + a_2\xi_2] = a_1E[\xi_1] + a_2E[\xi_2]. \tag{10.85}
$$

Proof: Suppose that ξ_1 and ξ_2 have uncertainty distributions Φ_1 and Φ_2, respectively. It follows from Theorem 10.26 that $a_1\xi_1 + a_2\xi_2$ has an inverse uncertainty distribution,

$$
\Phi^{-1}(\alpha) = a_1\Phi_1^{-1}(\alpha) + a_2\Phi_2^{-1}(\alpha)
$$

and

$$
\int_0^1 \Phi^{-1}(\alpha)\mathrm{d}\alpha = a_1\int_0^1 \Phi_1^{-1}(\alpha)\mathrm{d}\alpha + a_2\int_0^1 \Phi_2^{-1}(\alpha)\mathrm{d}\alpha.
$$

Then Theorem 10.30 tells us that

$$
E[a_1\xi_1 + a_2\xi_2] = \int_0^1 \Phi^{-1}(\alpha)\mathrm{d}\alpha = a_1E[\xi_1] + a_2E[\xi_2].
$$

The theorem is proved.

Theorem 10.33. *Assume $\xi_1, \xi_2, \cdots, \xi_n$ are independent uncertain sets with uncertainty distributions $\Phi_1, \Phi_2, \cdots, \Phi_n$, respectively. If $f : \Re^n \to \Re$ is a strictly monotone function, then the uncertain set $\xi = f(\xi_1, \xi_2, \cdots, \xi_n)$ has an expected value*

$$E[\xi] = \int_0^1 f(\Phi_1^{-1}(\alpha), \Phi_2^{-1}(\alpha), \cdots, \Phi_n^{-1}(\alpha)) \mathrm{d}\alpha \qquad (10.86)$$

provided that the expected value $E[\xi]$ exists.

Proof: Suppose that f is a strictly increasing function. It follows that the inverse uncertainty distribution of ξ is

$$\Phi^{-1}(\alpha) = f(\Phi_1^{-1}(\alpha), \Phi_2^{-1}(\alpha), \cdots, \Phi_n^{-1}(\alpha)).$$

Thus we obtain (10.86). When f is a strictly decreasing function, it follows that the inverse uncertainty distribution of ξ is

$$\Phi^{-1}(\alpha) = f(\Phi_1^{-1}(1-\alpha), \Phi_2^{-1}(1-\alpha), \cdots, \Phi_n^{-1}(1-\alpha)).$$

By using the change of variable of integral, we obtain (10.86). The theorem is proved.

Example 10.33: Let ξ and η be independent and nonnegative uncertain sets with uncertainty distributions Φ and Ψ, respectively. Then

$$E[\xi\eta] = \int_0^1 \Phi^{-1}(\alpha)\Psi^{-1}(\alpha)\mathrm{d}\alpha. \qquad (10.87)$$

Exercise 10.1: What is the expected value of an alternating monotone function of uncertain sets?

Exercise 10.2: Let ξ and η be independent and positive uncertain sets with uncertainty distributions Φ and Ψ, respectively. Prove

$$E\left[\frac{\xi}{\eta}\right] = \int_0^1 \frac{\Phi^{-1}(\alpha)}{\Psi^{-1}(1-\alpha)}\mathrm{d}\alpha. \qquad (10.88)$$

10.8   Critical Values

In order to rank uncertain sets, we may use two critical values: optimistic value and pessimistic value.

Definition 10.17 *(Liu [125]). Let ξ be an uncertain set, and $\alpha \in (0,1]$. Then*

$$\xi_{\sup}(\alpha) = \sup\left\{r \mid \mathcal{M}\left\{\xi \rhd [r, +\infty)\right\} \geq \alpha\right\} \qquad (10.89)$$

is called the α-optimistic value to ξ, and

$$\xi_{\inf}(\alpha) = \inf\left\{r \mid \mathcal{M}\left\{\xi \rhd (-\infty, r]\right\} \geq \alpha\right\} \qquad (10.90)$$

is called the α-pessimistic value to ξ.

Theorem 10.34. *Let ξ be an uncertain set with uncertainty distribution Φ. Then its α-optimistic value and α-pessimistic value are*

$$\xi_{\sup}(\alpha) = \Phi^{-1}(1-\alpha), \quad \xi_{\inf}(\alpha) = \Phi^{-1}(\alpha) \tag{10.91}$$

for any α with $0 < \alpha < 1$.

Proof: Since Φ is a strictly monotone function when $0 < \Phi(x) < 1$, we have

$$\xi_{\sup}(\alpha) = \sup\left\{r | \mathcal{M}\left\{\xi \triangleright [r, +\infty)\right\} \geq \alpha\right\} = \sup\{r | 1 - \Phi(r) \geq \alpha\} = \Phi^{-1}(1-\alpha),$$

$$\xi_{\inf}(\alpha) = \inf\left\{r | \mathcal{M}\left\{\xi \triangleright (-\infty, r]\right\} \geq \alpha\right\} = \inf\{r | \Phi(r) \geq \alpha\} = \Phi^{-1}(\alpha).$$

The theorem is proved.

Theorem 10.35. *Let $\xi_1, \xi_2, \cdots, \xi_n$ be independent uncertain sets with uncertainty distributions. If $f : \Re^n \to \Re$ is a continuous and strictly increasing function, then $\xi = f(\xi_1, \xi_2, \cdots, \xi_n)$ is an uncertain set, and*

$$\xi_{\sup}(\alpha) = f(\xi_{1\sup}(\alpha), \xi_{2\sup}(\alpha), \cdots, \xi_{n\sup}(\alpha)), \tag{10.92}$$

$$\xi_{\inf}(\alpha) = f(\xi_{1\inf}(\alpha), \xi_{2\inf}(\alpha), \cdots, \xi_{n\inf}(\alpha)). \tag{10.93}$$

Proof: Since f is a strictly increasing function, it follows that the inverse uncertainty distribution of ξ is

$$\Phi^{-1}(\alpha) = f(\Phi_1^{-1}(\alpha), \Phi_2^{-1}(\alpha), \cdots, \Phi_n^{-1}(\alpha))$$

where $\Phi_1, \Phi_2, \cdots, \Phi_n$ are uncertainty distributions of $\xi_1, \xi_2, \cdots, \xi_n$, respectively. Thus we get (10.92) and (10.93). The theorem is proved.

Example 10.34: Let ξ and η be independent uncertain sets with uncertainty distributions. Then

$$(\xi+\eta)_{\sup}(\alpha) = \xi_{\sup}(\alpha) + \eta_{\sup}(\alpha), \quad (\xi+\eta)_{\inf}(\alpha) = \xi_{\inf}(\alpha) + \eta_{\inf}(\alpha). \tag{10.94}$$

Example 10.35: Let ξ and η be independent and positive uncertain sets with uncertainty distributions. Then

$$(\xi\eta)_{\sup}(\alpha) = \xi_{\sup}(\alpha)\eta_{\sup}(\alpha), \quad (\xi\eta)_{\inf}(\alpha) = \xi_{\inf}(\alpha)\eta_{\inf}(\alpha). \tag{10.95}$$

Theorem 10.36. *Let $\xi_1, \xi_2, \cdots, \xi_n$ be independent uncertain sets with uncertainty distributions. If f is a continuous and strictly decreasing function, then*

$$\xi_{\sup}(\alpha) = f(\xi_{1\inf}(\alpha), \xi_{2\inf}(\alpha), \cdots, \xi_{n\inf}(\alpha)), \tag{10.96}$$

$$\xi_{\inf}(\alpha) = f(\xi_{1\sup}(\alpha), \xi_{2\sup}(\alpha), \cdots, \xi_{n\sup}(\alpha)). \tag{10.97}$$

Proof: Since f is a strictly decreasing function, it follows that the inverse uncertainty distribution of ξ is

$$\Phi^{-1}(\alpha) = f(\Phi_1^{-1}(1-\alpha), \Phi_2^{-1}(1-\alpha), \cdots, \Phi_n^{-1}(1-\alpha)).$$

Thus we get (10.96) and (10.97). The theorem is proved.

Exercise 10.3: What are the critical values to an alternating monotone function of uncertain sets?

Exercise 10.4: Let ξ and η be independent and positive uncertain sets. Prove

$$\left(\frac{\xi}{\eta}\right)_{\sup}(\alpha) = \frac{\xi_{\sup}(\alpha)}{\eta_{\inf}(\alpha)}, \qquad \left(\frac{\xi}{\eta}\right)_{\inf}(\alpha) = \frac{\xi_{\inf}(\alpha)}{\eta_{\sup}(\alpha)}. \tag{10.98}$$

10.9   Hausdorff Distance

Liu [125] generalized the Hausdorff distance to the domain of uncertain sets. Let ξ and η be two uncertain sets on the uncertainty space $(\Gamma, \mathcal{L}, \mathcal{M})$. For each $\gamma \in \Gamma$, it is clear that $\xi(\gamma)$ and $\eta(\gamma)$ are two sets of real numbers. Thus the Hausdorff distance between them is

$$\rho(\gamma) = \left(\sup_{a \in \xi(\gamma)} \inf_{b \in \eta(\gamma)} |a - b|\right) \vee \left(\sup_{b \in \eta(\gamma)} \inf_{a \in \xi(\gamma)} |a - b|\right). \tag{10.99}$$

Note that ρ is a function from $(\Gamma, \mathcal{L}, \mathcal{M})$ to the set of nonnegative numbers, and is just a nonnegative uncertain variable in the sense of Definition 1.5.

Definition 10.18 *(Liu [125]). Let ξ and η be two uncertain sets. Then the Hausdorff distance between ξ and η is*

$$d(\xi, \eta) = \int_0^{+\infty} \mathcal{M}\{\rho \geq r\}dr \tag{10.100}$$

where ρ is a nonnegative uncertain variable determined by (10.99).

If the uncertain sets degenerate to uncertain variables, then the Hausdorff distance between uncertain sets degenerates to the distance between uncertain variables in the sense of Definition 1.25.

Theorem 10.37. *Let ξ, η, τ be uncertain sets, and let $d(\cdot, \cdot)$ be the Hausdorff distance. Then we have*
(a) (Nonnegativity) $d(\xi, \eta) \geq 0$;
(b) (Identification) $d(\xi, \eta) = 0$ if and only if $\xi = \eta$;
(c) (Symmetry) $d(\xi, \eta) = d(\eta, \xi)$.

Proof: The theorem follows immediately from the definition.

10.10   Conditional Uncertainty

Let ξ be an uncertain set on $(\Gamma, \mathcal{L}, \mathcal{M})$. What is the conditional uncertain set of ξ after it has been learned that some event B has occurred? This section will answer this question.

Definition 10.19 *(Liu [125]). Let ξ be an uncertain set with membership function μ, and let B be an event with $\mathcal{M}\{B\} > 0$. Then the conditional membership function of ξ given B is defined by*

$$\mu(x|B) = \mathcal{M}\left\{\xi \in W_{\mu(x)} \mid B\right\} \tag{10.101}$$

where $W_{\mu(x)}$ is the $\mu(x)$-class of μ.

Definition 10.20 *(Liu [125]). Let ξ be an uncertain set and let B be an event with $\mathcal{M}\{B\} > 0$. Then the conditional uncertainty distribution Φ: $\Re \to [0, 1]$ of ξ given B is defined by*

$$\Phi(x|B) = \mathcal{M}\left\{\xi \rhd (-\infty, x] \mid B\right\}. \tag{10.102}$$

Definition 10.21 *(Liu [125]). Let ξ be an uncertain set and let B be an event with $\mathcal{M}\{B\} > 0$. Then the conditional expected value of ξ given B is defined by*

$$E[\xi|B] = \int_0^{+\infty} \mathcal{M}\left\{\xi \rhd [r, +\infty) \mid B\right\} \mathrm{d}r - \int_{-\infty}^0 \mathcal{M}\left\{\xi \rhd (-\infty, r] \mid B\right\} \mathrm{d}r$$

provided that at least one of the two integrals is finite.

Chapter 11

Uncertain Inference

Uncertain inference was proposed by Liu [125] in 2010 as a process of deriving consequences from uncertain knowledge or evidence via the tool of conditional uncertain set. Gao, Gao and Ralescu [42] extended the inference rule to the one with multiple antecedents and with multiple if-then rules.

This chapter will introduce an inference rule, and apply the tool to uncertain system and inference control. The technique of uncertain inference controller is also illustrated by an inverted pendulum system.

11.1 Inference Rule

Let $\mathbb{X}$ and $\mathbb{Y}$ be two concepts. It is assumed that we only have a rule "if $\mathbb{X}$ is ξ then $\mathbb{Y}$ is η" where ξ and η are two uncertain sets. We first have the following inference rule.

Inference Rule 11.1 *(Liu [125]). Let $\mathbb{X}$ and $\mathbb{Y}$ be two concepts. Assume a rule "if $\mathbb{X}$ is an uncertain set ξ then $\mathbb{Y}$ is an uncertain set η". From $\mathbb{X}$ is an uncertain set ξ^* we infer that $\mathbb{Y}$ is an uncertain set*

$$\eta^* = \eta|_{\xi^* \triangleright \xi} \tag{11.1}$$

which is the conditional uncertain set of η given $\xi^* \triangleright \xi$. The inference rule is represented by

$$
\begin{array}{l}
\text{Rule: If } \mathbb{X} \text{ is } \xi \text{ then } \mathbb{Y} \text{ is } \eta \\
\text{From: } \mathbb{X} \text{ is } \xi^* \\
\hline
\text{Infer: } \mathbb{Y} \text{ is } \eta^* = \eta|_{\xi^* \triangleright \xi}
\end{array}
\tag{11.2}
$$

Theorem 11.1. *Let ξ and η be independent uncertain sets with membership functions μ and ν, respectively. If ξ^* is a constant a, then inference rule 11.1 yields that η^* has a membership function*

$$
\nu^*(y) = \begin{cases}
\dfrac{\nu(y)}{\mu(a)}, & \text{if } \nu(y) < \mu(a)/2 \\[2mm]
\dfrac{\nu(y) + \mu(a) - 1}{\mu(a)}, & \text{if } \nu(y) > 1 - \mu(a)/2 \\[2mm]
0.5, & \text{otherwise.}
\end{cases}
\tag{11.3}
$$

B. Liu: Uncertainty Theory: A Branch of Mathematics, SCI 300, pp. 215–223.
springerlink.com   © Springer-Verlag Berlin Heidelberg 2010

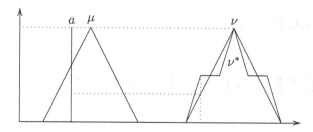

Figure 11.1: Graphical Illustration of Inference Rule

Proof: It follows from inference rule 11.1 that η^* has a membership function

$$\nu^*(y) = \mathcal{M}\{y \in \eta | a \triangleright \xi\}.$$

By using the definition of conditional uncertainty, we have

$$\mathcal{M}\{y \in \eta | a \triangleright \xi\} = \begin{cases} \dfrac{\mathcal{M}\{y \in \eta\}}{\mathcal{M}\{a \triangleright \xi\}}, & \text{if } \dfrac{\mathcal{M}\{y \in \eta\}}{\mathcal{M}\{a \triangleright \xi\}} < 0.5 \\[2ex] 1 - \dfrac{\mathcal{M}\{y \notin \eta\}}{\mathcal{M}\{a \triangleright \xi\}}, & \text{if } \dfrac{\mathcal{M}\{y \notin \eta\}}{\mathcal{M}\{a \triangleright \xi\}} < 0.5 \\[2ex] 0.5, & \text{otherwise.} \end{cases}$$

The equation (11.3) follows from $\mathcal{M}\{y \in \eta\} = \nu(y)$, $\mathcal{M}\{y \notin \eta\} = 1 - \nu(y)$ and $\mathcal{M}\{a \triangleright \xi\} = \mu(a)$ immediately. The theorem is proved.

Inference Rule 11.2 *(Gao, Gao and Ralescu [42]). Let $\mathbb{X}$, $\mathbb{Y}$ and $\mathbb{Z}$ be three concepts. Assume a rule "if $\mathbb{X}$ is an uncertain set ξ and $\mathbb{Y}$ is an uncertain set η then $\mathbb{Z}$ is an uncertain set τ". From $\mathbb{X}$ is an uncertain set ξ^* and $\mathbb{Y}$ is an uncertain set η^* we infer that $\mathbb{Z}$ is an uncertain set*

$$\tau^* = \tau|_{(\xi^* \triangleright \xi) \cap (\eta^* \triangleright \eta)} \tag{11.4}$$

which is the conditional uncertain set of τ given $\xi^* \triangleright \xi$ and $\eta^* \triangleright \eta$. The inference rule is represented by

$$\begin{array}{l} \text{Rule: If } \mathbb{X} \text{ is } \xi \text{ and } \mathbb{Y} \text{ is } \eta \text{ then } \mathbb{Z} \text{ is } \tau \\ \text{From: } \mathbb{X} \text{ is } \xi^* \text{ and } \mathbb{Y} \text{ is } \eta^* \\ \hline \text{Infer: } \mathbb{Z} \text{ is } \tau^* = \tau|_{(\xi^* \triangleright \xi) \cap (\eta^* \triangleright \eta)} \end{array} \tag{11.5}$$

Theorem 11.2. *Let ξ, η, τ be independent uncertain sets with membership functions μ, ν, λ, respectively. If ξ^* is a constant a and η^* is a constant b, then inference rule 11.2 yields that τ^* has a membership function*

$$\lambda^*(z) = \begin{cases} \dfrac{\lambda(z)}{\mu(a) \wedge \nu(b)}, & \text{if } \lambda(z) < \dfrac{\mu(a) \wedge \nu(b)}{2} \\[2ex] \dfrac{\lambda(z) + \mu(a) \wedge \nu(b) - 1}{\mu(a) \wedge \nu(b)}, & \text{if } \lambda(z) > 1 - \dfrac{\mu(a) \wedge \nu(b)}{2} \\[2ex] 0.5, & \text{otherwise.} \end{cases} \tag{11.6}$$

Proof: It follows from inference rule 11.2 that τ^* has a membership function

$$\lambda^*(z) = \mathcal{M}\{z \in \tau | (a \triangleright \xi) \cap (b \triangleright \eta)\}.$$

By using the definition of conditional uncertainty, $\mathcal{M}\{z \in \tau | (a \triangleright \xi) \cap (b \triangleright \eta)\}$ is

$$\begin{cases} \dfrac{\mathcal{M}\{z \in \tau\}}{\mathcal{M}\{(a \triangleright \xi) \cap (b \triangleright \eta)\}}, & \text{if } \dfrac{\mathcal{M}\{z \in \tau\}}{\mathcal{M}\{(a \triangleright \xi) \cap (b \triangleright \eta)\}} < 0.5 \\[4mm] 1 - \dfrac{\mathcal{M}\{z \notin \tau\}}{\mathcal{M}\{(a \triangleright \xi) \cap (b \triangleright \eta)\}}, & \text{if } \dfrac{\mathcal{M}\{z \notin \tau\}}{\mathcal{M}\{(a \triangleright \xi) \cap (b \triangleright \eta)\}} < 0.5 \\[4mm] 0.5, & \text{otherwise.} \end{cases}$$

The theorem follows from $\mathcal{M}\{z \in \tau\} = \lambda(z)$, $\mathcal{M}\{z \notin \tau\} = 1 - \lambda(z)$ and $\mathcal{M}\{(a \triangleright \xi) \cap (b \triangleright \eta)\} = \mu(a) \wedge \nu(b)$ immediately.

Inference Rule 11.3 *(Gao, Gao and Ralescu [42]). Let $\mathbb{X}$ and $\mathbb{Y}$ be two concepts. Assume two rules "if $\mathbb{X}$ is an uncertain set ξ_1 then $\mathbb{Y}$ is an uncertain set η_1" and "if $\mathbb{X}$ is an uncertain set ξ_2 then $\mathbb{Y}$ is an uncertain set η_2". From $\mathbb{X}$ is an uncertain set ξ^* we infer that $\mathbb{Y}$ is an uncertain set*

$$\eta^* = \frac{\mathcal{M}\{\xi^* \triangleright \xi_1\} \cdot \eta_1|_{\xi^* \triangleright \xi_1}}{\mathcal{M}\{\xi^* \triangleright \xi_1\} + \mathcal{M}\{\xi^* \triangleright \xi_2\}} + \frac{\mathcal{M}\{\xi^* \triangleright \xi_2\} \cdot \eta_2|_{\xi^* \triangleright \xi_2}}{\mathcal{M}\{\xi^* \triangleright \xi_1\} + \mathcal{M}\{\xi^* \triangleright \xi_2\}}. \tag{11.7}$$

The inference rule is represented by

$$\begin{array}{l} \text{Rule 1: If } \mathbb{X} \text{ is } \xi_1 \text{ then } \mathbb{Y} \text{ is } \eta_1 \\ \text{Rule 2: If } \mathbb{X} \text{ is } \xi_2 \text{ then } \mathbb{Y} \text{ is } \eta_2 \\ \underline{\text{From: } \mathbb{X} \text{ is } \xi^*} \\ \text{Infer: } \mathbb{Y} \text{ is } \eta^* \text{ determined by (11.7)} \end{array} \tag{11.8}$$

Theorem 11.3. *Let $\xi_1, \xi_2, \eta_1, \eta_2$ be independent uncertain sets with membership functions $\mu_1, \mu_2, \nu_1, \nu_2$, respectively. If ξ^* is a constant a, then inference rule 11.3 yields*

$$\eta^* = \frac{\mu_1(a)}{\mu_1(a) + \mu_2(a)} \eta_1^* + \frac{\mu_2(a)}{\mu_1(a) + \mu_2(a)} \eta_2^* \tag{11.9}$$

where η_1^* and η_2^* are uncertain sets whose membership functions are respectively given by

$$\nu_1^*(y) = \begin{cases} \dfrac{\nu_1(y)}{\mu_1(a)}, & \text{if } \nu_1(y) < \mu_1(a)/2 \\[4mm] \dfrac{\nu_1(y) + \mu_1(a) - 1}{\mu_1(a)}, & \text{if } \nu_1(y) > 1 - \mu_1(a)/2 \\[4mm] 0.5, & \text{otherwise,} \end{cases} \tag{11.10}$$

$$\nu_2^*(y) = \begin{cases} \dfrac{\nu_2(y)}{\mu_2(a)}, & \text{if } \nu_2(y) < \mu_2(a)/2 \\ \dfrac{\nu_2(y) + \mu_2(a) - 1}{\mu_2(a)}, & \text{if } \nu_2(y) > 1 - \mu_2(a)/2 \\ 0.5, & \text{otherwise.} \end{cases} \qquad (11.11)$$

Proof: It follows from inference rule 11.3 that the uncertain set η^* is just

$$\eta^* = \frac{\mathcal{M}\{a \triangleright \xi_1\} \cdot \eta_1|_{a \triangleright \xi_1}}{\mathcal{M}\{a \triangleright \xi_1\} + \mathcal{M}\{a \triangleright \xi_2\}} + \frac{\mathcal{M}\{a \triangleright \xi_2\} \cdot \eta_2|_{a \triangleright \xi_2}}{\mathcal{M}\{a \triangleright \xi_1\} + \mathcal{M}\{a \triangleright \xi_2\}}.$$

The theorem follows from $\mathcal{M}\{a \triangleright \xi_1\} = \mu_1(a)$ and $\mathcal{M}\{a \triangleright \xi_2\} = \mu_2(a)$ immediately.

Inference Rule 11.4. *Let* $\mathbb{X}_1, \mathbb{X}_2, \cdots, \mathbb{X}_m$ *be concepts. Assume rules "if* $\mathbb{X}_1$ *is* ξ_{i1} *and* $\cdots$ *and* $\mathbb{X}_m$ *is* ξ_{im} *then* $\mathbb{Y}$ *is* η_i *" for* $i = 1, 2, \cdots, k$. *From* $\mathbb{X}_1$ *is* ξ_1^* *and* $\cdots$ *and* $\mathbb{X}_m$ *is* ξ_m^* *we infer that* $\mathbb{Y}$ *is an uncertain set*

$$\eta^* = \sum_{i=1}^{k} \frac{c_i \cdot \eta_i|_{(\xi_1^* \triangleright \xi_{i1}) \cap (\xi_2^* \triangleright \xi_{i2}) \cap \cdots \cap (\xi_m^* \triangleright \xi_{im})}}{c_1 + c_2 + \cdots + c_k} \qquad (11.12)$$

where the coefficients are determined by

$$c_i = \mathcal{M}\left\{ (\xi_1^* \triangleright \xi_{i1}) \cap (\xi_2^* \triangleright \xi_{i2}) \cap \cdots \cap (\xi_m^* \triangleright \xi_{im}) \right\} \qquad (11.13)$$

for $i = 1, 2, \cdots, k$. *The inference rule is represented by*

Rule 1: If $\mathbb{X}_1$ is ξ_{11} and $\cdots$ and $\mathbb{X}_m$ is ξ_{1m} then $\mathbb{Y}$ is η_1
Rule 2: If $\mathbb{X}_1$ is ξ_{21} and $\cdots$ and $\mathbb{X}_m$ is ξ_{2m} then $\mathbb{Y}$ is η_2
$\cdots$
Rule k: If $\mathbb{X}_1$ is ξ_{k1} and $\cdots$ and $\mathbb{X}_m$ is ξ_{km} then $\mathbb{Y}$ is η_k $\qquad (11.14)$
From: $\mathbb{X}_1$ is ξ_1^* and $\cdots$ and $\mathbb{X}_m$ is ξ_m^*
———————————————————————————
Infer: $\mathbb{Y}$ is η^* determined by (11.12)

Theorem 11.4. *Assume* $\xi_{i1}, \xi_{i2}, \cdots, \xi_{im}, \eta_i$ *are independent uncertain sets with membership functions* $\mu_{i1}, \mu_{i2}, \cdots, \mu_{im}, \nu_i$, $i = 1, 2, \cdots, k$, *respectively. If* $\xi_1^*, \xi_2^*, \cdots, \xi_m^*$ *are constants* $a_1, a_2, \cdots, a_m$, *respectively, then inference rule 11.4 yields*

$$\eta^* = \sum_{i=1}^{k} \frac{c_i \cdot \eta_i^*}{c_1 + c_2 + \cdots + c_k} \qquad (11.15)$$

where η_i^* *are uncertain sets whose membership functions are given by*

$$\nu_i^*(y) = \begin{cases} \dfrac{\nu_i(y)}{c_i}, & \text{if } \nu_i(y) < c_i/2 \\ \dfrac{\nu_i(y) + c_i - 1}{c_i}, & \text{if } \nu_i(y) > 1 - c_i/2 \\ 0.5, & \text{otherwise} \end{cases} \qquad (11.16)$$

and c_i are constants determined by

$$c_i = \min_{1 \le l \le m} \mu_{il}(a_l) \qquad (11.17)$$

for $i = 1, 2, \cdots, k$, respectively.

Proof: For each i, since $a_1 \rhd \xi_{i1}, a_2 \rhd \xi_{i2}, \cdots, a_m \rhd \xi_{im}$ are independent events, we immediately have

$$\mathcal{M}\left\{ \bigcap_{j=1}^{m}(a_j \rhd \xi_{ij}) \right\} = \min_{1 \le j \le m} \mathcal{M}\{a_j \rhd \xi_{ij}\} = \min_{1 \le l \le m} \mu_{il}(a_l)$$

for $i = 1, 2, \cdots, k$. From those equations, we may prove the theorem by inference rule 11.4 immediately.

11.2   Uncertain System

An uncertain system, proposed by Liu [125], is a function from its inputs to outputs based on the inference rule. Now we consider a system in which there are m deterministic inputs $\alpha_1, \alpha_2, \cdots, \alpha_m$, and n deterministic outputs $\beta_1, \beta_2, \cdots, \beta_n$. At first, we infer n uncertain sets $\eta_1^*, \eta_2^*, \cdots, \eta_n^*$ from the m deterministic inputs by the rule-base (i.e., a set of if-then rules),

If ξ_{11} and ξ_{12} and$\cdots$and ξ_{1m} then η_{11} and η_{12} and$\cdots$and η_{1n}
If ξ_{21} and ξ_{22} and$\cdots$and ξ_{2m} then η_{21} and η_{22} and$\cdots$and η_{2n} $\qquad (11.18)$
$\cdots$
If ξ_{k1} and ξ_{k2} and$\cdots$and ξ_{km} then η_{k1} and η_{k2} and$\cdots$and η_{kn}

and the inference rule

$$\eta_j^* = \sum_{i=1}^{k} \frac{c_i \cdot \eta_{ij}\big|_{(\alpha_1 \rhd \xi_{i1}) \cap (\alpha_2 \rhd \xi_{i2}) \cap \cdots \cap (\alpha_m \rhd \xi_{im})}}{c_1 + c_2 + \cdots + c_k} \qquad (11.19)$$

for $j = 1, 2, \cdots, n$, where the coefficients are determined by

$$c_i = \mathcal{M}\left\{ (\alpha_1 \rhd \xi_{i1}) \cap (\alpha_2 \rhd \xi_{i2}) \cap \cdots \cap (\alpha_m \rhd \xi_{im}) \right\} \qquad (11.20)$$

for $i = 1, 2, \cdots, k$. Thus we obtain

$$\beta_j = E[\eta_j^*] \qquad (11.21)$$

for $j = 1, 2, \cdots, n$. Until now we have constructed a function from inputs $\alpha_1, \alpha_2, \cdots, \alpha_m$ to outputs $\beta_1, \beta_2, \cdots, \beta_n$. Write this function by f, i.e.,

$$(\beta_1, \beta_2, \cdots, \beta_n) = f(\alpha_1, \alpha_2, \cdots, \alpha_m). \qquad (11.22)$$

Then we get an uncertain system f.

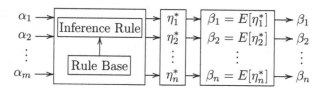

Figure 11.2: An Uncertain System

Theorem 11.5. *Assume $\xi_{i1}, \xi_{i2}, \cdots, \xi_{im}, \eta_{i1}, \eta_{i2}, \cdots, \eta_{in}$ are independent uncertain sets with membership functions $\mu_{i1}, \mu_{i2}, \cdots, \mu_{im}, \nu_{i1}, \nu_{i2}, \cdots, \nu_{in}$, $i = 1, 2, \cdots, k$, respectively. Then the uncertain system from $(\alpha_1, \alpha_2, \cdots, \alpha_m)$ to $(\beta_1, \beta_2, \cdots, \beta_n)$ is*

$$\beta_j = \sum_{i=1}^{k} \frac{c_i \cdot E[\eta_{ij}^*]}{c_1 + c_2 + \cdots + c_k} \tag{11.23}$$

for $j = 1, 2, \cdots, n$, where η_{ij}^* are uncertain sets whose membership functions are given by

$$\nu_{ij}^*(y) = \begin{cases} \dfrac{\nu_{ij}(y)}{c_i}, & \text{if } \nu_{ij}(y) < c_i/2 \\[2mm] \dfrac{\nu_{ij}(y) + c_i - 1}{c_i}, & \text{if } \nu_{ij}(y) > 1 - c_i/2 \\[2mm] 0.5, & \text{otherwise} \end{cases} \tag{11.24}$$

and c_i are constants determined by

$$c_i = \min_{1 \le l \le m} \mu_{il}(\alpha_l) \tag{11.25}$$

for $i = 1, 2, \cdots, k$, $j = 1, 2, \cdots, n$, respectively.

Proof: It follows from inference rule 11.4 that the uncertain sets η_j^* are

$$\eta_j^* = \sum_{i=1}^{k} \frac{c_i \cdot \eta_{ij}^*}{c_1 + c_2 + \cdots + c_k}$$

for $j = 1, 2, \cdots, n$. Since $\eta_{ij}^*, i = 1, 2, \cdots, k, j = 1, 2, \cdots, n$ are independent uncertain sets, we get the theorem immediately by the linearity of expected value operator.

Remark 11.1: The uncertain system allows the uncertain sets η_{ij} in the rule-base (11.18) become constants b_{ij}, i.e.,

$$\eta_{ij} = b_{ij} \tag{11.26}$$

for $i = 1, 2, \cdots, k$ and $j = 1, 2, \cdots, n$. For this case, the uncertain system (11.23) becomes

$$\beta_j = \sum_{i=1}^{k} \frac{c_i \cdot b_{ij}}{c_1 + c_2 + \cdots + c_k} \qquad (11.27)$$

for $j = 1, 2, \cdots, n$.

Remark 11.2: The uncertain system allows the uncertain sets η_{ij} in the rule-base (11.18) become functions h_{ij} of inputs $\alpha_1, \alpha_2, \cdots, \alpha_m$, i.e.,

$$\eta_{ij} = h_{ij}(\alpha_1, \alpha_2, \cdots, \alpha_m) \qquad (11.28)$$

for $i = 1, 2, \cdots, k$ and $j = 1, 2, \cdots, n$. For this case, the uncertain system (11.23) becomes

$$\beta_j = \sum_{i=1}^{k} \frac{c_i \cdot h_{ij}(\alpha_1, \alpha_2, \cdots, \alpha_m)}{c_1 + c_2 + \cdots + c_k} \qquad (11.29)$$

for $j = 1, 2, \cdots, n$.

Uncertain Systems are Universal Approximators

Uncertain systems are capable of approximating any continuous function on a compact set (i.e., bounded and closed set) to arbitrary accuracy. This is the reason why uncertain systems may play a controller. The following theorem shows this fact.

Theorem 11.6 *(Peng [178]). For any given continuous function g on a compact set $D \subset \Re^m$ and any given $\varepsilon > 0$, there exists an uncertain system f such that*

$$\sup_{(\alpha_1, \alpha_2, \cdots, \alpha_m) \in D} \| f(\alpha_1, \alpha_2, \cdots, \alpha_m) - g(\alpha_1, \alpha_2, \cdots, \alpha_m) \| < \varepsilon. \qquad (11.30)$$

Proof: Without loss of generality, we assume that the function g is a real-valued function with only two variables α_1 and α_2, and the compact set is a unit rectangle $D = [0, 1] \times [0, 1]$. Since g is continuous on D and then is uniformly continuous, for any given number $\varepsilon > 0$, there is a number $\delta > 0$ such that

$$|g(\alpha_1, \alpha_2) - g(\alpha'_1, \alpha'_2)| < \varepsilon \qquad (11.31)$$

whenever $\|(\alpha_1, \alpha_2) - (\alpha'_1, \alpha'_2)\| < \delta$. Let k be an integer larger than $1/(\sqrt{2}\delta)$, and write

$$D_{ij} = \left\{ (\alpha_1, \alpha_2) \mid \frac{i-1}{k} < \alpha_1 \le \frac{i}{k}, \frac{j-1}{k} < \alpha_2 \le \frac{j}{k} \right\} \qquad (11.32)$$

for $i, j = 1, 2, \cdots, k$. Note that $\{D_{ij}\}$ is a sequence of disjoint rectangles whose "diameter" is less than δ. Define rectangular uncertain sets

$$\xi_i = \left(\frac{i-1}{k}, \frac{i}{k}\right), \quad i = 1, 2, \cdots, k, \tag{11.33}$$

$$\eta_j = \left(\frac{j-1}{k}, \frac{j}{k}\right), \quad j = 1, 2, \cdots, k. \tag{11.34}$$

Then we assume a rule-base with $k \times k$ if-then rules,

$$\text{Rule } ij: \text{ If } \xi_i \text{ and } \eta_j \text{ then } g(i/k, j/k), \quad i, j = 1, 2, \cdots, k. \tag{11.35}$$

According to the inference rule, the corresponding uncertain system from D to $\Re$ is

$$f(\alpha_1, \alpha_2) = g(i/k, j/k), \quad \text{if } (\alpha_1, \alpha_2) \in D_{ij}, \, i, j = 1, 2, \cdots, k. \tag{11.36}$$

It follows from (11.31) that

$$\sup_{(\alpha_1,\alpha_2)\in D} |f(\alpha_1, \alpha_2) - g(\alpha_1, \alpha_2)| = \max_{1\le i,j\le k} \sup_{(\alpha_1,\alpha_2)\in D_{ij}} |f(\alpha_1, \alpha_2) - g(\alpha_1, \alpha_2)|$$

$$= \max_{1\le i,j\le k} \sup_{(\alpha_1,\alpha_2)\in D_{ij}} |g(i/k, j/k) - g(\alpha_1, \alpha_2)| < \max_{1\le i,j\le k} \varepsilon = \varepsilon.$$

The theorem is thus verified.

11.3   Inference Control

An inference controller is a controller based on the inference rule. Figure 11.3 shows an inference control system consisting of an inference controller and a process. Note that t represents time, $\alpha_1(t), \alpha_2(t), \cdots, \alpha_m(t)$ are not only the inputs of inference controller but also the outputs of process, and $\beta_1(t), \beta_2(t), \cdots, \beta_n(t)$ are not only the outputs of inference controller but also the inputs of process.

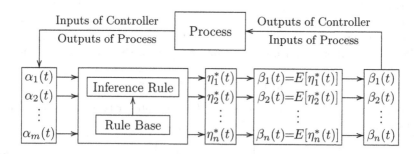

Figure 11.3: An Inference Control System

11.4   Inverted Pendulum

Inverted pendulum system is a nonlinear unstable system that is widely used as a benchmark for testing control algorithms. Many good techniques already exist for balancing inverted pendulum. Especially, Gao [43] successfully balanced an inverted pendulum by the inference controller with 5×5 if-then rules.

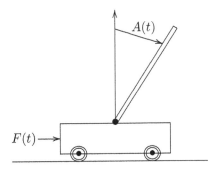

Figure 11.4: An Inverted Pendulum in which $A(t)$ represents the angular position and $F(t)$ represents the force that moves the cart at time t.

Appendix A

Supplements

This appendix introduces a law of truth conservation and a maximum uncertainty principle that have been used in this book. This appendix also provides a brief history of evolution of measures and discusses why uncertainty theory is reasonable. In addition, this appendix proposes a way to determine uncertainty distributions via expert's experimental data. Finally, we answer the question "what is uncertainty".

A.1 Law of Truth Conservation

The law of excluded middle tells us that a proposition is either true or false, and the law of contradiction tells us that a proposition cannot be both true and false. In the state of uncertainty, some people said, the law of excluded middle and the law of contradiction are no longer valid because the truth degree of a proposition is no longer 0 or 1. I cannot gainsay this viewpoint to a certain extent.

But it does not mean that you might "go as you please". At least, I think, the law of truth conservation should be valid in the state of uncertainty. In other words, the sum of truth values of a proposition and its negative proposition is identical to 1. That is, we always have

$$\mathcal{M}\{\Lambda\} + \mathcal{M}\{\Lambda^c\} = 1 \tag{A.1}$$

for each proposition Λ. This means that the law of truth conservation is nothing but the self-duality property. The law of truth conservation is weaker than the law of excluded middle and the law of contradiction. Furthermore, the law of truth conservation agrees with the law of excluded middle and the law of contradiction when the uncertainty vanishes, i.e., when $\mathcal{M}\{\Lambda\}$ tends to either 0 or 1.

A.2 Maximum Uncertainty Principle

An event has no uncertainty if its measure is 1 (or 0) because we may believe that the event occurs (or not). An event is the most uncertain if its measure is 0.5 because the event and its complement may be regarded as "equally likely".

In practice, if there is no information about the measure of an event, we should assign 0.5 to it. Sometimes, only partial information is available.

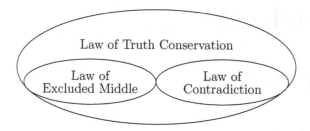

Figure A.1: Relationship among Three Laws

For this case, the value of measure may be specified in some range. What value does the measure take? For the safety purpose, we should assign it the value as close to 0.5 as possible. This is the maximum uncertainty principle proposed by Liu [120].

Maximum Uncertainty Principle: *For any event, if there are multiple reasonable values that an uncertain measure may take, then the value as close to 0.5 as possible is assigned to the event.*

Example A.1: Let Λ be an event. Based on some given information, the measure value $\mathcal{M}\{\Lambda\}$ is on the interval $[a, b]$. By using the maximum uncertainty principle, we should assign

$$
\mathcal{M}\{\Lambda\} = \begin{cases} a, & \text{if } 0.5 < a \le b \\ 0.5, & \text{if } a \le 0.5 \le b \\ b, & \text{if } a \le b < 0.5. \end{cases} \tag{A.2}
$$

Especially, if $\mathcal{M}\{\Lambda\}$ is known to be less than a, then we assign

$$
\mathcal{M}\{\Lambda\} = \begin{cases} a, & \text{if } a < 0.5 \\ 0.5, & \text{if } a \ge 0.5; \end{cases} \tag{A.3}
$$

if $\mathcal{M}\{\Lambda\}$ is known to be greater than b, then we assign

$$
\mathcal{M}\{\Lambda\} = \begin{cases} 0.5, & \text{if } b < 0.5 \\ b, & \text{if } b \ge 0.5. \end{cases} \tag{A.4}
$$

A.3   How to Determine Distribution?

How do we determine the uncertainty distribution for an uncertain variable like "about 100km"? Personally I think uncertainty distribution determination is based on expert's experimental data rather than historical data!

How do we obtain expert's experimental data? The starting point is to invite one or more domain experts who are asked to complete a questionnaire about the meaning of an uncertain variable ξ like "about 100km".

We may ask the domain expert to choose a possible value x (say 110km) that the uncertain variable ξ may take, and then quiz him

"How likely is ξ less than x?"

Denote the expert's belief degree by α (say 0.6). An experimental data (x, α) is thus acquired from the domain expert.

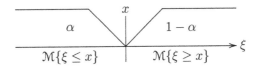

Figure A.2: An Experimental Data (x, α)

Assume that the following experimental data are obtained by the questionnaire,

$$(x_1, \alpha_1), \ (x_2, \alpha_2), \ \cdots, \ (x_n, \alpha_n). \tag{A.5}$$

We will accept them as the expert's experimental data if (perhaps after a rearrangement)

$$x_1 < x_2 < \cdots < x_n, \quad 0 \le \alpha_1 \le \alpha_2 \le \cdots \le \alpha_n \le 1. \tag{A.6}$$

Otherwise, the experimental data are inconsistent and rejected.

Empirical Uncertainty Distribution

Based on the expert's experimental data $(x_1, \alpha_1), (x_2, \alpha_2), \cdots, (x_n, \alpha_n)$, we obtain an empirical uncertainty distribution

$$\Phi(x) = \begin{cases} 0, & \text{if } x < x_1 \\ \alpha_i + \dfrac{(\alpha_{i+1} - \alpha_i)(x - x_i)}{x_{i+1} - x_i}, & \text{if } x_i \le x \le x_{i+1}, 1 \le i < n \\ 1, & \text{if } x > x_n \end{cases} \tag{A.7}$$

Multiple Domain Experts

Assume there are m domain experts and each produces an uncertainty distribution. Then we may get m uncertainty distributions $\Phi_1(x), \Phi_2(x), \cdots, \Phi_m(x)$ that are aggregated to an uncertainty distribution

$$\Phi(x) = w_1 \Phi_1(x) + w_2 \Phi_2(x) + \cdots + w_m \Phi_m(x) \tag{A.8}$$

where $w_1, w_2, \cdots, w_m$ are convex combination coefficients representing weights of the domain experts.

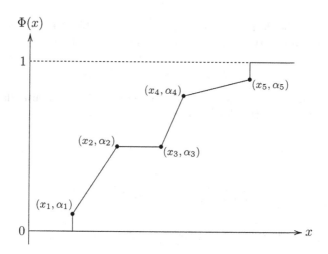

Figure A.3: Empirical Uncertainty Distribution $\Phi(x)$

Principle of Least Squares

Assume that the uncertainty distribution to be determined has a known functional form with one or more unknown parameters, say, $\Phi(x|a, b)$. Based on the expert's experimental data $(x_1, \alpha_1), (x_2, \alpha_2), \cdots, (x_n, \alpha_n)$, the unknown parameters a and b should solve the optimization problem,

$$\min_{a,b} \sum_{i=1}^{n} (\Phi(x_i|a, b) - \alpha_i)^2. \qquad (A.9)$$

For example, assume that the uncertainty distribution has a linear form with two unknown parameters, i.e.,

$$\Phi(x) = ax + b \qquad (A.10)$$

where the unknown parameters a and b should solve

$$\min_{a,b} \sum_{i=1}^{n} (ax_i + b - \alpha_i)^2 \qquad (A.11)$$

whose optimal solution tells us that the linear uncertainty distribution is (not rigorous)

$$\Phi(x) = a^*x + b^* \qquad (A.12)$$

where

$$a^* = \frac{n\widehat{x}\widehat{\alpha} - \sum_{i=1}^{n} x_i\alpha_i}{n\widehat{x}^2 - \sum_{i=1}^{n} x_i^2}, \qquad (A.13)$$

$$b^* = \widehat{\alpha} - a^* \widehat{x}, \tag{A.14}$$

$$\widehat{x} = (x_1 + x_2 + \cdots + x_n)/n, \tag{A.15}$$

$$\widehat{\alpha} = (\alpha_1 + \alpha_2 + \cdots + \alpha_n)/n. \tag{A.16}$$

A.4   Evolution of Measures

Perhaps we are accustomed to assigning to an event two state symbols, true or false, to represent the truth degree of the event. However, starting with Aristotle's epoch, it was observed that such a two-valued assignment has its shortages. Today, it is well-known that the truth degree may be assigned any values between 0 and 1, where 0 represents "completely false" and 1 represents "completely true". The higher the truth degree is, the more true the event is. It is clear that there are multiple assignment ways. This fact has resulted in several types of measure, for example,

> 1933: Probability Measure (A.N. Kolmogoroff);
> 1954: Capacity (G. Choquet);
> 1974: Fuzzy Measure (M. Sugeno);
> 1978: Possibility Measure (L.A. Zadeh);
> 2002: Credibility Measure (B. Liu and Y. Liu);
> 2007: Uncertain Measure (B. Liu).

Probability Measure

A classical measure is essentially a set function satisfying nonnegativity and countable additivity axioms. In order to deal with randomness, Kolmogoroff (1933) defined a probability measure as a special classical measure with normality axiom. In other words, the following three axioms must be satisfied:

Axiom 1. *(Normality)* $\Pr\{\Omega\} = 1$ *for the universal set* Ω.

Axiom 2. *(Nonnegativity)* $\Pr\{A\} \geq 0$ *for any event* A.

Axiom 3. *(Countable Additivity)* *For every countable sequence of mutually disjoint events* $\{A_i\}$, *we have*

$$\Pr\left\{\bigcup_{i=1}^{\infty} A_i\right\} = \sum_{i=1}^{\infty} \Pr\{A_i\}. \tag{A.17}$$

It is clear that probability measure obeys the law of truth conservation and is consistent with the law of excluded middle and the law of contradiction.

Capacity

In order to deal with human systems, the additivity axiom seems too strong. The earliest challenge was from the theory of capacities by Choquet (1954) in which the following axioms are assumed:

Axiom 1. $\pi\{\emptyset\} = 0$.

Axiom 2. $\pi\{A\} \leq \pi\{B\}$ whenever $A \subset B$.

Axiom 3. $\pi\left\{\lim_{i\to\infty} A_i\right\} = \lim_{i\to\infty} \pi\{A_i\}$.

One disadvantage is that capacity does not obey the law of truth conservation and is inconsistent with the law of excluded middle and the law of contradiction.

Fuzzy Measure

Sugeno (1974) generalized classical measure theory to fuzzy measure theory by replacing additivity axiom with weaker axioms of monotonicity and continuity:

Axiom 1. $\pi\{\emptyset\} = 0$.

Axiom 2. $\pi\{A\} \leq \pi\{B\}$ whenever $A \subset B$.

Axiom 3. $\pi\left\{\lim_{i\to\infty} A_i\right\} = \lim_{i\to\infty} \pi\{A_i\}$.

This version of fuzzy measure seems identical with Choquet's capacity. The continuity axiom was replaced with semicontinuity axiom by Sugeno in 1977. However, every version of fuzzy measure does not obey the law of truth conservation and is inconsistent with the law of excluded middle and the law of contradiction.

Possibility Measure

In order to deal with fuzziness, Zadeh (1978) proposed a possibility measure that satisfies the following axioms:

Axiom 1. *(Normality)* $\mathrm{Pos}\{\Theta\} = 1$ *for the universal set* Θ.

Axiom 1. *(Nonnegativity)* $\mathrm{Pos}\{\emptyset\} = 0$ *for the empty set* $\emptyset$.

Axiom 3. *(Maximality) For every sequence of events* $\{A_i\}$, *we have*

$$\mathrm{Pos}\left\{\bigcup_{i=1}^{\infty} A_i\right\} = \bigvee_{i=1}^{\infty} \mathrm{Pos}\{A_i\}. \tag{A.18}$$

Unfortunately, possibility measure does not obey the law of truth conservation and is inconsistent with the law of excluded middle and the law of contradiction.

Credibility Measure

In order to overcome the shortage of possibility measure, Liu and Liu (2002) presented a credibility measure that may be defined by the following four axioms:

Axiom 1. *(Normality)* $\text{Cr}\{\Theta\} = 1$ *for the universal set* Θ.

Axiom 2. *(Monotonicity)* $\text{Cr}\{A\} \leq \text{Cr}\{B\}$ *whenever* $A \subset B$.

Axiom 3. *(Self-Duality)* $\text{Cr}\{A\} + \text{Cr}\{A^c\} = 1$ *for any event* A.

Axiom 4. *(Maximality)* *For every sequence of events* $\{A_i\}$, *we have*

$$\text{Cr}\left\{\bigcup_{i=1}^{\infty} A_i\right\} = \bigvee_{i=1}^{\infty} \text{Cr}\{A_i\}, \quad \text{if} \quad \bigvee_{i=1}^{\infty} \text{Cr}\{A_i\} < 0.5. \qquad (A.19)$$

Note that credibility measure and possibility measure are uniquely determined by each other via the following two equations,

$$\text{Cr}\{A\} = \frac{1}{2}\left(\text{Pos}\{A\} + 1 - \text{Pos}\{A^c\}\right), \qquad (A.20)$$

$$\text{Pos}\{A\} = (2\text{Cr}\{A\}) \wedge 1. \qquad (A.21)$$

Credibility measure obeys the law of truth conservation and is consistent with the law of excluded middle and the law of contradiction. For exploring the credibility theory, the reader may consult the book [120].

Uncertain Measure

In order to deal with uncertainty in human systems, Liu (2007) proposed an uncertain measure based on the following five axioms:

Axiom 1. *(Normality)* $\mathcal{M}\{\Gamma\} = 1$ *for the universal set* Γ.

Axiom 2. *(Monotonicity)* $\mathcal{M}\{\Lambda_1\} \leq \mathcal{M}\{\Lambda_2\}$ *whenever* $\Lambda_1 \subset \Lambda_2$.

Axiom 3. *(Self-Duality)* $\mathcal{M}\{\Lambda\} + \mathcal{M}\{\Lambda^c\} = 1$ *for any event* Λ.

Axiom 4. *(Countable Subadditivity)* *For every countable sequence of events* $\{\Lambda_i\}$, *we have*

$$\mathcal{M}\left\{\bigcup_{i=1}^{\infty} \Lambda_i\right\} \leq \sum_{i=1}^{\infty} \mathcal{M}\{\Lambda_i\}. \qquad (A.22)$$

Axiom 5. *(Product Measure Axiom)* *Let* $(\Gamma_k, \mathcal{L}_k, \mathcal{M}_k)$ *be uncertainty spaces for* $k = 1, 2, \cdots, n$. *Then the product uncertain measure* $\mathcal{M}$ *is an uncertain measure on the product* σ-*algebra* $\mathcal{L}_1 \times \mathcal{L}_2 \times \cdots \times \mathcal{L}_n$ *satisfying*

$$\mathcal{M}\left\{\prod_{k=1}^{n} \Lambda_k\right\} = \min_{1 \leq k \leq n} \mathcal{M}_k\{\Lambda_k\}. \qquad (A.23)$$

Uncertain measure is neither a completely additive measure nor a completely nonadditive measure. In fact, uncertain measure is a "partially additive measure" because of its self-duality. Uncertain measure obeys the law of truth conservation and is consistent with the law of excluded middle and the law of contradiction.

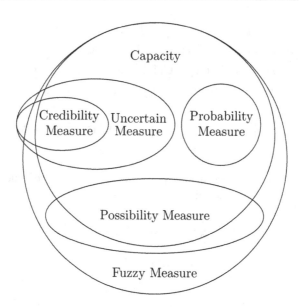

Figure A.4: Relationship among Various Measures. Randomness belongs to the domain of probability measure and uncertainty belongs to the domain of uncertain measure undoubtedly. Fuzziness belongs to the domain of possibility/credibility measure. Some scholars deal with roughness by probability measure and then roughness is an alternative explanation of randomness; some scholars deal with roughness by possibility/credibility measure and then roughness is an alternative explanation of fuzziness; and some scholars deal with roughness by uncertain measure and then roughness is an alternative explanation of uncertainty. Greyness uses probability measure and then is an alternative explanation of randomness.

A.5   Uncertainty vs. Randomness

Probability theory is a branch of mathematics based on Kolmogoroff's axioms. In fact, probability theory may be equivalently reconstructed based on the following 5 axioms:

Axiom 1. *(Normality)* $\Pr\{\Omega\} = 1$ *for the universal set Ω.*

Axiom 2. *(Monotonicity)* $\Pr\{A_1\} \leq \Pr\{A_2\}$ *whenever $A_1 \subset A_2$.*

Axiom 3. *(Self-Duality)* $\Pr\{A\} + \Pr\{A^c\} = 1$ *for any event A.*

Axiom 4. *(Countable Additivity) For every countable sequence of mutually disjoint events $\{A_i\}$, we have*

$$\Pr\left\{\bigcup_{i=1}^{\infty} A_i\right\} = \sum_{i=1}^{\infty} \Pr\{A_i\}. \tag{A.24}$$

Axiom 5. *(Product Probability Axiom) Let $(\Omega_k, \mathcal{A}_k, \Pr_k)$ be probability spaces for $k = 1, 2, \cdots, n$. Then the product probability measure $\Pr$ is a probability measure on the product σ-algebra $\mathcal{A}_1 \times \mathcal{A}_2 \times \cdots \times \mathcal{A}_n$ satisfying*

$$\Pr\left\{\prod_{k=1}^{n} A_k\right\} = \prod_{k=1}^{n} \Pr_k\{A_k\}. \tag{A.25}$$

It is clear that uncertain measure and probability measure share first three axioms. The first differentia is that uncertain measure assumes countable subadditivity axiom and probability measure assumes countable additivity axiom. The second differentia is that product uncertain measure is the minimum of uncertain measures of independent uncertain events and product probability measure is the product of probability measures of independent random events.

Probability theory and uncertainty theory are complementary mathematical systems that provide two acceptable mathematical models to deal with imprecise quantities. Probability model usually simulates objective randomness, and uncertainty model usually simulates human uncertainty.

A.6   Uncertainty + Randomness

In many cases, uncertainty and randomness simultaneously appear in a system. For example, what is an uncertain variable plus a random variable? Actually, you will find that the answer is quite simple.

Suppose $(\Gamma, \mathcal{L}, \mathcal{M})$ is an uncertainty space and $(\Omega, \mathcal{A}, \Pr)$ is a probability space. Then their product is $(\Gamma \times \Omega, \mathcal{L} \times \mathcal{A}, \mathcal{M} \wedge \Pr)$ in which the universal $\Gamma \times \Omega$ is clearly the set of all ordered pairs of (γ, ω) whenever $\gamma \in \Gamma$ and $\omega \in \Omega$. Also, the product σ-algebra $\mathcal{L} \times \mathcal{A}$ is unambiguous too. What is the product measure $\mathcal{M} \wedge \Pr$? In fact, since a probability measure satisfies the first four axioms of uncertain measure, the product measure $\mathcal{M} \wedge \Pr$ satisfying

$$(\mathcal{M} \wedge \Pr)\{\Lambda \times A\} = \mathcal{M}\{\Lambda\} \wedge \Pr\{A\} \tag{A.26}$$

is just an uncertain measure on $\Gamma \times \Omega$, where Λ is an uncertain event and A is a random event.

Recall that an uncertain variable ξ is a measurable function from an uncertainty space to the set of real numbers, and a random variable η is a measurable function from a probability space to the set of real numbers. Since

$$(\Gamma \times \Omega, \mathcal{L} \times \mathcal{A}, \mathcal{M} \wedge \Pr)$$

is an uncertainty space, the functions of uncertain variable and random variable, say $\xi + \eta$ or $\xi \times \eta$, are uncertain variables because they are measurable functions on the uncertainty space $(\Gamma \times \Omega, \mathcal{L} \times \mathcal{A}, \mathcal{M} \wedge \Pr)$.

Theorem A.1. *Let ξ be an uncertain variable with uncertainty distribution Φ, and let η be a random variable with probability distribution Ψ. If $f(x, y)$ is* *a strictly increasing function, then $\tau = f(\xi, \eta)$ is an uncertain variable with inverse uncertainty distribution*

$$\Upsilon^{-1}(\alpha) = f(\Phi^{-1}(\alpha), \Psi^{-1}(\alpha)). \tag{A.27}$$

Proof: It follows from Theorem 1.20 immediately.

Example A.2: Let ξ be an uncertain variable with uncertainty distribution Φ, and let η be a random variable with probability distribution Ψ. Then $\xi + \eta$ is an uncertain variable with inverse uncertainty distribution

$$\Upsilon^{-1}(\alpha) = \Phi^{-1}(\alpha) + \Psi^{-1}(\alpha). \tag{A.28}$$

Example A.3: Let ξ be a nonnegative uncertain variable with uncertainty distribution Φ, and let η be a nonnegative random variable with probability distribution Ψ. Then $\xi \times \eta$ is an uncertain variable with inverse uncertainty distribution

$$\Upsilon^{-1}(\alpha) = \Phi^{-1}(\alpha) \times \Psi^{-1}(\alpha). \tag{A.29}$$

Theorem A.2. *Let ξ be an uncertain variable with uncertainty distribution Φ, and let η be a random variable with probability distribution Ψ. If $f(x, y)$ is a strictly decreasing function, then $\tau = f(\xi, \eta)$ is an uncertain variable with inverse uncertainty distribution*

$$\Upsilon^{-1}(\alpha) = f(\Phi^{-1}(1 - \alpha), \Psi^{-1}(1 - \alpha)). \tag{A.30}$$

Proof: It follows from Theorem 1.25 immediately.

Theorem A.3. *Let ξ be an uncertain variable with uncertainty distribution Φ, and let η be a random variable with probability distribution Ψ. If $f(x, y)$ is strictly increasing with respect to x and strictly decreasing with respect to y, then $\tau = f(\xi, \eta)$ is an uncertain variable with inverse uncertainty distribution*

$$\Upsilon^{-1}(\alpha) = f(\Phi^{-1}(\alpha), \Psi^{-1}(1 - \alpha)). \tag{A.31}$$

Proof: It follows from Theorem 1.26 immediately.

Example A.4: Let ξ be an uncertain variable with uncertainty distribution Φ, and let η be a random variable with probability distribution Ψ. Then $\xi - \eta$ is an uncertain variable with inverse uncertainty distribution

$$\Upsilon^{-1}(\alpha) = \Phi^{-1}(\alpha) - \Psi^{-1}(1 - \alpha). \tag{A.32}$$

Example A.5: Let ξ be a positive uncertain variable with uncertainty distribution Φ, and let η be a positive random variable with probability distribution Ψ. Then ξ/η is an uncertain variable with inverse uncertainty distribution

$$\Upsilon^{-1}(\alpha) = \Phi^{-1}(\alpha)/\Psi^{-1}(1 - \alpha). \tag{A.33}$$

Exercise A.1: Let $\xi_1, \xi_2, \cdots, \xi_m$ be uncertain variables with uncertainty distributions $\Phi_1, \Phi_2, \cdots, \Phi_m$, and let $\eta_1, \eta_2, \cdots, \eta_n$ be random variables with probability distributions $\Psi_1, \Psi_2, \cdots, \Psi_n$, respectively. Assume f is an alternating monotone function. Please determine the uncertainty distribution of

$$\tau = f(\xi_1, \xi_2, \cdots, \xi_m, \eta_1, \eta_2, \cdots, \eta_n). \tag{A.34}$$

Theorem A.4. *Assume that ξ is a Boolean uncertain variable and η is a Boolean random variable, i.e.,*

$$\xi = \begin{cases} 1 & \text{with uncertain measure } a \\ 0 & \text{with uncertain measure } 1-a, \end{cases} \tag{A.35}$$

$$\eta = \begin{cases} 1 & \text{with probability measure } b \\ 0 & \text{with probability measure } 1-b. \end{cases} \tag{A.36}$$

For any Boolean function f, the uncertain variable $\tau = f(\xi, \eta)$ is also Boolean and

$$\mathcal{M}\{\tau = 1\} = \begin{cases} \sup_{f(x,y)=1} \mu(x) \wedge \nu(y), & \text{if } \sup_{f(x,y)=1} \mu(x) \wedge \nu(y) < 0.5 \\ 1 - \sup_{f(x,y)=0} \mu(x) \wedge \nu(y), & \text{if } \sup_{f(x,y)=1} \mu(x) \wedge \nu(y) \geq 0.5 \end{cases}$$

and

$$\mathcal{M}\{\tau = 0\} = \begin{cases} \sup_{f(x,y)=0} \mu(x) \wedge \nu(y), & \text{if } \sup_{f(x,y)=0} \mu(x) \wedge \nu(y) < 0.5 \\ 1 - \sup_{f(x,y)=1} \mu(x) \wedge \nu(y), & \text{if } \sup_{f(x,y)=0} \mu(x) \wedge \nu(y) \geq 0.5 \end{cases}$$

where x and y take values either 0 or 1, and μ and ν are defined by

$$\mu(x) = \begin{cases} a, & \text{if } x = 1 \\ 1-a, & \text{if } x = 0, \end{cases} \tag{A.37}$$

$$\nu(y) = \begin{cases} b, & \text{if } y = 1 \\ 1-b, & \text{if } y = 0. \end{cases} \tag{A.38}$$

Proof: It follows from Theorem 1.27 immediately.

Example A.6: Let ξ be an uncertain proposition with truth value a in uncertain measure, and let η be a random proposition with truth value b in probability measure. Then

$$T(\xi \wedge \eta) = \mathcal{M}\{\xi = 1\} \wedge \Pr\{\eta = 1\} = a \wedge b, \tag{A.39}$$

$$T(\xi \vee \eta) = \mathcal{M}\{\xi = 1\} \vee \Pr\{\eta = 1\} = a \vee b, \tag{A.40}$$

$$T(\xi \to \eta) = (1 - \mathcal{M}\{\xi = 1\}) \vee \Pr\{\eta = 1\} = (1 - a) \vee b. \tag{A.41}$$

Exercise A.2: Let $\xi_1, \xi_2, \cdots, \xi_m$ be Boolean uncertain variables, and let $\eta_1, \eta_2, \cdots, \eta_n$ be Boolean random variables. Assume f is a Boolean function. Please determine the Boolean uncertain variable

$$\tau = f(\xi_1, \xi_2, \cdots, \xi_m, \eta_1, \eta_2, \cdots, \eta_n). \tag{A.42}$$

A.7 Uncertainty vs. Fuzziness

The essential differentia between fuzziness and uncertainty is that the former assumes

$$\mathrm{Cr}\{A \cup B\} = \mathrm{Cr}\{A\} \vee \mathrm{Cr}\{B\}, \quad \text{if } \mathrm{Cr}\{A\} \vee \mathrm{Cr}\{B\} < 0.5$$

for any events A and B no matter if they are independent or not, and the latter assumes

$$\mathcal{M}\{A \cup B\} = \mathcal{M}\{A\} \vee \mathcal{M}\{B\}$$

only for independent events A and B. However, a lot of surveys showed that the measure of union of events is not necessarily maxitive, i.e.,

$$\mathcal{M}\{A \cup B\} \neq \mathcal{M}\{A\} \vee \mathcal{M}\{B\}$$

when the events A and B are not independent. This fact states that human systems do not behave fuzziness.

For example, it is assumed that the distance between Beijing and Tianjin is "about 100km". If "about 100km" is regarded as a fuzzy concept, then we may assign it a membership function, say

$$\mu(x) = \begin{cases} (x - 80)/20, & \text{if } 80 \le x \le 100 \\ (120 - x)/20, & \text{if } 100 \le x \le 120. \end{cases}$$

This membership function represents a triangular fuzzy variable $(80, 100, 120)$. Please do not argue why I choose such a membership function because it is not important for the focus of debate. Based on this membership function, possibility theory (or credibility theory) will conclude the following proposition:

> *The distance between Beijing and Tianjin is "exactly 100km" with belief degree 1 in possibility measure (or 0.5 in credibility measure).*

However, it is doubtless that the belief degree of "exactly 100km" is almost zero. Nobody is so naive to expect that "exactly 100km" is the true distance between Beijing and Tianjin. This paradox shows that those imprecise quantities like "about 100km" cannot be quantified by possibility measure (or credibility measure) and then they are not fuzzy concepts.

If those imprecise quantities are understood as uncertain variables, then the paradox will disappear immediately. Furthermore, uncertainty theory is competent to do almost all jobs of fuzzy theory. This is the main reason why we need the uncertainty theory.

A.8   What Is Uncertainty?

Now we are arriving at the end of this book. Perhaps some readers may complain that I never clarify what uncertainty is. In fact, I really have no idea how to use natural language to define the concept of uncertainty clearly, and I think all existing definitions by natural language are specious just like a riddle. A very personal and ultra viewpoint is that the words like randomness, fuzziness, roughness, greyness, and uncertainty are nothing but ambiguity of human language!

However, fortunately, some "mathematical scales" have been invented to measure the truth degree of an event, for example, probability measure, capacity, fuzzy measure, possibility measure, credibility measure as well as uncertain measure. All of those measures may be defined clearly and precisely by axiomatic methods.

Let us go back to the first question "what is uncertainty". Perhaps we can answer it this way. If it happened that some phenomena can be quantified by uncertain measure, then we call the phenomena "uncertainty". In other words, uncertainty is any concept that satisfies the axioms of uncertainty theory. Thus there are various valid possibilities (*e.g.*, a personal belief degree) to interpret uncertainty theory. Could you agree with me? I hope that uncertainty theory may play a mathematical model of uncertainty in your own problem.

Appendix B

Probability Theory

Probability theory is a branch of mathematics for studying the behavior of random phenomena. The emphasis in this appendix is mainly on probability space, random variable, probability distribution, independence, expected value, variance, moments, critical values, entropy and conditional probability. The main results in this appendix are well-known. For this reason the credit references are not provided.

B.1   Probability Space

Let Ω be a nonempty set, and $\mathcal{A}$ a σ-algebra over Ω. If Ω is countable, usually $\mathcal{A}$ is the power set of Ω. If Ω is uncountable, for example $\Omega = [0, 1]$, usually $\mathcal{A}$ is the Borel algebra of Ω. Each element in $\mathcal{A}$ is called an event. In order to present an axiomatic definition of probability, it is necessary to assign to each event A a number $\Pr\{A\}$ which indicates the probability that A will occur. In order to ensure that the number $\Pr\{A\}$ has certain mathematical properties which we intuitively expect a probability to have, the following three axioms must be satisfied:

Axiom 1. *(Normality)* $\Pr\{\Omega\} = 1$.

Axiom 2. *(Nonnegativity)* $\Pr\{A\} \geq 0$ *for any event A.*

Axiom 3. *(Countable Additivity) For every countable sequence of mutually disjoint events $\{A_i\}$, we have*

$$\Pr\left\{\bigcup_{i=1}^{\infty} A_i\right\} = \sum_{i=1}^{\infty} \Pr\{A_i\}. \tag{B.1}$$

Definition B.1. *The set function $\Pr$ is called a probability measure if it satisfies the normality, nonnegativity, and countable additivity axioms.*

Example B.1: Let $\Omega = \{\omega_1, \omega_2, \cdots\}$, and let $\mathcal{A}$ be the power set of Ω. Assume that $p_1, p_2, \cdots$ are nonnegative numbers such that $p_1 + p_2 + \cdots = 1$. Define a set function on $\mathcal{A}$ as

$$\Pr\{A\} = \sum_{\omega_i \in A} p_i, \quad A \in \mathcal{A}. \tag{B.2}$$

Then $\Pr$ is a probability measure.

Example B.2: Let ϕ be a nonnegative and integrable function on $\Re$ (the set of real numbers) such that

$$\int_{\Re} \phi(x)\mathrm{d}x = 1. \tag{B.3}$$

Then for any Borel set A, the set function

$$\Pr\{A\} = \int_A \phi(x)\mathrm{d}x \tag{B.4}$$

is a probability measure on $\Re$.

Theorem B.1. *Let Ω be a nonempty set, $\mathcal{A}$ a σ-algebra over Ω, and $\Pr$ a probability measure. Then we have*
(a) $\Pr$ is self-dual, i.e., $\Pr\{A\} + \Pr\{A^c\} = 1$ for any $A \in \mathcal{A}$;
(b) $\Pr$ is increasing, i.e., $\Pr\{A\} \le \Pr\{B\}$ whenever $A \subset B$.

Proof: (a) Since A and A^c are disjoint events and $A \cup A^c = \Omega$, we have $\Pr\{A\} + \Pr\{A^c\} = \Pr\{\Omega\} = 1$. (b) Since $A \subset B$, we have $B = A \cup (B \cap A^c)$, where A and $B \cap A^c$ are disjoint events. Therefore $\Pr\{B\} = \Pr\{A\} + \Pr\{B \cap A^c\} \ge \Pr\{A\}$.

Probability Continuity Theorem

Theorem B.2 *(Probability Continuity Theorem). Let Ω be a nonempty set, $\mathcal{A}$ a σ-algebra over Ω, and $\Pr$ a probability measure. If $A_1, A_2, \cdots \in \mathcal{A}$ and $\lim_{i\to\infty} A_i$ exists, then*

$$\lim_{i\to\infty} \Pr\{A_i\} = \Pr\left\{\lim_{i\to\infty} A_i\right\}. \tag{B.5}$$

Proof: STEP 1: Suppose $\{A_i\}$ is an increasing sequence. Write $A_i \to A$ and $A_0 = \emptyset$. Then $\{A_i \backslash A_{i-1}\}$ is a sequence of disjoint events and

$$\bigcup_{i=1}^{\infty}(A_i \backslash A_{i-1}) = A, \quad \bigcup_{i=1}^{k}(A_i \backslash A_{i-1}) = A_k$$

for $k = 1, 2, \cdots$ Thus we have

$$\Pr\{A\} = \Pr\left\{\bigcup_{i=1}^{\infty}(A_i \backslash A_{i-1})\right\} = \sum_{i=1}^{\infty}\Pr\{A_i \backslash A_{i-1}\}$$

$$= \lim_{k\to\infty}\sum_{i=1}^{k}\Pr\{A_i \backslash A_{i-1}\} = \lim_{k\to\infty}\Pr\left\{\bigcup_{i=1}^{k}(A_i \backslash A_{i-1})\right\}$$

$$= \lim_{k\to\infty}\Pr\{A_k\}.$$

STEP 2: If $\{A_i\}$ is a decreasing sequence, then the sequence $\{A_1 \setminus A_i\}$ is clearly increasing. It follows that

$$\Pr\{A_1\} - \Pr\{A\} = \Pr\left\{\lim_{i \to \infty} (A_1 \setminus A_i)\right\} = \lim_{i \to \infty} \Pr\{A_1 \setminus A_i\}$$

$$= \Pr\{A_1\} - \lim_{i \to \infty} \Pr\{A_i\}$$

which implies that $\Pr\{A_i\} \to \Pr\{A\}$.

STEP 3: If $\{A_i\}$ is a sequence of events such that $A_i \to A$, then for each k, we have

$$\bigcap_{i=k}^{\infty} A_i \subset A_k \subset \bigcup_{i=k}^{\infty} A_i.$$

Since Pr is increasing, we have

$$\Pr\left\{\bigcap_{i=k}^{\infty} A_i\right\} \leq \Pr\{A_k\} \leq \Pr\left\{\bigcup_{i=k}^{\infty} A_i\right\}.$$

Note that

$$\bigcap_{i=k}^{\infty} A_i \uparrow A, \quad \bigcup_{i=k}^{\infty} A_i \downarrow A.$$

It follows from Steps 1 and 2 that $\Pr\{A_i\} \to \Pr\{A\}$.

Probability Space

Definition B.2. *Let Ω be a nonempty set, $\mathcal{A}$ a σ-algebra over Ω, and Pr a probability measure. Then the triplet $(\Omega, \mathcal{A}, \Pr)$ is called a probability space.*

Example B.3: Let $\Omega = \{\omega_1, \omega_2, \cdots\}$, $\mathcal{A}$ the power set of Ω, and Pr a probability measure defined by (B.2). Then $(\Omega, \mathcal{A}, \Pr)$ is a probability space.

Example B.4: Let $\Omega = [0, 1]$, $\mathcal{A}$ the Borel algebra over Ω, and Pr the Lebesgue measure. Then $([0, 1], \mathcal{A}, \Pr)$ is a probability space, and sometimes is called *Lebesgue unit interval*. For many purposes it is sufficient to use it as the basic probability space.

Product Probability Space

Let $(\Omega_i, \mathcal{A}_i, \Pr_i)$, $i = 1, 2, \cdots, n$ be probability spaces, and $\Omega = \Omega_1 \times \Omega_2 \times \cdots \times \Omega_n$, $\mathcal{A} = \mathcal{A}_1 \times \mathcal{A}_2 \times \cdots \times \mathcal{A}_n$. Note that the probability measures $\Pr_i, i = 1, 2, \cdots, n$ are finite. It follows from the classical measure theory that there is a unique measure Pr on $\mathcal{A}$ such that

$$\Pr\{A_1 \times A_2 \times \cdots \times A_n\} = \Pr_1\{A_1\} \times \Pr_2\{A_2\} \times \cdots \times \Pr_n\{A_n\}$$

for any $A_i \in \mathcal{A}_i$, $i = 1, 2, \cdots, n$. This conclusion is called the *product probability theorem*. The measure Pr is also a probability measure since

$$\Pr\{\Omega\} = \Pr_1\{\Omega_1\} \times \Pr_2\{\Omega_2\} \times \cdots \times \Pr_n\{\Omega_n\} = 1.$$

Such a probability measure is called the product probability measure, denoted by $\Pr = \Pr_1 \times \Pr_2 \times \cdots \times \Pr_n$.

Definition B.3. *Let* $(\Omega_i, \mathcal{A}_i, \Pr_i)$, $i = 1, 2, \cdots, n$ *be probability spaces, and* $\Omega = \Omega_1 \times \Omega_2 \times \cdots \times \Omega_n$, $\mathcal{A} = \mathcal{A}_1 \times \mathcal{A}_2 \times \cdots \times \mathcal{A}_n$, $\Pr = \Pr_1 \times \Pr_2 \times \cdots \times \Pr_n$. *Then the triplet* $(\Omega, \mathcal{A}, \Pr)$ *is called the product probability space.*

B.2   Random Variable

Definition B.4. *A random variable is a measurable function from a probability space* $(\Omega, \mathcal{A}, \Pr)$ *to the set of real numbers, i.e., for any Borel set B of real numbers, the set*

$$\{\xi \in B\} = \{\omega \in \Omega \mid \xi(\omega) \in B\} \tag{B.6}$$

is an event.

Example B.5: Take $(\Omega, \mathcal{A}, \Pr)$ to be $\{\omega_1, \omega_2\}$ with $\Pr\{\omega_1\} = \Pr\{\omega_2\} = 0.5$. Then the function

$$\xi(\omega) = \begin{cases} 0, & \text{if } \omega = \omega_1 \\ 1, & \text{if } \omega = \omega_2 \end{cases}$$

is a random variable.

Example B.6: Take $(\Omega, \mathcal{A}, \Pr)$ to be the interval $[0, 1]$ with Borel algebra and Lebesgue measure. We define ξ as an identity function from Ω to $[0,1]$. Since ξ is a measurable function, it is a random variable.

Example B.7: A deterministic number c may be regarded as a special random variable. In fact, it is the constant function $\xi(\omega) \equiv c$ on the probability space $(\Omega, \mathcal{A}, \Pr)$.

Definition B.5. *Let ξ_1 and ξ_2 be random variables defined on the probability space* $(\Omega, \mathcal{A}, \Pr)$. *We say $\xi_1 = \xi_2$ if $\xi_1(\omega) = \xi_2(\omega)$ for almost all $\omega \in \Omega$.*

Random Vector

Definition B.6. *An n-dimensional random vector is a measurable function from a probability space* $(\Omega, \mathcal{A}, \Pr)$ *to the set of n-dimensional real vectors, i.e., for any Borel set B of $\Re^n$, the set*

$$\{\boldsymbol{\xi} \in B\} = \{\omega \in \Omega \mid \boldsymbol{\xi}(\omega) \in B\} \tag{B.7}$$

is an event.

Theorem B.3. *The vector* $(\xi_1, \xi_2, \cdots, \xi_n)$ *is a random vector if and only if* $\xi_1, \xi_2, \cdots, \xi_n$ *are random variables.*

Proof: Write $\boldsymbol{\xi} = (\xi_1, \xi_2, \cdots, \xi_n)$. Suppose that $\boldsymbol{\xi}$ is a random vector on the probability space $(\Omega, \mathcal{A}, \mathrm{Pr})$. For any Borel set B of $\Re$, the set $B \times \Re^{n-1}$ is also a Borel set of $\Re^n$. Thus we have

$$\{\xi_1 \in B\} = \{\xi_1 \in B, \xi_2 \in \Re, \cdots, \xi_n \in \Re\} = \{\boldsymbol{\xi} \in B \times \Re^{n-1}\} \in \mathcal{A}$$

which implies that ξ_1 is a random variable. A similar process may prove that $\xi_2, \xi_3, \cdots, \xi_n$ are random variables. Conversely, suppose that all $\xi_1, \xi_2, \cdots, \xi_n$ are random variables on the probability space $(\Omega, \mathcal{A}, \mathrm{Pr})$. We define

$$\mathcal{B} = \left\{ B \subset \Re^n \mid \{\boldsymbol{\xi} \in B\} \in \mathcal{A} \right\}.$$

The vector $\boldsymbol{\xi} = (\xi_1, \xi_2, \cdots, \xi_n)$ is proved to be a random vector if we can prove that $\mathcal{B}$ contains all Borel sets of $\Re^n$. First, the class $\mathcal{B}$ contains all open intervals of $\Re^n$ because

$$\left\{ \boldsymbol{\xi} \in \prod_{i=1}^{n} (a_i, b_i) \right\} = \bigcap_{i=1}^{n} \{\xi_i \in (a_i, b_i)\} \in \mathcal{A}.$$

Next, the class $\mathcal{B}$ is a σ-algebra of $\Re^n$ because (i) we have $\Re^n \in \mathcal{B}$ since $\{\boldsymbol{\xi} \in \Re^n\} = \Omega \in \mathcal{A}$; (ii) if $B \in \mathcal{B}$, then $\{\boldsymbol{\xi} \in B\} \in \mathcal{A}$, and

$$\{\boldsymbol{\xi} \in B^c\} = \{\boldsymbol{\xi} \in B\}^c \in \mathcal{A}$$

which implies that $B^c \in \mathcal{B}$; (iii) if $B_i \in \mathcal{B}$ for $i = 1, 2, \cdots$, then $\{\boldsymbol{\xi} \in B_i\} \in \mathcal{A}$ and

$$\left\{ \boldsymbol{\xi} \in \bigcup_{i=1}^{\infty} B_i \right\} = \bigcup_{i=1}^{\infty} \{\boldsymbol{\xi} \in B_i\} \in \mathcal{A}$$

which implies that $\cup_i B_i \in \mathcal{B}$. Since the smallest σ-algebra containing all open intervals of $\Re^n$ is just the Borel algebra of $\Re^n$, the class $\mathcal{B}$ contains all Borel sets of $\Re^n$. The theorem is proved.

Random Arithmetic

In this subsections, we will suppose that all random variables are defined on a common probability space. Otherwise, we may embed them into the product probability space.

Definition B.7. *Let* $f : \Re^n \to \Re$ *be a measurable function, and* $\xi_1, \xi_2, \cdots, \xi_n$ *random variables defined on the probability space* $(\Omega, \mathcal{A}, \mathrm{Pr})$. *Then* $\xi = f(\xi_1, \xi_2, \cdots, \xi_n)$ *is a random variable defined by*

$$\xi(\omega) = f(\xi_1(\omega), \xi_2(\omega), \cdots, \xi_n(\omega)), \quad \forall \omega \in \Omega. \tag{B.8}$$

Example B.8: Let ξ_1 and ξ_2 be random variables on the probability space $(\Omega, \mathcal{A}, \mathrm{Pr})$. Then their sum is

$$(\xi_1 + \xi_2)(\omega) = \xi_1(\omega) + \xi_2(\omega), \quad \forall \omega \in \Omega$$

and their product is

$$(\xi_1 \times \xi_2)(\omega) = \xi_1(\omega) \times \xi_2(\omega), \quad \forall \omega \in \Omega.$$

The reader may wonder whether $\xi(\omega_1, \omega_2, \cdots, \omega_n)$ defined by (B.7) is a random variable. The following theorem answers this question.

Theorem B.4. *Let $\boldsymbol{\xi}$ be an n-dimensional random vector, and $f : \Re^n \to \Re$ a measurable function. Then $f(\boldsymbol{\xi})$ is a random variable.*

Proof: Assume that $\boldsymbol{\xi}$ is a random vector on the probability space $(\Omega, \mathcal{A}, \mathrm{Pr})$. For any Borel set B of $\Re$, since f is a measurable function, $f^{-1}(B)$ is also a Borel set of $\Re^n$. Thus we have

$$\{f(\boldsymbol{\xi}) \in B\} = \left\{\boldsymbol{\xi} \in f^{-1}(B)\right\} \in \mathcal{A}$$

which implies that $f(\boldsymbol{\xi})$ is a random variable.

B.3   Probability Distribution

Definition B.8. *The probability distribution Φ: $\Re \to [0, 1]$ of a random variable ξ is defined by*

$$\Phi(x) = \mathrm{Pr}\{\xi \leq x\}. \tag{B.9}$$

That is, $\Phi(x)$ is the probability that the random variable ξ takes a value less than or equal to x.

Example B.9: Take $(\Omega, \mathcal{A}, \mathrm{Pr})$ to be $\{\omega_1, \omega_2\}$ with $\mathrm{Pr}\{\omega_1\} = \mathrm{Pr}\{\omega_2\} = 0.5$. We now define a random variable as follows,

$$\xi(\omega) = \begin{cases} -1, & \text{if } \omega = \omega_1 \\ 1, & \text{if } \omega = \omega_2. \end{cases}$$

Then ξ has a probability distribution

$$\Phi(x) = \begin{cases} 0, & \text{if } x < -1 \\ 0.5, & \text{if } -1 \leq x < 1 \\ 1, & \text{if } x \geq 1. \end{cases}$$

Theorem B.5 *(Sufficient and Necessary Condition for Probability Distribution). A function $\Phi : \Re \to [0, 1]$ is a probability distribution if and only if it is an increasing and right-continuous function with*

$$\lim_{x \to -\infty} \Phi(x) = 0; \qquad \lim_{x \to +\infty} \Phi(x) = 1. \tag{B.10}$$

Proof: For any $x, y \in \Re$ with $x < y$, we have

$$\Phi(y) - \Phi(x) = \Pr\{x < \xi \leq y\} \geq 0.$$

Thus the probability distribution Φ is increasing. Next, let $\{\varepsilon_i\}$ be a sequence of positive numbers such that $\varepsilon_i \to 0$ as $i \to \infty$. Then, for every $i \geq 1$, we have

$$\Phi(x + \varepsilon_i) - \Phi(x) = \Pr\{x < \xi \leq x + \varepsilon_i\}.$$

It follows from the probability continuity theorem that

$$\lim_{i \to \infty} \Phi(x + \varepsilon_i) - \Phi(x) = \Pr\{\emptyset\} = 0.$$

Hence Φ is a right-continuous function. Finally,

$$\lim_{x \to -\infty} \Phi(x) = \lim_{x \to -\infty} \Pr\{\xi \leq x\} = \Pr\{\emptyset\} = 0,$$

$$\lim_{x \to +\infty} \Phi(x) = \lim_{x \to +\infty} \Pr\{\xi \leq x\} = \Pr\{\Omega\} = 1.$$

Conversely, it is known there is a unique probability measure Pr on the Borel algebra over $\Re$ such that $\Pr\{(-\infty, x]\} = \Phi(x)$ for all $x \in \Re$. Furthermore, it is easy to verify that the random variable defined by $\xi(x) = x$ from the probability space $(\Re, \mathcal{A}, \Pr)$ to $\Re$ has the probability distribution Φ.

Probability Density Function

Definition B.9 . *The probability density function $\phi \colon \Re \to [0, +\infty)$ of a random variable ξ is a function such that*

$$\Phi(x) = \int_{-\infty}^{x} \phi(y)\mathrm{d}y \tag{B.11}$$

holds for all $x \in \Re$, where Φ is the probability distribution of the random variable ξ.

Theorem B.6 *(Probability Inversion Theorem). Let ξ be a random variable whose probability density function ϕ exists. Then for any Borel set B of $\Re$, we have*

$$\Pr\{\xi \in B\} = \int_{B} \phi(y)\mathrm{d}y. \tag{B.12}$$

Proof: Let $\mathcal{C}$ be the class of all subsets C of $\Re$ for which the relation

$$\Pr\{\xi \in C\} = \int_{C} \phi(y)\mathrm{d}y \tag{B.13}$$

holds. We will show that $\mathcal{C}$ contains all Borel sets of $\Re$. It follows from the probability continuity theorem and relation (B.13) that $\mathcal{C}$ is a monotone class. It is also clear that $\mathcal{C}$ contains all intervals of the form $(-\infty, a]$, $(a, b]$, (b, ∞) and $\emptyset$ since

$$\Pr\{\xi \in (-\infty, a]\} = \Phi(a) = \int_{-\infty}^{a} \phi(y)\mathrm{d}y,$$

$$\Pr\{\xi \in (b, +\infty)\} = \Phi(+\infty) - \Phi(b) = \int_{b}^{+\infty} \phi(y)\mathrm{d}y,$$

$$\Pr\{\xi \in (a, b]\} = \Phi(b) - \Phi(a) = \int_{a}^{b} \phi(y)\mathrm{d}y,$$

$$\Pr\{\xi \in \emptyset\} = 0 = \int_{\emptyset} \phi(y)\mathrm{d}y$$

where Φ is the probability distribution of ξ. Let $\mathcal{F}$ be the algebra consisting of all finite unions of disjoint sets of the form $(-\infty, a]$, $(a, b]$, (b, ∞) and $\emptyset$. Note that for any disjoint sets $C_1, C_2, \cdots, C_m$ of $\mathcal{F}$ and $C = C_1 \cup C_2 \cup \cdots \cup C_m$, we have

$$\Pr\{\xi \in C\} = \sum_{j=1}^{m} \Pr\{\xi \in C_j\} = \sum_{j=1}^{m} \int_{C_j} \phi(y)\mathrm{d}y = \int_{C} \phi(y)\mathrm{d}y.$$

That is, $C \in \mathcal{C}$. Hence we have $\mathcal{F} \subset \mathcal{C}$. Since the smallest σ-algebra containing $\mathcal{F}$ is just the Borel algebra of $\Re$, the monotone class theorem implies that $\mathcal{C}$ contains all Borel sets of $\Re$.

Some Special Distributions

Uniform Distribution: A random variable ξ has a uniform distribution if its probability density function is defined by

$$\phi(x) = \frac{1}{b - a}, \quad a \leq x \leq b. \tag{B.14}$$

where a and b are given real numbers with $a < b$.

Exponential Distribution: A random variable ξ has an exponential distribution if its probability density function is defined by

$$\phi(x) = \frac{1}{\beta} \exp\left(-\frac{x}{\beta}\right), \quad x \geq 0 \tag{B.15}$$

where β is a positive number.

Normal Distribution: A random variable ξ has a normal distribution if its probability density function is defined by

$$\phi(x) = \frac{1}{\sigma\sqrt{2\pi}} \exp\left(-\frac{(x - \mu)^2}{2\sigma^2}\right), \quad x \in \Re \tag{B.16}$$

where μ and σ are real numbers.

B.4 Independence

Definition B.10. *The random variables $\xi_1, \xi_2, \cdots, \xi_m$ are said to be independent if*

$$\Pr\left\{\bigcap_{i=1}^{m}\{\xi_i \in B_i\}\right\} = \prod_{i=1}^{m}\Pr\{\xi_i \in B_i\} \tag{B.17}$$

for any Borel sets $B_1, B_2, \cdots, B_m$ of $\Re$.

Theorem B.7. *Let ξ_i be random variables with probability distributions Φ_i, $i = 1, 2, \cdots, m$, respectively, and Φ the probability distribution of the random vector $(\xi_1, \xi_2, \cdots, \xi_m)$. Then $\xi_1, \xi_2, \cdots, \xi_m$ are independent if and only if*

$$\Phi(x_1, x_2, \cdots, x_m) = \Phi_1(x_1)\Phi_2(x_2)\cdots\Phi_m(x_m) \tag{B.18}$$

for all $(x_1, x_2, \cdots, x_m) \in \Re^m$.

Proof: If $\xi_1, \xi_2, \cdots, \xi_m$ are independent random variables, then we have

$$
\begin{aligned}
\Phi(x_1, x_2, \cdots, x_m) &= \Pr\{\xi_1 \le x_1, \xi_2 \le x_2, \cdots, \xi_m \le x_m\} \\
&= \Pr\{\xi_1 \le x_1\}\Pr\{\xi_2 \le x_2\}\cdots\Pr\{\xi_m \le x_m\} \\
&= \Phi_1(x_1)\Phi_2(x_2)\cdots\Phi_m(x_m)
\end{aligned}
$$

for all $(x_1, x_2, \cdots, x_m) \in \Re^m$. Conversely, assume that (B.18) holds. Let $x_2, x_3, \cdots, x_m$ be fixed real numbers, and $\mathcal{C}$ the class of all subsets C of $\Re$ for which the relation

$$\Pr\{\xi_1 \in C, \xi_2 \le x_2, \cdots, \xi_m \le x_m\} = \Pr\{\xi_1 \in C\}\prod_{i=2}^{m}\Pr\{\xi_i \le x_i\} \tag{B.19}$$

holds. We will show that $\mathcal{C}$ contains all Borel sets of $\Re$. It follows from the probability continuity theorem and relation (B.19) that $\mathcal{C}$ is a monotone class. It is also clear that $\mathcal{C}$ contains all intervals of the form $(-\infty, a]$, $(a, b]$, (b, ∞) and $\emptyset$. Let $\mathcal{F}$ be the algebra consisting of all finite unions of disjoint sets of the form $(-\infty, a]$, $(a, b]$, (b, ∞) and $\emptyset$. Note that for any disjoint sets $C_1, C_2, \cdots, C_k$ of $\mathcal{F}$ and $C = C_1 \cup C_2 \cup \cdots \cup C_k$, we have

$$
\begin{aligned}
&\Pr\{\xi_1 \in C, \xi_2 \le x_2, \cdots, \xi_m \le x_m\} \\
&= \sum_{j=1}^{m}\Pr\{\xi_1 \in C_j, \xi_2 \le x_2, \cdots, \xi_m \le x_m\} \\
&= \Pr\{\xi_1 \in C\}\Pr\{\xi_2 \le x_2\}\cdots\Pr\{\xi_m \le x_m\}.
\end{aligned}
$$

That is, $C \in \mathcal{C}$. Hence we have $\mathcal{F} \subset \mathcal{C}$. Since the smallest σ-algebra containing $\mathcal{F}$ is just the Borel algebra of $\Re$, the monotone class theorem implies that $\mathcal{C}$ contains all Borel sets of $\Re$. Applying the same reasoning to each ξ_i in turn, we obtain the independence of the random variables.

Theorem B.8. *Let ξ_i be random variables with probability density functions ϕ_i, $i = 1, 2, \cdots, m$, respectively, and ϕ the probability density function of the random vector $(\xi_1, \xi_2, \cdots, \xi_m)$. Then $\xi_1, \xi_2, \cdots, \xi_m$ are independent if and only if*

$$\phi(x_1, x_2, \cdots, x_m) = \phi_1(x_1)\phi_2(x_2)\cdots\phi_m(x_m) \qquad (B.20)$$

for almost all $(x_1, x_2, \cdots, x_m) \in \Re^m$.

Proof: If $\phi(x_1, x_2, \cdots, x_m) = \phi_1(x_1)\phi_2(x_2)\cdots\phi_m(x_m)$ a.e., then we have

$$\begin{aligned}
\Phi(x_1, x_2, \cdots, x_m) &= \int_{-\infty}^{x_1} \int_{-\infty}^{x_2} \cdots \int_{-\infty}^{x_m} \phi(t_1, t_2, \cdots, t_m) dt_1 dt_2 \cdots dt_m \\
&= \int_{-\infty}^{x_1} \int_{-\infty}^{x_2} \cdots \int_{-\infty}^{x_m} \phi_1(t_1)\phi_2(t_2)\cdots\phi_m(t_m) dt_1 dt_2 \cdots dt_m \\
&= \int_{-\infty}^{x_1} \phi_1(t_1) dt_1 \int_{-\infty}^{x_2} \phi_2(t_2) dt_2 \cdots \int_{-\infty}^{x_m} \phi_m(t_m) dt_m \\
&= \Phi_1(x_1)\Phi_2(x_2)\cdots\Phi_m(x_m)
\end{aligned}$$

for all $(x_1, x_2, \cdots, x_m) \in \Re^m$. Thus $\xi_1, \xi_2, \cdots, \xi_m$ are independent. Conversely, if $\xi_1, \xi_2, \cdots, \xi_m$ are independent, then for any $(x_1, x_2, \cdots, x_m) \in \Re^m$, we have $\Phi(x_1, x_2, \cdots, x_m) = \Phi_1(x_1)\Phi_2(x_2)\cdots\Phi_m(x_m)$. Hence

$$\Phi(x_1, x_2, \cdots, x_m) = \int_{-\infty}^{x_1} \int_{-\infty}^{x_2} \cdots \int_{-\infty}^{x_m} \phi_1(t_1)\phi_2(t_2)\cdots\phi_m(t_m) dt_1 dt_2 \cdots dt_m$$

which implies that $\phi(x_1, x_2, \cdots, x_m) = \phi_1(x_1)\phi_2(x_2)\cdots\phi_m(x_m)$ a.e.

Example B.10: Let $\xi_1, \xi_2, \cdots, \xi_m$ be independent random variables with probability density functions $\phi_1, \phi_2, \cdots, \phi_m$, respectively, and $f : \Re^m \to \Re$ a measurable function. Then for any Borel set B of real numbers, the probability $\Pr\{f(\xi_1, \xi_2, \cdots, \xi_m) \in B\}$ is

$$\iint \cdots \int_{f(x_1, x_2, \cdots, x_m) \in B} \phi_1(x_1)\phi_2(x_2)\cdots\phi_m(x_m) dx_1 dx_2 \cdots dx_m.$$

B.5   Expected Value

Definition B.11. *Let ξ be a random variable. Then the expected value of ξ is defined by*

$$E[\xi] = \int_0^{+\infty} \Pr\{\xi \geq r\} dr - \int_{-\infty}^0 \Pr\{\xi \leq r\} dr \qquad (B.21)$$

provided that at least one of the two integrals is finite.

Example B.11: Assume that ξ is a discrete random variable taking values x_i with probabilities p_i, $i = 1, 2, \cdots, m$, respectively. It follows from the definition of expected value operator that

$$E[\xi] = \sum_{i=1}^{m} p_i x_i.$$

Theorem B.9. *Let ξ be a random variable whose probability density function ϕ exists. If the Lebesgue integral*

$$\int_{-\infty}^{+\infty} x\phi(x)\mathrm{d}x$$

is finite, then we have

$$E[\xi] = \int_{-\infty}^{+\infty} x\phi(x)\mathrm{d}x. \tag{B.22}$$

Proof: It follows from Definition B.11 and Fubini Theorem that

$$
\begin{aligned}
E[\xi] &= \int_{0}^{+\infty} \Pr\{\xi \geq r\}\mathrm{d}r - \int_{-\infty}^{0} \Pr\{\xi \leq r\}\mathrm{d}r \\
&= \int_{0}^{+\infty} \left[\int_{r}^{+\infty} \phi(x)\mathrm{d}x\right]\mathrm{d}r - \int_{-\infty}^{0} \left[\int_{-\infty}^{r} \phi(x)\mathrm{d}x\right]\mathrm{d}r \\
&= \int_{0}^{+\infty} \left[\int_{0}^{x} \phi(x)\mathrm{d}r\right]\mathrm{d}x - \int_{-\infty}^{0} \left[\int_{x}^{0} \phi(x)\mathrm{d}r\right]\mathrm{d}x \\
&= \int_{0}^{+\infty} x\phi(x)\mathrm{d}x + \int_{-\infty}^{0} x\phi(x)\mathrm{d}x \\
&= \int_{-\infty}^{+\infty} x\phi(x)\mathrm{d}x.
\end{aligned}
$$

The theorem is proved.

Theorem B.10. *Let ξ be a random variable with probability distribution Φ. If the Lebesgue-Stieltjes integral*

$$\int_{-\infty}^{+\infty} x\mathrm{d}\Phi(x)$$

is finite, then we have

$$E[\xi] = \int_{-\infty}^{+\infty} x\mathrm{d}\Phi(x). \tag{B.23}$$

Proof: Since the Lebesgue-Stieltjes integral $\int_{-\infty}^{+\infty} x\mathrm{d}\Phi(x)$ is finite, we immediately have

$$\lim_{y\to+\infty} \int_{0}^{y} x\mathrm{d}\Phi(x) = \int_{0}^{+\infty} x\mathrm{d}\Phi(x), \quad \lim_{y\to-\infty} \int_{y}^{0} x\mathrm{d}\Phi(x) = \int_{-\infty}^{0} x\mathrm{d}\Phi(x)$$

and

$$\lim_{y \to +\infty} \int_y^{+\infty} x d\Phi(x) = 0, \quad \lim_{y \to -\infty} \int_{-\infty}^y x d\Phi(x) = 0.$$

It follows from

$$\int_y^{+\infty} x d\Phi(x) \geq y \left(\lim_{z \to +\infty} \Phi(z) - \Phi(y) \right) = y(1 - \Phi(y)) \geq 0, \quad \text{if } y > 0,$$

$$\int_{-\infty}^y x d\Phi(x) \leq y \left(\Phi(y) - \lim_{z \to -\infty} \Phi(z) \right) = y\Phi(y) \leq 0, \quad \text{if } y < 0$$

that

$$\lim_{y \to +\infty} y \left(1 - \Phi(y) \right) = 0, \quad \lim_{y \to -\infty} y\Phi(y) = 0.$$

Let $0 = x_0 < x_1 < x_2 < \cdots < x_n = y$ be a partition of $[0, y]$. Then we have

$$\sum_{i=0}^{n-1} x_i \left(\Phi(x_{i+1}) - \Phi(x_i) \right) \to \int_0^y x d\Phi(x)$$

and

$$\sum_{i=0}^{n-1} (1 - \Phi(x_{i+1}))(x_{i+1} - x_i) \to \int_0^y \Pr\{\xi \geq r\} dr$$

as $\max\{|x_{i+1} - x_i| : i = 0, 1, \cdots, n-1\} \to 0$. Since

$$\sum_{i=0}^{n-1} x_i \left(\Phi(x_{i+1}) - \Phi(x_i) \right) - \sum_{i=0}^{n-1} (1 - \Phi(x_{i+1}))(x_{i+1} - x_i) = y(\Phi(y) - 1) \to 0$$

as $y \to +\infty$. This fact implies that

$$\int_0^{+\infty} \Pr\{\xi \geq r\} dr = \int_0^{+\infty} x d\Phi(x).$$

A similar way may prove that

$$-\int_{-\infty}^0 \Pr\{\xi \leq r\} dr = \int_{-\infty}^0 x d\Phi(x).$$

Thus (B.23) is verified by the above two equations.

Linearity of Expected Value Operator

Theorem B.11. *Let ξ and η be random variables with finite expected values. Then for any numbers a and b, we have*

$$E[a\xi + b\eta] = aE[\xi] + bE[\eta]. \tag{B.24}$$

Proof: STEP 1: We first prove that $E[\xi + b] = E[\xi] + b$ for any real number b. When $b \geq 0$, we have

$$
\begin{aligned}
E[\xi + b] &= \int_0^\infty \Pr\{\xi + b \geq r\} \mathrm{d}r - \int_{-\infty}^0 \Pr\{\xi + b \leq r\} \mathrm{d}r \\
&= \int_0^\infty \Pr\{\xi \geq r - b\} \mathrm{d}r - \int_{-\infty}^0 \Pr\{\xi \leq r - b\} \mathrm{d}r \\
&= E[\xi] + \int_0^b \left(\Pr\{\xi \geq r - b\} + \Pr\{\xi < r - b\} \right) \mathrm{d}r \\
&= E[\xi] + b.
\end{aligned}
$$

If $b < 0$, then we have

$$
E[\xi + b] = E[\xi] - \int_b^0 \left(\Pr\{\xi \geq r - b\} + \Pr\{\xi < r - b\} \right) \mathrm{d}r = E[\xi] + b.
$$

STEP 2: We prove that $E[a\xi] = aE[\xi]$ for any real number a. If $a = 0$, then the equation $E[a\xi] = aE[\xi]$ holds trivially. If $a > 0$, we have

$$
\begin{aligned}
E[a\xi] &= \int_0^\infty \Pr\{a\xi \geq r\} \mathrm{d}r - \int_{-\infty}^0 \Pr\{a\xi \leq r\} \mathrm{d}r \\
&= \int_0^\infty \Pr\left\{\xi \geq \frac{r}{a}\right\} \mathrm{d}r - \int_{-\infty}^0 \Pr\left\{\xi \leq \frac{r}{a}\right\} \mathrm{d}r \\
&= a \int_0^\infty \Pr\left\{\xi \geq \frac{r}{a}\right\} \mathrm{d}\left(\frac{r}{a}\right) - a \int_{-\infty}^0 \Pr\left\{\xi \leq \frac{r}{a}\right\} \mathrm{d}\left(\frac{r}{a}\right) \\
&= aE[\xi].
\end{aligned}
$$

If $a < 0$, we have

$$
\begin{aligned}
E[a\xi] &= \int_0^\infty \Pr\{a\xi \geq r\} \mathrm{d}r - \int_{-\infty}^0 \Pr\{a\xi \leq r\} \mathrm{d}r \\
&= \int_0^\infty \Pr\left\{\xi \leq \frac{r}{a}\right\} \mathrm{d}r - \int_{-\infty}^0 \Pr\left\{\xi \geq \frac{r}{a}\right\} \mathrm{d}r \\
&= a \int_0^\infty \Pr\left\{\xi \geq \frac{r}{a}\right\} \mathrm{d}\left(\frac{r}{a}\right) - a \int_{-\infty}^0 \Pr\left\{\xi \leq \frac{r}{a}\right\} \mathrm{d}\left(\frac{r}{a}\right) \\
&= aE[\xi].
\end{aligned}
$$

STEP 3: We prove that $E[\xi + \eta] = E[\xi] + E[\eta]$ when both ξ and η are nonnegative simple random variables taking values $a_1, a_2, \cdots, a_m$ and $b_1, b_2, \cdots, b_n$, respectively. Then $\xi + \eta$ is also a nonnegative simple random variable taking values $a_i + b_j$, $i = 1, 2, \cdots, m$, $j = 1, 2, \cdots, n$. Thus we have

$$E[\xi + \eta] = \sum_{i=1}^{m} \sum_{j=1}^{n} (a_i + b_j) \Pr\{\xi = a_i, \eta = b_j\}$$

$$= \sum_{i=1}^{m} \sum_{j=1}^{n} a_i \Pr\{\xi = a_i, \eta = b_j\} + \sum_{i=1}^{m} \sum_{j=1}^{n} b_j \Pr\{\xi = a_i, \eta = b_j\}$$

$$= \sum_{i=1}^{m} a_i \Pr\{\xi = a_i\} + \sum_{j=1}^{n} b_j \Pr\{\eta = b_j\}$$

$$= E[\xi] + E[\eta].$$

STEP 4: We prove that $E[\xi + \eta] = E[\xi] + E[\eta]$ when both ξ and η are nonnegative random variables. For every $i \geq 1$ and every $\omega \in \Omega$, we define

$$\xi_i(\omega) = \begin{cases} \dfrac{k-1}{2^i}, & \text{if } \dfrac{k-1}{2^i} \leq \xi(\omega) < \dfrac{k}{2^i}, \ k = 1, 2, \cdots, i2^i \\ i, & \text{if } i \leq \xi(\omega), \end{cases}$$

$$\eta_i(\omega) = \begin{cases} \dfrac{k-1}{2^i}, & \text{if } \dfrac{k-1}{2^i} \leq \eta(\omega) < \dfrac{k}{2^i}, \ k = 1, 2, \cdots, i2^i \\ i, & \text{if } i \leq \eta(\omega). \end{cases}$$

Then $\{\xi_i\}$, $\{\eta_i\}$ and $\{\xi_i + \eta_i\}$ are three sequences of nonnegative simple random variables such that $\xi_i \uparrow \xi$, $\eta_i \uparrow \eta$ and $\xi_i + \eta_i \uparrow \xi + \eta$ as $i \to \infty$. Note that the functions $\Pr\{\xi_i > r\}$, $\Pr\{\eta_i > r\}$, $\Pr\{\xi_i + \eta_i > r\}$, $i = 1, 2, \cdots$ are also simple. It follows from the probability continuity theorem that

$$\Pr\{\xi_i > r\} \uparrow \Pr\{\xi > r\}, \ \forall r \geq 0$$

as $i \to \infty$. Since the expected value $E[\xi]$ exists, we have

$$E[\xi_i] = \int_0^{+\infty} \Pr\{\xi_i > r\} dr \to \int_0^{+\infty} \Pr\{\xi > r\} dr = E[\xi]$$

as $i \to \infty$. Similarly, we may prove that $E[\eta_i] \to E[\eta]$ and $E[\xi_i + \eta_i] \to E[\xi + \eta]$ as $i \to \infty$. It follows from Step 3 that $E[\xi + \eta] = E[\xi] + E[\eta]$.

STEP 5: We prove that $E[\xi + \eta] = E[\xi] + E[\eta]$ when ξ and η are arbitrary random variables. Define

$$\xi_i(\omega) = \begin{cases} \xi(\omega), & \text{if } \xi(\omega) \geq -i \\ -i, & \text{otherwise,} \end{cases} \qquad \eta_i(\omega) = \begin{cases} \eta(\omega), & \text{if } \eta(\omega) \geq -i \\ -i, & \text{otherwise.} \end{cases}$$

Since the expected values $E[\xi]$ and $E[\eta]$ are finite, we have

$$\lim_{i \to \infty} E[\xi_i] = E[\xi], \quad \lim_{i \to \infty} E[\eta_i] = E[\eta], \quad \lim_{i \to \infty} E[\xi_i + \eta_i] = E[\xi + \eta].$$

Note that $(\xi_i + i)$ and $(\eta_i + i)$ are nonnegative random variables. It follows from Steps 1 and 4 that

$$
\begin{aligned}
E[\xi + \eta] &= \lim_{i \to \infty} E[\xi_i + \eta_i] \\
&= \lim_{i \to \infty} \left(E[(\xi_i + i) + (\eta_i + i)] - 2i \right) \\
&= \lim_{i \to \infty} \left(E[\xi_i + i] + E[\eta_i + i] - 2i \right) \\
&= \lim_{i \to \infty} \left(E[\xi_i] + i + E[\eta_i] + i - 2i \right) \\
&= \lim_{i \to \infty} E[\xi_i] + \lim_{i \to \infty} E[\eta_i] \\
&= E[\xi] + E[\eta].
\end{aligned}
$$

STEP 6: The linearity $E[a\xi + b\eta] = aE[\xi] + bE[\eta]$ follows immediately from Steps 2 and 5. The theorem is proved.

Product of Independent Random Variables

Theorem B.12. *Let ξ and η be independent random variables with finite expected values. Then the expected value of $\xi\eta$ exists and*

$$
E[\xi\eta] = E[\xi]E[\eta]. \tag{B.25}
$$

Proof: STEP 1: We first prove the case where both ξ and η are nonnegative simple random variables taking values $a_1, a_2, \cdots, a_m$ and $b_1, b_2, \cdots, b_n$, respectively. Then $\xi\eta$ is also a nonnegative simple random variable taking values $a_i b_j$, $i = 1, 2, \cdots, m$, $j = 1, 2, \cdots, n$. It follows from the independence of ξ and η that

$$
\begin{aligned}
E[\xi\eta] &= \sum_{i=1}^{m} \sum_{j=1}^{n} a_i b_j \Pr\{\xi = a_i, \eta = b_j\} \\
&= \sum_{i=1}^{m} \sum_{j=1}^{n} a_i b_j \Pr\{\xi = a_i\} \Pr\{\eta = b_j\} \\
&= \left(\sum_{i=1}^{m} a_i \Pr\{\xi = a_i\} \right) \left(\sum_{j=1}^{n} b_j \Pr\{\eta = b_j\} \right) \\
&= E[\xi]E[\eta].
\end{aligned}
$$

STEP 2: Next we prove the case where ξ and η are nonnegative random variables. For every $i \geq 1$ and every $\omega \in \Omega$, we define

$$
\xi_i(\omega) = \begin{cases} \dfrac{k-1}{2^i}, & \text{if } \dfrac{k-1}{2^i} \leq \xi(\omega) < \dfrac{k}{2^i}, \ k = 1, 2, \cdots, i2^i \\[2ex] i, & \text{if } i \leq \xi(\omega), \end{cases}
$$

$$
\eta_i(\omega) = \begin{cases} \dfrac{k-1}{2^i}, & \text{if } \dfrac{k-1}{2^i} \leq \eta(\omega) < \dfrac{k}{2^i}, \ k = 1, 2, \cdots, i2^i \\[2ex] i, & \text{if } i \leq \eta(\omega). \end{cases}
$$

Then $\{\xi_i\}$, $\{\eta_i\}$ and $\{\xi_i\eta_i\}$ are three sequences of nonnegative simple random variables such that $\xi_i \uparrow \xi$, $\eta_i \uparrow \eta$ and $\xi_i\eta_i \uparrow \xi\eta$ as $i \to \infty$. It follows from the independence of ξ and η that ξ_i and η_i are independent. Hence we have $E[\xi_i\eta_i] = E[\xi_i]E[\eta_i]$ for $i = 1, 2, \cdots$ It follows from the probability continuity theorem that $\Pr\{\xi_i > r\}, i = 1, 2, \cdots$ are simple functions such that

$$\Pr\{\xi_i > r\} \uparrow \Pr\{\xi > r\}, \quad \text{for all } r \geq 0$$

as $i \to \infty$. Since the expected value $E[\xi]$ exists, we have

$$E[\xi_i] = \int_0^{+\infty} \Pr\{\xi_i > r\}\mathrm{d}r \to \int_0^{+\infty} \Pr\{\xi > r\}\mathrm{d}r = E[\xi]$$

as $i \to \infty$. Similarly, we may prove that $E[\eta_i] \to E[\eta]$ and $E[\xi_i\eta_i] \to E[\xi\eta]$ as $i \to \infty$. Therefore $E[\xi\eta] = E[\xi]E[\eta]$.

STEP 3: Finally, if ξ and η are arbitrary independent random variables, then the nonnegative random variables ξ^+ and η^+ are independent and so are ξ^+ and η^-, ξ^- and η^+, ξ^- and η^-. Thus we have

$$E[\xi^+\eta^+] = E[\xi^+]E[\eta^+], \quad E[\xi^+\eta^-] = E[\xi^+]E[\eta^-],$$
$$E[\xi^-\eta^+] = E[\xi^-]E[\eta^+], \quad E[\xi^-\eta^-] = E[\xi^-]E[\eta^-].$$

It follows that

$$
\begin{aligned}
E[\xi\eta] &= E[(\xi^+ - \xi^-)(\eta^+ - \eta^-)] \\
&= E[\xi^+\eta^+] - E[\xi^+\eta^-] - E[\xi^-\eta^+] + E[\xi^-\eta^-] \\
&= E[\xi^+]E[\eta^+] - E[\xi^+]E[\eta^-] - E[\xi^-]E[\eta^+] + E[\xi^-]E[\eta^-] \\
&= (E[\xi^+] - E[\xi^-])(E[\eta^+] - E[\eta^-]) \\
&= E[\xi^+ - \xi^-]E[\eta^+ - \eta^-] \\
&= E[\xi]E[\eta]
\end{aligned}
$$

which proves the theorem.

B.6   Variance

Definition B.12. *Let ξ be a random variable with finite expected value e. Then the variance of ξ is defined by $V[\xi] = E[(\xi - e)^2]$.*

The variance of a random variable provides a measure of the spread of the distribution around its expected value. A small value of variance indicates that the random variable is tightly concentrated around its expected value; and a large value of variance indicates that the random variable has a wide spread around its expected value.

Theorem B.13. *If ξ is a random variable whose variance exists, a and b are real numbers, then $V[a\xi + b] = a^2 V[\xi]$.*

Proof: It follows from the definition of variance that

$$V[a\xi + b] = E\left[(a\xi + b - aE[\xi] - b)^2\right] = a^2 E[(\xi - E[\xi])^2] = a^2 V[\xi].$$

Theorem B.14. *Let ξ be a random variable with expected value e. Then $V[\xi] = 0$ if and only if $\Pr\{\xi = e\} = 1$.*

Proof: If $V[\xi] = 0$, then $E[(\xi - e)^2] = 0$. Thus we have

$$\int_0^{+\infty} \Pr\{(\xi - e)^2 \geq r\} \mathrm{d}r = 0$$

which implies $\Pr\{(\xi - e)^2 \geq r\} = 0$ for any $r > 0$. Hence we have $\Pr\{(\xi - e)^2 = 0\} = 1$, i.e., $\Pr\{\xi = e\} = 1$. Conversely, if $\Pr\{\xi = e\} = 1$, then we have $\Pr\{(\xi - e)^2 = 0\} = 1$ and $\Pr\{(\xi - e)^2 \geq r\} = 0$ for any $r > 0$. Thus

$$V[\xi] = \int_0^{+\infty} \Pr\{(\xi - e)^2 \geq r\} \mathrm{d}r = 0.$$

Theorem B.15. *If $\xi_1, \xi_2, \cdots, \xi_n$ are independent random variables with finite expected values, then*

$$V[\xi_1 + \xi_2 + \cdots + \xi_n] = V[\xi_1] + V[\xi_2] + \cdots + V[\xi_n]. \tag{B.26}$$

Proof: It follows from the definition of variance that

$$V\left[\sum_{i=1}^n \xi_i\right] = E\left[(\xi_1 + \xi_2 + \cdots + \xi_n - E[\xi_1] - E[\xi_2] - \cdots - E[\xi_n])^2\right]$$

$$= \sum_{i=1}^n E\left[(\xi_i - E[\xi_i])^2\right] + 2\sum_{i=1}^{n-1} \sum_{j=i+1}^n E\left[(\xi_i - E[\xi_i])(\xi_j - E[\xi_j])\right].$$

Since $\xi_1, \xi_2, \cdots, \xi_n$ are independent, $E\left[(\xi_i - E[\xi_i])(\xi_j - E[\xi_j])\right] = 0$ for all i, j with $i \neq j$. Thus (B.26) holds.

B.7 Moments

Definition B.13. *Let ξ be a random variable, and k a positive number. Then*
(a) the expected value $E[\xi^k]$ is called the kth moment;
(b) the expected value $E[|\xi|^k]$ is called the kth absolute moment;
(c) the expected value $E[(\xi - E[\xi])^k]$ is called the kth central moment;
(d) the expected value $E[|\xi - E[\xi]|^k]$ is called the kth absolute central moment.

Note that the first central moment is always 0, the first moment is just the expected value, and the second central moment is just the variance.

Theorem B.16. *Let ξ be a nonnegative random variable, and k a positive number. Then the k-th moment*

$$E[\xi^k] = k \int_0^{+\infty} r^{k-1} \Pr\{\xi \geq r\} \mathrm{d}r. \tag{B.27}$$

Proof: It follows from the nonnegativity of ξ that

$$E[\xi^k] = \int_0^\infty \Pr\{\xi^k \geq x\}\mathrm{d}x = \int_0^\infty \Pr\{\xi \geq r\}\mathrm{d}r^k = k\int_0^\infty r^{k-1}\Pr\{\xi \geq r\}\mathrm{d}r.$$

The theorem is proved.

B.8   Critical Values

Let ξ be a random variable. In order to measure it, we may use its expected value. Alternately, we may employ α-optimistic value and α-pessimistic value as a ranking measure.

Definition B.14. *Let ξ be a random variable, and $\alpha \in (0,1]$. Then*

$$\xi_{\mathrm{sup}}(\alpha) = \sup\left\{r \mid \Pr\left\{\xi \geq r\right\} \geq \alpha\right\} \tag{B.28}$$

is called the α-optimistic value of ξ, and

$$\xi_{\mathrm{inf}}(\alpha) = \inf\left\{r \mid \Pr\left\{\xi \leq r\right\} \geq \alpha\right\} \tag{B.29}$$

is called the α-pessimistic value of ξ.

This means that the random variable ξ will reach upwards of the α-optimistic value $\xi_{\mathrm{sup}}(\alpha)$ at least α of time, and will be below the α-pessimistic value $\xi_{\mathrm{inf}}(\alpha)$ at least α of time. The optimistic value is also called percentile.

Theorem B.17. *Let ξ be a random variable, and $\alpha \in (0,1]$. Then we have*

$$\Pr\{\xi \geq \xi_{\mathrm{sup}}(\alpha)\} \geq \alpha, \quad \Pr\{\xi \leq \xi_{\mathrm{inf}}(\alpha)\} \geq \alpha \tag{B.30}$$

where $\xi_{\mathrm{sup}}(\alpha)$ and $\xi_{\mathrm{inf}}(\alpha)$ are the α-optimistic and α-pessimistic values of the random variable ξ, respectively.

Proof: It follows from the definition of the optimistic value that there exists an increasing sequence $\{r_i\}$ such that $\Pr\{\xi \geq r_i\} \geq \alpha$ and $r_i \uparrow \xi_{\mathrm{sup}}(\alpha)$ as $i \to \infty$. Since $\{\xi \geq r_i\} \downarrow \{\xi \geq \xi_{\mathrm{sup}}(\alpha)\}$, it follows from the probability continuity theorem that

$$\Pr\{\xi \geq \xi_{\mathrm{sup}}(\alpha)\} = \lim_{i\to\infty} \Pr\{\xi \geq r_i\} \geq \alpha.$$

The inequality $\Pr\{\xi \leq \xi_{\mathrm{inf}}(\alpha)\} \geq \alpha$ may be proved similarly.

Theorem B.18. *Let ξ be a random variable, and $\alpha \in (0,1]$. Then we have*
(a) $\xi_{\mathrm{inf}}(\alpha)$ is an increasing and left-continuous function of α;
(b) $\xi_{\mathrm{sup}}(\alpha)$ is a decreasing and left-continuous function of α.

Proof: (a) Let α_1 and α_2 be two numbers with $0 < \alpha_1 < \alpha_2 \leq 1$. Then for any number $r < \xi_{\sup}(\alpha_2)$, we have $\Pr\{\xi \geq r\} \geq \alpha_2 > \alpha_1$. Thus, by the definition of optimistic value, we obtain $\xi_{\sup}(\alpha_1) \geq \xi_{\sup}(\alpha_2)$. That is, the value $\xi_{\sup}(\alpha)$ is a decreasing function of α. Next, we prove the left-continuity of $\xi_{\inf}(\alpha)$ with respect to α. Let $\{\alpha_i\}$ be an arbitrary sequence of positive numbers such that $\alpha_i \uparrow \alpha$. Then $\{\xi_{\inf}(\alpha_i)\}$ is an increasing sequence. If the limitation is equal to $\xi_{\inf}(\alpha)$, then the left-continuity is proved. Otherwise, there exists a number z^* such that

$$\lim_{i\to\infty} \xi_{\inf}(\alpha_i) < z^* < \xi_{\inf}(\alpha).$$

Thus $\Pr\{\xi \leq z^*\} \geq \alpha_i$ for each i. Letting $i \to \infty$, we get $\Pr\{\xi \leq z^*\} \geq \alpha$. Hence $z^* \geq \xi_{\inf}(\alpha)$. A contradiction proves the left-continuity of $\xi_{\inf}(\alpha)$ with respect to α. The part (b) may be proved similarly.

B.9   Entropy

Given a random variable, what is the degree of difficulty of predicting the specified value that the random variable will take? In order to answer this question, Shannon [191] defined a concept of entropy as a measure of uncertainty.

Entropy of Discrete Random Variables

Definition B.15. *Let ξ be a discrete random variable taking values x_i with probabilities p_i, $i = 1, 2, \cdots$, respectively. Then its entropy is defined by*

$$H[\xi] = -\sum_{i=1}^{\infty} p_i \ln p_i. \tag{B.31}$$

It should be noticed that the entropy depends only on the number of values and their probabilities and does not depend on the actual values that the random variable takes.

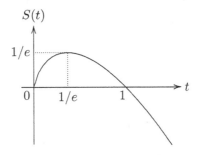

Figure B.1: Function $S(t) = -t \ln t$ is concave

Theorem B.19. *Let ξ be a discrete random variable taking values x_i with probabilities p_i, $i = 1, 2, \cdots$, respectively. Then*

$$H[\xi] \geq 0 \tag{B.32}$$

and equality holds if and only if there exists an index k such that $p_k = 1$, i.e., ξ is essentially a deterministic number.

Proof: The nonnegativity is clear. In addition, $H[\xi] = 0$ if and only if $p_i = 0$ or 1 for each i. That is, there exists one and only one index k such that $p_k = 1$. The theorem is proved.

This theorem states that the entropy of a discrete random variable reaches its minimum 0 when the random variable degenerates to a deterministic number. In this case, there is no uncertainty.

Theorem B.20. *Let ξ be a simple random variable taking values x_i with probabilities p_i, $i = 1, 2, \cdots, n$, respectively. Then*

$$H[\xi] \leq \ln n \tag{B.33}$$

and equality holds if and only if $p_i \equiv 1/n$ for all $i = 1, 2, \cdots, n$.

Proof: Since the function $S(t)$ is a concave function of t and $p_1 + p_2 + \cdots + p_n = 1$, we have

$$-\sum_{i=1}^{n} p_i \ln p_i \leq -n \left(\frac{1}{n} \sum_{i=1}^{n} p_i \right) \ln \left(\frac{1}{n} \sum_{i=1}^{n} p_i \right) = \ln n$$

which implies that $H[\xi] \leq \ln n$ and equality holds if and only if $p_1 = p_2 = \cdots = p_n$, i.e., $p_i \equiv 1/n$ for all $i = 1, 2, \cdots, n$.

This theorem states that the entropy of a simple random variable reaches its maximum $\ln n$ when all outcomes are equiprobable. In this case, there is no preference among all the values that the random variable will take.

Entropy of Absolutely Continuous Random Variables

Definition B.16. *Let ξ be a random variable with probability density function ϕ. Then its entropy is defined by*

$$H[\xi] = -\int_{-\infty}^{+\infty} \phi(x) \ln \phi(x) \mathrm{d}x. \tag{B.34}$$

Example B.12: Let ξ be a uniformly distributed random variable on $[a, b]$. Then its entropy is $H[\xi] = \ln(b - a)$. This example shows that the entropy of absolutely continuous random variable may assume both positive and negative values since $\ln(b - a) < 0$ if $b - a < 1$; and $\ln(b - a) > 0$ if $b - a > 1$.

Example B.13: Let ξ be an exponentially distributed random variable with expected value β. Then its entropy is $H[\xi] = 1 + \ln \beta$.

Example B.14: Let ξ be a normally distributed random variable with expected value e and variance σ^2. Then its entropy is $H[\xi] = 1/2 + \ln \sqrt{2\pi}\sigma$.

Maximum Entropy Principle

Given some constraints, for example, expected value and variance, there are usually multiple compatible probability distributions. For this case, we would like to select the distribution that maximizes the value of entropy and satisfies the prescribed constraints. This method is often referred to as the *maximum entropy principle* (Jaynes [61]).

Example B.15: Let ξ be an absolutely continuous random variable on $[a, b]$. The maximum entropy principle attempts to find the probability density function $\phi(x)$ that maximizes the entropy

$$- \int_a^b \phi(x) \ln \phi(x) \mathrm{d}x$$

subject to the natural constraint $\int_a^b \phi(x)\mathrm{d}x = 1$. The Lagrangian is

$$L = - \int_a^b \phi(x) \ln \phi(x) \mathrm{d}x - \lambda \left(\int_a^b \phi(x)\mathrm{d}x - 1 \right).$$

It follows from the Euler-Lagrange equation that the maximum entropy probability density function meets

$$\ln \phi(x) + 1 + \lambda = 0$$

and has the form $\phi(x) = \exp(-1 - \lambda)$. Substituting it into the natural constraint, we get

$$\phi^*(x) = \frac{1}{b - a}, \quad a \le x \le b$$

which is just the uniformly distributed random variable, and the maximum entropy is $H[\xi^*] = \ln(b - a)$.

Example B.16: Let ξ be an absolutely continuous random variable on $[0, \infty)$. Assume that the expected value of ξ is prescribed to be β. The maximum entropy probability density function $\phi(x)$ should maximize the entropy

$$- \int_0^{+\infty} \phi(x) \ln \phi(x) \mathrm{d}x$$

subject to the constraints

$$\int_0^{+\infty} \phi(x)\mathrm{d}x = 1, \quad \int_0^{+\infty} x\phi(x)\mathrm{d}x = \beta.$$

The Lagrangian is

$$L = -\int_0^\infty \phi(x) \ln \phi(x) dx - \lambda_1 \left(\int_0^\infty \phi(x) dx - 1 \right) - \lambda_2 \left(\int_0^\infty x\phi(x) dx - \beta \right).$$

The maximum entropy probability density function meets Euler-Lagrange equation

$$\ln \phi(x) + 1 + \lambda_1 + \lambda_2 x = 0$$

and has the form $\phi(x) = \exp(-1 - \lambda_1 - \lambda_2 x)$. Substituting it into the constraints, we get

$$\phi^*(x) = \frac{1}{\beta} \exp\left(-\frac{x}{\beta}\right), \quad x \geq 0$$

which is just the exponentially distributed random variable, and the maximum entropy is $H[\xi^*] = 1 + \ln \beta$.

Example B.17: Let ξ be an absolutely continuous random variable on $(-\infty, +\infty)$. Assume that the expected value and variance of ξ are prescribed to be μ and σ^2, respectively. The maximum entropy probability density function $\phi(x)$ should maximize the entropy

$$-\int_{-\infty}^{+\infty} \phi(x) \ln \phi(x) dx$$

subject to the constraints

$$\int_{-\infty}^{+\infty} \phi(x) dx = 1, \quad \int_{-\infty}^{+\infty} x\phi(x) dx = \mu, \quad \int_{-\infty}^{+\infty} (x - \mu)^2 \phi(x) dx = \sigma^2.$$

The Lagrangian is

$$L = -\int_{-\infty}^{+\infty} \phi(x) \ln \phi(x) dx - \lambda_1 \left(\int_{-\infty}^{+\infty} \phi(x) dx - 1 \right)$$

$$- \lambda_2 \left(\int_{-\infty}^{+\infty} x\phi(x) dx - \mu \right) - \lambda_3 \left(\int_{-\infty}^{+\infty} (x - \mu)^2 \phi(x) dx - \sigma^2 \right).$$

The maximum entropy probability density function meets Euler-Lagrange equation

$$\ln \phi(x) + 1 + \lambda_1 + \lambda_2 x + \lambda_3 (x - \mu)^2 = 0$$

and has the form $\phi(x) = \exp(-1 - \lambda_1 - \lambda_2 x - \lambda_3 (x - \mu)^2)$. Substituting it into the constraints, we get

$$\phi^*(x) = \frac{1}{\sigma\sqrt{2\pi}} \exp\left(-\frac{(x - \mu)^2}{2\sigma^2}\right), \quad x \in \Re$$

which is just the normally distributed random variable, and the maximum entropy is $H[\xi^*] = 1/2 + \ln \sqrt{2\pi}\sigma$.

B.10   Conditional Probability

We consider the probability of an event A after it has been learned that some other event B has occurred. This new probability of A is called the conditional probability of A given B.

Definition B.17. *Let $(\Omega, \mathcal{A}, \mathrm{Pr})$ be a probability space, and $A, B \in \mathcal{A}$. Then the conditional probability of A given B is defined by*

$$\mathrm{Pr}\{A|B\} = \frac{\mathrm{Pr}\{A \cap B\}}{\mathrm{Pr}\{B\}} \tag{B.35}$$

provided that $\mathrm{Pr}\{B\} > 0$.

Theorem B.21. *Let $(\Omega, \mathcal{A}, \mathrm{Pr})$ be a probability space, and B an event with $\mathrm{Pr}\{B\} > 0$. Then $\mathrm{Pr}\{\cdot|B\}$ defined by (B.35) is a probability measure, and $(\Omega, \mathcal{A}, \mathrm{Pr}\{\cdot|B\})$ is a probability space.*

Proof: It is sufficient to prove that $\mathrm{Pr}\{\cdot|B\}$ satisfies the normality, nonnegativity and countable additivity axioms. At first, we have

$$\mathrm{Pr}\{\Omega|B\} = \frac{\mathrm{Pr}\{\Omega \cap B\}}{\mathrm{Pr}\{B\}} = \frac{\mathrm{Pr}\{B\}}{\mathrm{Pr}\{B\}} = 1.$$

Secondly, for any $A \in \mathcal{A}$, the set function $\mathrm{Pr}\{A|B\}$ is nonnegative. Finally, for any countable sequence $\{A_i\}$ of mutually disjoint events, we have

$$\mathrm{Pr}\left\{\bigcup_{i=1}^{\infty} A_i|B\right\} = \frac{\mathrm{Pr}\left\{\left(\bigcup_{i=1}^{\infty} A_i\right) \cap B\right\}}{\mathrm{Pr}\{B\}} = \frac{\sum_{i=1}^{\infty}\mathrm{Pr}\{A_i \cap B\}}{\mathrm{Pr}\{B\}} = \sum_{i=1}^{\infty}\mathrm{Pr}\{A_i|B\}.$$

Thus $\mathrm{Pr}\{\cdot|B\}$ is a probability measure. Furthermore, $(\Omega, \mathcal{A}, \mathrm{Pr}\{\cdot|B\})$ is a probability space.

Theorem B.22 *(Bayes Formula).* *Let the events $A_1, A_2, \cdots, A_n$ form a partition of the space Ω such that $\mathrm{Pr}\{A_i\} > 0$ for $i = 1, 2, \cdots, n$, and let B be an event with $\mathrm{Pr}\{B\} > 0$. Then we have*

$$\mathrm{Pr}\{A_k|B\} = \frac{\mathrm{Pr}\{A_k\}\,\mathrm{Pr}\{B|A_k\}}{\sum_{i=1}^{n}\mathrm{Pr}\{A_i\}\,\mathrm{Pr}\{B|A_i\}} \tag{B.36}$$

for $k = 1, 2, \cdots, n$.

Proof: Since $A_1, A_2, \cdots, A_n$ form a partition of the space Ω, we have

$$\mathrm{Pr}\{B\} = \sum_{i=1}^{n}\mathrm{Pr}\{A_i \cap B\} = \sum_{i=1}^{n}\mathrm{Pr}\{A_i\}\,\mathrm{Pr}\{B|A_i\}$$

which is also called the *formula for total probability*. Thus, for any k, we have

$$\Pr\{A_k|B\} = \frac{\Pr\{A_k \cap B\}}{\Pr\{B\}} = \frac{\Pr\{A_k\}\Pr\{B|A_k\}}{\sum_{i=1}^{n}\Pr\{A_i\}\Pr\{B|A_i\}}.$$

The theorem is proved.

Remark B.1: Especially, let A and B be two events with $\Pr\{A\} > 0$ and $\Pr\{B\} > 0$. Then A and A^c form a partition of the space Ω, and the *Bayes formula* is

$$\Pr\{A|B\} = \frac{\Pr\{A\}\Pr\{B|A\}}{\Pr\{B\}}. \tag{B.37}$$

Remark B.2: In statistical applications, the events $A_1, A_2, \cdots, A_n$ are often called *hypotheses*. Furthermore, for each i, the $\Pr\{A_i\}$ is called the *prior* probability of A_i, and $\Pr\{A_i|B\}$ is called the *posterior* probability of A_i after the occurrence of event B.

Example B.18: Let ξ be an exponentially distributed random variable with expected value β. Then for any real numbers $a > 0$ and $x > 0$, the conditional probability of $\xi \geq a + x$ given $\xi \geq a$ is

$$\Pr\{\xi \geq a + x | \xi \geq a\} = \exp(-x/\beta) = \Pr\{\xi \geq x\}$$

which means that the conditional probability is identical to the original probability. This is the so-called memoryless property of exponential distribution. In other words, it is as good as new if it is functioning on inspection.

Definition B.18. *The conditional probability distribution* $\Phi\colon \Re \to [0,1]$ *of a random variable* ξ *given* B *is defined by*

$$\Phi(x|B) = \Pr\{\xi \leq x|B\} \tag{B.38}$$

provided that $\Pr\{B\} > 0$.

Definition B.19. *The conditional probability density function* ϕ *of a random variable* ξ *given* B *is a nonnegative function such that*

$$\Phi(x|B) = \int_{-\infty}^{x} \phi(y|B)\mathrm{d}y, \quad \forall x \in \Re \tag{B.39}$$

where $\Phi(x|B)$ *is the conditional probability distribution of* ξ *given* B.

Example B.19: Let (ξ, η) be a random vector with joint probability density function ψ. Then the marginal probability density functions of ξ and η are

$$f(x) = \int_{-\infty}^{+\infty} \psi(x,y)\mathrm{d}y, \quad g(y) = \int_{-\infty}^{+\infty} \psi(x,y)\mathrm{d}x,$$

respectively. Furthermore, we have

$$\Pr\{\xi \le x, \eta \le y\} = \int_{-\infty}^{x} \int_{-\infty}^{y} \psi(r,t) \mathrm{d}r \mathrm{d}t = \int_{-\infty}^{y} \left[\int_{-\infty}^{x} \frac{\psi(r,t)}{g(t)} \mathrm{d}r \right] g(t) \mathrm{d}t$$

which implies that the conditional probability distribution of ξ given $\eta = y$ is

$$\Phi(x|\eta = y) = \int_{-\infty}^{x} \frac{\psi(r,y)}{g(y)} \mathrm{d}r, \quad \text{a.s.} \tag{B.40}$$

and the conditional probability density function of ξ given $\eta = y$ is

$$\phi(x|\eta = y) = \frac{\psi(x,y)}{g(y)} = \frac{\psi(x,y)}{\displaystyle\int_{-\infty}^{+\infty} \psi(x,y) \mathrm{d}x}, \quad \text{a.s.} \tag{B.41}$$

Note that (B.40) and (B.41) are defined only for $g(y) \ne 0$. In fact, the set $\{y|g(y) = 0\}$ has probability 0. Especially, if ξ and η are independent random variables, then $\psi(x,y) = f(x)g(y)$ and $\phi(x|\eta = y) = f(x)$.

Definition B.20. *Let ξ be a random variable. Then the conditional expected value of ξ given B is defined by*

$$E[\xi|B] = \int_{0}^{+\infty} \Pr\{\xi \ge r|B\} \mathrm{d}r - \int_{-\infty}^{0} \Pr\{\xi \le r|B\} \mathrm{d}r \tag{B.42}$$

provided that at least one of the two integrals is finite.

B.11   Random Set

Random set is a well-known concept in probability theory, and widely applied in science and engineering. Here we deal with random set after the fashion of uncertain set.

Definition B.21. *A random set is a measurable function ξ from a probability space $(\Omega, \mathcal{A}, \Pr)$ to a collection of sets of real numbers, i.e., for any Borel set B, the set*

$$\{\xi \subset B\} = \{\omega \in \Omega \mid \xi(\omega) \subset B\} \tag{B.43}$$

is an event.

Let ξ and η be two nonempty random sets. Then the strong membership degree of η to ξ is defined as the probability measure that η is strongly included in ξ, i.e., $\Pr\{\eta \subset \xi\}$. The weak membership degree of η to ξ is defined as the probability measure that η is weakly included in ξ, i.e., $\Pr\{\eta \not\subset \xi^c\}$.

Definition B.22 . *Let ξ and η be two nonempty random sets. Then the membership degree of η to ξ is defined as the average of strong and weak membership degrees, i.e.,*

$$\Pr\{\eta \rhd \xi\} = \frac{1}{2} \left(\Pr\{\eta \subset \xi\} + \Pr\{\eta \not\subset \xi^c\} \right). \tag{B.44}$$

The membership degree is understood as the probability measure that η is imaginarily included in ξ.

Note that if η degenerates to a single point a, then the strong inclusion is identical with the weak inclusion, and $\Pr\{a \triangleright \xi\} = \Pr\{a \in \xi\} = \Pr\{a \notin \xi^c\}$.

Definition B.23. *Let ξ be a nonempty random set. Then the function*

$$\Phi(x) = \Pr\{\xi \triangleright (-\infty, x]\}, \quad \forall x \in \Re \tag{B.45}$$

is called the probability distribution of ξ.

The concept of membership function is also applicable to random set except that the membership degree takes values in probability measure.

Definition B.24. *A random set ξ is said to have a membership function μ if the range of ξ is just the total class of μ, and*

$$\Pr\{\xi \in W_\alpha\} = \alpha, \quad \forall \alpha \in [0, 1] \tag{B.46}$$

where W_α is the α-class of μ.

A representation theorem states that, if ξ is a random set with membership function μ, then ξ may be represented by

$$\xi = \bigcup_{0 \le \alpha \le 1} \alpha \cdot \mu_\alpha \tag{B.47}$$

where μ_α is the α-cut of membership function μ.

Warning: The complement ξ^c, union $\xi \cup \eta$, intersection $\xi \cap \eta$, sum $\xi + \eta$ and product $\xi \times \eta$ of random sets have no membership functions even though the original random sets have their own membership functions.

Definition B.25. *Let ξ be a nonempty random set. Then the expected value of ξ is defined by*

$$E[\xi] = \int_0^{+\infty} \Pr\{\xi \triangleright [r, +\infty)\}\mathrm{d}r - \int_{-\infty}^0 \Pr\{\xi \triangleright (-\infty, r]\}\mathrm{d}r \tag{B.48}$$

provided that at least one of the two integrals is finite.

Let ξ be a nonempty random set with probability distribution Φ. If ξ has a finite expected value, then

$$E[\xi] = \int_0^{+\infty} (1 - \Phi(x))\mathrm{d}x - \int_{-\infty}^0 \Phi(x)\mathrm{d}x. \tag{B.49}$$

Definition B.26. *Let ξ be a random set, and $\alpha \in (0, 1]$. Then*

$$\xi_{\text{sup}}(\alpha) = \sup\left\{r \mid \Pr\{\xi \triangleright [r, +\infty)\} \geq \alpha\right\} \tag{B.50}$$

is called the α-optimistic value to ξ, and

$$\xi_{\text{inf}}(\alpha) = \inf\left\{r \mid \Pr\{\xi \triangleright (-\infty, r]\} \geq \alpha\right\} \tag{B.51}$$

is called the α-pessimistic value to ξ.

Let ξ be a random set with probability distribution Φ. Then its α-optimistic value and α-pessimistic value are

$$\xi_{\text{sup}}(\alpha) = \Phi^{-1}(1 - \alpha), \quad \xi_{\text{inf}}(\alpha) = \Phi^{-1}(\alpha) \tag{B.52}$$

for any α with $0 < \alpha < 1$.

Appendix C

Credibility Theory

The concept of fuzzy set was initiated by Zadeh [222] via membership function in 1965. In order to measure a fuzzy event, Zadeh [225] proposed the concept of possibility measure. Although possibility measure has been widely used, it does not obey the law of truth conservation and is inconsistent with the law of excluded middle and the law of contradiction. The main reason is that possibility measure has no self-duality property. However, a self-dual measure is absolutely needed in both theory and practice. In order to define a self-dual measure, Liu and Liu [114] presented the concept of credibility measure. In addition, a sufficient and necessary condition for credibility measure was given by Li and Liu [91]. Credibility theory, founded by Liu [117] in 2004 and refined by Liu [120] in 2007, is a branch of mathematics for studying the behavior of fuzzy phenomena.

The emphasis in this appendix is mainly on credibility measure, credibility space, fuzzy variable, membership function, credibility distribution, independence, expected value, variance, moments, critical values, entropy and conditional credibility.

C.1 Credibility Space

Let Θ be a nonempty set, and $\mathcal{P}$ the power set of Θ (i.e., the larggest σ-algebra over Θ). Each element in $\mathcal{P}$ is called an *event*. In order to present an axiomatic definition of credibility, it is necessary to assign to each event A a number $\mathrm{Cr}\{A\}$ which indicates the credibility that A will occur. In order to ensure that the number $\mathrm{Cr}\{A\}$ has certain mathematical properties which we intuitively expect a credibility to have, we accept the following four axioms:

Axiom 1. *(Normality)* $\mathrm{Cr}\{\Theta\} = 1$.

Axiom 2. *(Monotonicity)* $\mathrm{Cr}\{A\} \leq \mathrm{Cr}\{B\}$ *whenever* $A \subset B$.

Axiom 3. *(Self-Duality)* $\mathrm{Cr}\{A\} + \mathrm{Cr}\{A^c\} = 1$ *for any event* A.

Axiom 4. *(Maximality)* $\mathrm{Cr}\{\cup_i A_i\} = \sup_i \mathrm{Cr}\{A_i\}$ *for any events* $\{A_i\}$ *with* $\sup_i \mathrm{Cr}\{A_i\} < 0.5$.

Definition C.1 *(Liu and Liu [114]). The set function Cr is called a credibility measure if it satisfies the normality, monotonicity, self-duality, and maximality axioms.*

Example C.1: Let $\Theta = \{\theta_1, \theta_2\}$. For this case, there are only four events: $\emptyset, \{\theta_1\}, \{\theta_2\}, \Theta$. Define $\text{Cr}\{\emptyset\} = 0$, $\text{Cr}\{\theta_1\} = 0.7$, $\text{Cr}\{\theta_2\} = 0.3$, and $\text{Cr}\{\Theta\} = 1$. Then the set function Cr is a credibility measure because it satisfies the four axioms.

Example C.2: Let Θ be a nonempty set. Define $\text{Cr}\{\emptyset\} = 0$, $\text{Cr}\{\Theta\} = 1$ and $\text{Cr}\{A\} = 1/2$ for any subset A (excluding $\emptyset$ and Θ). Then the set function Cr is a credibility measure.

Example C.3: Let μ be a nonnegative function on Θ (for example, the set of real numbers) such that

$$\sup_{x \in \Theta} \mu(x) = 1. \tag{C.1}$$

Then the set function

$$\text{Cr}\{A\} = \frac{1}{2}\left(\sup_{x \in A} \mu(x) + 1 - \sup_{x \in A^c} \mu(x)\right) \tag{C.2}$$

is a credibility measure on Θ.

Theorem C.1. *Let Θ be a nonempty set, $\mathcal{P}$ the power set of Θ, and Cr the credibility measure. Then $\text{Cr}\{\emptyset\} = 0$ and $0 \leq \text{Cr}\{A\} \leq 1$ for any $A \in \mathcal{P}$.*

Proof: It follows from Axioms 1 and 3 that $\text{Cr}\{\emptyset\} = 1 - \text{Cr}\{\Theta\} = 1 - 1 = 0$. Since $\emptyset \subset A \subset \Theta$, we have $0 \leq \text{Cr}\{A\} \leq 1$ by using Axiom 2.

Theorem C.2. *Let Θ be a nonempty set, $\mathcal{P}$ the power set of Θ, and Cr the credibility measure. Then for any $A, B \in \mathcal{P}$, we have*

$$\text{Cr}\{A \cup B\} = \text{Cr}\{A\} \vee \text{Cr}\{B\} \quad \text{if } \text{Cr}\{A \cup B\} \leq 0.5, \tag{C.3}$$

$$\text{Cr}\{A \cap B\} = \text{Cr}\{A\} \wedge \text{Cr}\{B\} \quad \text{if } \text{Cr}\{A \cap B\} \geq 0.5. \tag{C.4}$$

The above equations hold for not only finite number of events but also infinite number of events.

Proof: If $\text{Cr}\{A \cup B\} < 0.5$, then $\text{Cr}\{A\} \vee \text{Cr}\{B\} < 0.5$ by using Axiom 2. Thus the equation (C.3) follows immediately from Axiom 4. If $\text{Cr}\{A \cup B\} = 0.5$ and (C.3) does not hold, then we have $\text{Cr}\{A\} \vee \text{Cr}\{B\} < 0.5$. It follows from Axiom 4 that

$$\text{Cr}\{A \cup B\} = \text{Cr}\{A\} \vee \text{Cr}\{B\} < 0.5.$$

A contradiction proves (C.3). Next we prove (C.4). Since $\text{Cr}\{A \cap B\} \geq 0.5$, we have $\text{Cr}\{A^c \cup B^c\} \leq 0.5$ by the self-duality. Thus

$$\text{Cr}\{A \cap B\} = 1 - \text{Cr}\{A^c \cup B^c\} = 1 - \text{Cr}\{A^c\} \vee \text{Cr}\{B^c\}$$

$$= (1 - \text{Cr}\{A^c\}) \wedge (1 - \text{Cr}\{B^c\}) = \text{Cr}\{A\} \wedge \text{Cr}\{B\}.$$

The theorem is proved.

Theorem C.3. *Let* Θ *be a nonempty set,* $\mathcal{P}$ *the power set of* Θ, *and* Cr *the credibility measure. Then for any* $A, B \in \mathcal{P}$, *we have*

$$\text{Cr}\{A \cup B\} = \text{Cr}\{A\} \vee \text{Cr}\{B\} \quad \textit{if } \text{Cr}\{A\} + \text{Cr}\{B\} < 1, \qquad (C.5)$$

$$\text{Cr}\{A \cap B\} = \text{Cr}\{A\} \wedge \text{Cr}\{B\} \quad \textit{if } \text{Cr}\{A\} + \text{Cr}\{B\} > 1. \qquad (C.6)$$

Proof: Suppose $\text{Cr}\{A\} + \text{Cr}\{B\} < 1$. Then there exists at least one term less than 0.5, say $\text{Cr}\{B\} < 0.5$. If $\text{Cr}\{A\} < 0.5$ also holds, then the equation (C.3) follows immediately from Axiom 4. If $\text{Cr}\{A\} \geq 0.5$, then by using Theorem C.2, we obtain

$$\text{Cr}\{A\} = \text{Cr}\{A \cup (B \cap B^c)\} = \text{Cr}\{(A \cup B) \cap (A \cup B^c)\} = \text{Cr}\{A \cup B\} \wedge \text{Cr}\{A \cup B^c\}.$$

On the other hand, we have

$$\text{Cr}\{A\} < 1 - \text{Cr}\{B\} = \text{Cr}\{B^c\} \leq \text{Cr}\{A \cup B^c\}.$$

Hence we must have $\text{Cr}\{A \cup B\} = \text{Cr}\{A\} = \text{Cr}\{A\} \vee \text{Cr}\{B\}$. The equation (C.5) is proved. Next we suppose $\text{Cr}\{A\} + \text{Cr}\{B\} > 1$. Then $\text{Cr}\{A^c\} + \text{Cr}\{B^c\} < 1$. It follows from (C.5) that

$$\text{Cr}\{A \cap B\} = 1 - \text{Cr}\{A^c \cup B^c\} = 1 - \text{Cr}\{A^c\} \vee \text{Cr}\{B^c\}$$
$$= (1 - \text{Cr}\{A^c\}) \wedge (1 - \text{Cr}\{B^c\}) = \text{Cr}\{A\} \wedge \text{Cr}\{B\}.$$

The theorem is proved.

Credibility Subadditivity Theorem

Theorem C.4 *(Liu [117], Credibility Subadditivity Theorem). The credibility measure is subadditive. That is,*

$$\text{Cr}\{A \cup B\} \leq \text{Cr}\{A\} + \text{Cr}\{B\} \qquad (C.7)$$

for any events A and B. In fact, credibility measure is not only finitely subadditive but also countably subadditive.

Proof: The argument breaks down into three cases. Case 1: $\text{Cr}\{A\} < 0.5$ and $\text{Cr}\{B\} < 0.5$. It follows from Axiom 4 that

$$\text{Cr}\{A \cup B\} = \text{Cr}\{A\} \vee \text{Cr}\{B\} \leq \text{Cr}\{A\} + \text{Cr}\{B\}.$$

Case 2: $\text{Cr}\{A\} \geq 0.5$. For this case, by using Axioms 2 and 3, we have $\text{Cr}\{A^c\} \leq 0.5$ and $\text{Cr}\{A \cup B\} \geq \text{Cr}\{A\} \geq 0.5$. Then

$$\text{Cr}\{A^c\} = \text{Cr}\{A^c \cap B\} \vee \text{Cr}\{A^c \cap B^c\}$$
$$\leq \text{Cr}\{A^c \cap B\} + \text{Cr}\{A^c \cap B^c\}$$
$$\leq \text{Cr}\{B\} + \text{Cr}\{A^c \cap B^c\}.$$

Applying this inequality, we obtain

$$\begin{aligned}
\mathrm{Cr}\{A\} + \mathrm{Cr}\{B\} &= 1 - \mathrm{Cr}\{A^c\} + \mathrm{Cr}\{B\} \\
&\geq 1 - \mathrm{Cr}\{B\} - \mathrm{Cr}\{A^c \cap B^c\} + \mathrm{Cr}\{B\} \\
&= 1 - \mathrm{Cr}\{A^c \cap B^c\} \\
&= \mathrm{Cr}\{A \cup B\}.
\end{aligned}$$

Case 3: $\mathrm{Cr}\{B\} \geq 0.5$. This case may be proved by a similar process of Case 2. The theorem is proved.

Remark C.1: For any events A and B, it follows from the credibility sub-additivity theorem that the credibility measure is null-additive, i.e., $\mathrm{Cr}\{A \cup B\} = \mathrm{Cr}\{A\} + \mathrm{Cr}\{B\}$ if either $\mathrm{Cr}\{A\} = 0$ or $\mathrm{Cr}\{B\} = 0$.

Theorem C.5. *Let $\{B_i\}$ be a decreasing sequence of events with $\mathrm{Cr}\{B_i\} \to 0$ as $i \to \infty$. Then for any event A, we have*

$$\lim_{i \to \infty} \mathrm{Cr}\{A \cup B_i\} = \lim_{i \to \infty} \mathrm{Cr}\{A \backslash B_i\} = \mathrm{Cr}\{A\}. \tag{C.8}$$

Proof: It follows from the monotonicity axiom and credibility subadditivity theorem that

$$\mathrm{Cr}\{A\} \leq \mathrm{Cr}\{A \cup B_i\} \leq \mathrm{Cr}\{A\} + \mathrm{Cr}\{B_i\}$$

for each i. Thus we get $\mathrm{Cr}\{A \cup B_i\} \to \mathrm{Cr}\{A\}$ by using $\mathrm{Cr}\{B_i\} \to 0$. Since $(A \backslash B_i) \subset A \subset ((A \backslash B_i) \cup B_i)$, we have

$$\mathrm{Cr}\{A \backslash B_i\} \leq \mathrm{Cr}\{A\} \leq \mathrm{Cr}\{A \backslash B_i\} + \mathrm{Cr}\{B_i\}.$$

Hence $\mathrm{Cr}\{A \backslash B_i\} \to \mathrm{Cr}\{A\}$ by using $\mathrm{Cr}\{B_i\} \to 0$.

Credibility Semicontinuity Law

Generally speaking, the credibility measure is neither lower semicontinuous nor upper semicontinuous. However, we have the following credibility semi-continuity law.

Theorem C.6 *(Liu [117], Credibility Semicontinuity Law). For any events $A_1, A_2, \cdots$, we have*

$$\lim_{i \to \infty} \mathrm{Cr}\{A_i\} = \mathrm{Cr}\left\{\lim_{i \to \infty} A_i\right\} \tag{C.9}$$

if one of the following conditions is satisfied:
(a) $\mathrm{Cr}\{A\} \leq 0.5$ and $A_i \uparrow A$;   (b) $\lim_{i \to \infty} \mathrm{Cr}\{A_i\} < 0.5$ and $A_i \uparrow A$;
(c) $\mathrm{Cr}\{A\} \geq 0.5$ and $A_i \downarrow A$;   (d) $\lim_{i \to \infty} \mathrm{Cr}\{A_i\} > 0.5$ and $A_i \downarrow A$.

Proof: (a) Since $\mathrm{Cr}\{A\} \leq 0.5$, we have $\mathrm{Cr}\{A_i\} \leq 0.5$ for each i. It follows from Axiom 4 that

$$\mathrm{Cr}\{A\} = \mathrm{Cr}\{\cup_i A_i\} = \sup_i \mathrm{Cr}\{A_i\} = \lim_{i \to \infty} \mathrm{Cr}\{A_i\}.$$

(b) Since $\lim_{i \to \infty} \mathrm{Cr}\{A_i\} < 0.5$, we have $\sup_i \mathrm{Cr}\{A_i\} < 0.5$. It follows from Axiom 4 that

$$\mathrm{Cr}\{A\} = \mathrm{Cr}\{\cup_i A_i\} = \sup_i \mathrm{Cr}\{A_i\} = \lim_{i \to \infty} \mathrm{Cr}\{A_i\}.$$

(c) Since $\mathrm{Cr}\{A\} \geq 0.5$ and $A_i \downarrow A$, it follows from the self-duality of credibility measure that $\mathrm{Cr}\{A^c\} \leq 0.5$ and $A_i^c \uparrow A^c$. Thus

$$\lim_{i \to \infty} \mathrm{Cr}\{A_i\} = 1 - \lim_{i \to \infty} \mathrm{Cr}\{A_i^c\} = 1 - \mathrm{Cr}\{A^c\} = \mathrm{Cr}\{A\}.$$

(d) Since $\lim_{i \to \infty} \mathrm{Cr}\{A_i\} > 0.5$ and $A_i \downarrow A$, it follows from the self-duality of credibility measure that

$$\lim_{i \to \infty} \mathrm{Cr}\{A_i^c\} = \lim_{i \to \infty} (1 - \mathrm{Cr}\{A_i\}) < 0.5$$

and $A_i^c \uparrow A^c$. Thus $\mathrm{Cr}\{A_i\} = 1 - \mathrm{Cr}\{A_i^c\} \to 1 - \mathrm{Cr}\{A^c\} = \mathrm{Cr}\{A\}$ as $i \to \infty$. The theorem is proved.

Credibility Asymptotic Theorem

Theorem C.7 *(Credibility Asymptotic Theorem). For any events $A_1, A_2, \cdots$, we have*

$$\lim_{i \to \infty} \mathrm{Cr}\{A_i\} \geq 0.5, \quad \textit{if } A_i \uparrow \Theta, \tag{C.10}$$

$$\lim_{i \to \infty} \mathrm{Cr}\{A_i\} \leq 0.5, \quad \textit{if } A_i \downarrow \emptyset. \tag{C.11}$$

Proof: Assume $A_i \uparrow \Theta$. If $\lim_{i \to \infty} \mathrm{Cr}\{A_i\} < 0.5$, it follows from the credibility semicontinuity law that

$$\mathrm{Cr}\{\Theta\} = \lim_{i \to \infty} \mathrm{Cr}\{A_i\} < 0.5$$

which is in contradiction with $\mathrm{Cr}\{\Theta\} = 1$. The first inequality is proved. The second one may be verified similarly.

Credibility Extension Theorem

Suppose that the credibility of each singleton is given. Is the credibility measure fully and uniquely determined? This subsection will answer the question.

Theorem C.8. *Suppose that* Θ *is a nonempty set. If* Cr *is a credibility measure, then we have*

$$\sup_{\theta \in \Theta} \mathrm{Cr}\{\theta\} \geq 0.5,$$

$$\mathrm{Cr}\{\theta^*\} + \sup_{\theta \neq \theta^*} \mathrm{Cr}\{\theta\} = 1 \ if \ \mathrm{Cr}\{\theta^*\} \geq 0.5. \tag{C.12}$$

We will call (C.12) the credibility extension condition.

Proof: If $\sup \mathrm{Cr}\{\theta\} < 0.5$, then by using Axiom 4, we have

$$1 = \mathrm{Cr}\{\Theta\} = \sup_{\theta \in \Theta} \mathrm{Cr}\{\theta\} < 0.5.$$

This contradiction proves $\sup \mathrm{Cr}\{\theta\} \geq 0.5$. We suppose that $\theta^* \in \Theta$ is a point with $\mathrm{Cr}\{\theta^*\} \geq 0.5$. It follows from Axioms 3 and 4 that $\mathrm{Cr}\{\Theta \setminus \{\theta^*\}\} \leq 0.5$, and

$$\mathrm{Cr}\{\Theta \setminus \{\theta^*\}\} = \sup_{\theta \neq \theta^*} \mathrm{Cr}\{\theta\}.$$

Hence the second formula of (C.12) is true by the self-duality of credibility measure.

Theorem C.9 *(Li and Liu [91], Credibility Extension Theorem). Suppose that* Θ *is a nonempty set, and* $\mathrm{Cr}\{\theta\}$ *is a nonnegative function on* Θ *satisfying the credibility extension condition (C.12). Then* $\mathrm{Cr}\{\theta\}$ *has a unique extension to a credibility measure as follows,*

$$\mathrm{Cr}\{A\} = \begin{cases} \sup_{\theta \in A} \mathrm{Cr}\{\theta\}, & if \ \sup_{\theta \in A} \mathrm{Cr}\{\theta\} < 0.5 \\ 1 - \sup_{\theta \in A^c} \mathrm{Cr}\{\theta\}, & if \ \sup_{\theta \in A} \mathrm{Cr}\{\theta\} \geq 0.5. \end{cases} \tag{C.13}$$

Proof: We first prove that the set function $\mathrm{Cr}\{A\}$ defined by (C.13) is a credibility measure.

STEP 1: By the credibility extension condition $\sup_{\theta \in \Theta} \mathrm{Cr}\{\theta\} \geq 0.5$, we have

$$\mathrm{Cr}\{\Theta\} = 1 - \sup_{\theta \in \emptyset} \mathrm{Cr}\{\theta\} = 1 - 0 = 1.$$

STEP 2: If $A \subset B$, then $B^c \subset A^c$. The proof breaks down into two cases. Case 1: $\sup_{\theta \in A} \mathrm{Cr}\{\theta\} < 0.5$. For this case, we have

$$\mathrm{Cr}\{A\} = \sup_{\theta \in A} \mathrm{Cr}\{\theta\} \leq \sup_{\theta \in B} \mathrm{Cr}\{\theta\} \leq \mathrm{Cr}\{B\}.$$

Case 2: $\sup_{\theta \in A} \mathrm{Cr}\{\theta\} \geq 0.5$. For this case, we have $\sup_{\theta \in B} \mathrm{Cr}\{\theta\} \geq 0.5$, and

$$\mathrm{Cr}\{A\} = 1 - \sup_{\theta \in A^c} \mathrm{Cr}\{\theta\} \leq 1 - \sup_{\theta \in B^c} \mathrm{Cr}\{\theta\} = \mathrm{Cr}\{B\}.$$

Step 3: In order to prove $\mathrm{Cr}\{A\} + \mathrm{Cr}\{A^c\} = 1$, the argument breaks down into two cases. Case 1: $\sup_{\theta \in A} \mathrm{Cr}\{\theta\} < 0.5$. For this case, we have $\sup_{\theta \in A^c} \mathrm{Cr}\{\theta\} \geq 0.5$. Thus,

$$\mathrm{Cr}\{A\} + \mathrm{Cr}\{A^c\} = \sup_{\theta \in A} \mathrm{Cr}\{\theta\} + 1 - \sup_{\theta \in A} \mathrm{Cr}\{\theta\} = 1.$$

Case 2: $\sup_{\theta \in A} \mathrm{Cr}\{\theta\} \geq 0.5$. For this case, we have $\sup_{\theta \in A^c} \mathrm{Cr}\{\theta\} \leq 0.5$, and

$$\mathrm{Cr}\{A\} + \mathrm{Cr}\{A^c\} = 1 - \sup_{\theta \in A^c} \mathrm{Cr}\{\theta\} + \sup_{\theta \in A^c} \mathrm{Cr}\{\theta\} = 1.$$

Step 4: For any collection $\{A_i\}$ with $\sup_i \mathrm{Cr}\{A_i\} < 0.5$, we have

$$\mathrm{Cr}\{\cup_i A_i\} = \sup_{\theta \in \cup_i A_i} \mathrm{Cr}\{\theta\} = \sup_i \sup_{\theta \in A_i} \mathrm{Cr}\{\theta\} = \sup_i \mathrm{Cr}\{A_i\}.$$

Thus Cr is a credibility measure because it satisfies the four axioms.

Finally, let us prove the uniqueness. Assume that Cr_1 and Cr_2 are two credibility measures such that $\mathrm{Cr}_1\{\theta\} = \mathrm{Cr}_2\{\theta\}$ for each $\theta \in \Theta$. Let us prove that $\mathrm{Cr}_1\{A\} = \mathrm{Cr}_2\{A\}$ for any event A. The argument breaks down into three cases. Case 1: $\mathrm{Cr}_1\{A\} < 0.5$. For this case, it follows from Axiom 4 that

$$\mathrm{Cr}_1\{A\} = \sup_{\theta \in A} \mathrm{Cr}_1\{\theta\} = \sup_{\theta \in A} \mathrm{Cr}_2\{\theta\} = \mathrm{Cr}_2\{A\}.$$

Case 2: $\mathrm{Cr}_1\{A\} > 0.5$. For this case, we have $\mathrm{Cr}_1\{A^c\} < 0.5$. It follows from the first case that $\mathrm{Cr}_1\{A^c\} = \mathrm{Cr}_2\{A^c\}$ which implies $\mathrm{Cr}_1\{A\} = \mathrm{Cr}_2\{A\}$. Case 3: $\mathrm{Cr}_1\{A\} = 0.5$. For this case, we have $\mathrm{Cr}_1\{A^c\} = 0.5$, and

$$\mathrm{Cr}_2\{A\} \geq \sup_{\theta \in A} \mathrm{Cr}_2\{\theta\} = \sup_{\theta \in A} \mathrm{Cr}_1\{\theta\} = \mathrm{Cr}_1\{A\} = 0.5,$$

$$\mathrm{Cr}_2\{A^c\} \geq \sup_{\theta \in A^c} \mathrm{Cr}_2\{\theta\} = \sup_{\theta \in A^c} \mathrm{Cr}_1\{\theta\} = \mathrm{Cr}_1\{A^c\} = 0.5.$$

Hence $\mathrm{Cr}_2\{A\} = 0.5 = \mathrm{Cr}_1\{A\}$. The uniqueness is proved.

Credibility Space

Definition C.2. *Let Θ be a nonempty set, $\mathcal{P}$ the power set of Θ, and Cr a credibility measure. Then the triplet $(\Theta, \mathcal{P}, \mathrm{Cr})$ is called a credibility space.*

Example C.4: The triplet $(\Theta, \mathcal{P}, \mathrm{Cr})$ is a credibility space if

$$\Theta = \{\theta_1, \theta_2, \cdots\}, \ \mathrm{Cr}\{\theta_i\} \equiv 1/2 \ \text{for} \ i = 1, 2, \cdots \tag{C.14}$$

Note that the credibility measure is produced by the credibility extension theorem as follows,

$$\mathrm{Cr}\{A\} = \begin{cases} 0, & \text{if } A = \emptyset \\ 1, & \text{if } A = \Theta \\ 1/2, & \text{otherwise.} \end{cases}$$

Example C.5: The triplet $(\Theta, \mathcal{P}, \mathrm{Cr})$ is a credibility space if

$$\Theta = \{\theta_1, \theta_2, \cdots\}, \; \mathrm{Cr}\{\theta_i\} = i/(2i+1) \; \text{ for } i = 1, 2, \cdots \qquad \text{(C.15)}$$

By using the credibility extension theorem, we obtain the following credibility measure,

$$\mathrm{Cr}\{A\} = \begin{cases} \sup_{\theta_i \in A} \dfrac{i}{2i+1}, & \text{if } A \text{ is finite} \\[2ex] 1 - \sup_{\theta_i \in A^c} \dfrac{i}{2i+1}, & \text{if } A \text{ is infinite.} \end{cases}$$

Example C.6: The triplet $(\Theta, \mathcal{P}, \mathrm{Cr})$ is a credibility space if

$$\Theta = \{\theta_1, \theta_2, \cdots\}, \; \mathrm{Cr}\{\theta_1\} = 1/2, \; \mathrm{Cr}\{\theta_i\} = 1/i \; \text{ for } i = 2, 3, \cdots \qquad \text{(C.16)}$$

For this case, the credibility measure is

$$\mathrm{Cr}\{A\} = \begin{cases} \sup_{\theta_i \in A} 1/i, & \text{if } A \text{ contains neither } \theta_1 \text{ nor } \theta_2 \\[1ex] 1/2, & \text{if } A \text{ contains only one of } \theta_1 \text{ and } \theta_2 \\[1ex] 1 - \sup_{\theta_i \in A^c} 1/i, & \text{if } A \text{ contains both } \theta_1 \text{ and } \theta_2. \end{cases}$$

Example C.7: The triplet $(\Theta, \mathcal{P}, \mathrm{Cr})$ is a credibility space if

$$\Theta = [0, 1], \quad \mathrm{Cr}\{\theta\} = \theta/2 \; \text{ for } \theta \in \Theta. \qquad \text{(C.17)}$$

For this case, the credibility measure is

$$\mathrm{Cr}\{A\} = \begin{cases} \dfrac{1}{2} \sup_{\theta \in A} \theta, & \text{if } \sup_{\theta \in A} \theta < 1 \\[2ex] 1 - \dfrac{1}{2} \sup_{\theta \in A^c} \theta, & \text{if } \sup_{\theta \in A} \theta = 1. \end{cases}$$

Product Credibility Measure

Product credibility measure may be defined in multiple ways. This book accepts the following axiom.

Axiom 5. *(Product Credibility Axiom) Let Θ_k be nonempty sets on which Cr_k are credibility measures, $k = 1, 2, \cdots, n$, respectively, and $\Theta = \Theta_1 \times \Theta_2 \times \cdots \times \Theta_n$. Then*

$$\mathrm{Cr}\{(\theta_1, \theta_2, \cdots, \theta_n)\} = \mathrm{Cr}_1\{\theta_1\} \wedge \mathrm{Cr}_2\{\theta_2\} \wedge \cdots \wedge \mathrm{Cr}_n\{\theta_n\} \qquad \text{(C.18)}$$

for each $(\theta_1, \theta_2, \cdots, \theta_n) \in \Theta$.

Theorem C.10 *(Product Credibility Theorem).* *Let* Θ_k *be nonempty sets on which* Cr_k *are the credibility measures,* $k = 1, 2, \cdots, n$, *respectively, and* $\Theta = \Theta_1 \times \Theta_2 \times \cdots \times \Theta_n$. *Then* $\mathrm{Cr} = \mathrm{Cr}_1 \wedge \mathrm{Cr}_2 \wedge \cdots \wedge \mathrm{Cr}_n$ *defined by Axiom 5 has a unique extension to a credibility measure on* Θ *as follows,*

$$\mathrm{Cr}\{A\} = \begin{cases} \sup_{(\theta_1, \theta_2 \cdots, \theta_n) \in A} \min_{1 \le k \le n} \mathrm{Cr}_k\{\theta_k\}, \\ \quad if \sup_{(\theta_1, \theta_2, \cdots, \theta_n) \in A} \min_{1 \le k \le n} \mathrm{Cr}_k\{\theta_k\} < 0.5 \\ 1 - \sup_{(\theta_1, \theta_2, \cdots, \theta_n) \in A^c} \min_{1 \le k \le n} \mathrm{Cr}_k\{\theta_k\}, \\ \quad if \sup_{(\theta_1, \theta_2, \cdots, \theta_n) \in A} \min_{1 \le k \le n} \mathrm{Cr}_k\{\theta_k\} \ge 0.5. \end{cases} \quad (C.19)$$

Proof: For each $\boldsymbol{\theta} = (\theta_1, \theta_2, \cdots, \theta_n) \in \Theta$, we have $\mathrm{Cr}\{\boldsymbol{\theta}\} = \mathrm{Cr}_1\{\theta_1\} \wedge \mathrm{Cr}_2\{\theta_2\} \wedge \cdots \wedge \mathrm{Cr}_n\{\theta_n\}$. Let us prove that $\mathrm{Cr}\{\boldsymbol{\theta}\}$ satisfies the credibility extension condition. Since $\sup_{\theta_k \in \Theta_k} \mathrm{Cr}\{\theta_k\} \ge 0.5$ for each k, we have

$$\sup_{\boldsymbol{\theta} \in \Theta} \mathrm{Cr}\{\boldsymbol{\theta}\} = \sup_{(\theta_1, \theta_2, \cdots, \theta_n) \in \Theta} \min_{1 \le k \le n} \mathrm{Cr}_k\{\theta_k\} \ge 0.5.$$

Now we suppose that $\boldsymbol{\theta}^* = (\theta_1^*, \theta_2^*, \cdots, \theta_n^*)$ is a point with $\mathrm{Cr}\{\boldsymbol{\theta}^*\} \ge 0.5$. Without loss of generality, let i be the index such that

$$\mathrm{Cr}\{\boldsymbol{\theta}^*\} = \min_{1 \le k \le n} \mathrm{Cr}_k\{\theta_k^*\} = \mathrm{Cr}_i\{\theta_i^*\}. \quad (C.20)$$

We also immediately have

$$\mathrm{Cr}_k\{\theta_k^*\} \ge 0.5, \quad k = 1, 2, \cdots, n; \quad (C.21)$$

$$\mathrm{Cr}_k\{\theta_k^*\} + \sup_{\theta_k \ne \theta_k^*} \mathrm{Cr}_k\{\theta_k\} = 1, \quad k = 1, 2, \cdots, n; \quad (C.22)$$

$$\sup_{\theta_i \ne \theta_i^*} \mathrm{Cr}_i\{\theta_i\} \ge \sup_{\theta_k \ne \theta_k^*} \mathrm{Cr}_k\{\theta_k\}, \quad k = 1, 2, \cdots, n; \quad (C.23)$$

$$\sup_{\theta_k \ne \theta_k^*} \mathrm{Cr}_k\{\theta_k\} \le 0.5, \quad k = 1, \cdots, n. \quad (C.24)$$

It follows from (C.21) and (C.24) that

$$\sup_{\boldsymbol{\theta} \ne \boldsymbol{\theta}^*} \mathrm{Cr}\{\boldsymbol{\theta}\} = \sup_{(\theta_1, \theta_2, \cdots, \theta_n) \ne (\theta_1^*, \theta_2^*, \cdots, \theta_n^*)} \min_{1 \le k \le n} \mathrm{Cr}_k\{\theta_k\}$$

$$\ge \sup_{\theta_i \ne \theta_i^*} \min_{1 \le k \le i-1} \mathrm{Cr}_k\{\theta_k^*\} \wedge \mathrm{Cr}_i\{\theta_i\} \wedge \min_{i+1 \le k \le n} \mathrm{Cr}_k\{\theta_k^*\}$$

$$= \sup_{\theta_i \ne \theta_i^*} \mathrm{Cr}_i\{\theta_i\}.$$

We next suppose that

$$\sup_{\boldsymbol{\theta} \ne \boldsymbol{\theta}^*} \mathrm{Cr}\{\boldsymbol{\theta}\} > \sup_{\theta_i \ne \theta_i^*} \mathrm{Cr}_i\{\theta_i\}.$$

Then there is a point $(\theta'_1, \theta'_2, \cdots, \theta'_n) \neq (\theta^*_1, \theta^*_2, \cdots, \theta^*_n)$ such that

$$\min_{1 \leq k \leq n} \mathrm{Cr}_k\{\theta'_k\} > \sup_{\theta_i \neq \theta^*_i} \mathrm{Cr}_i\{\theta_i\}.$$

Let j be one of the index such that $\theta'_j \neq \theta^*_j$. Then

$$\mathrm{Cr}_j\{\theta'_j\} > \sup_{\theta_i \neq \theta^*_i} \mathrm{Cr}_i\{\theta_i\}.$$

That is,

$$\sup_{\theta_j \neq \theta^*_j} \mathrm{Cr}_j\{\theta_j\} > \sup_{\theta_i \neq \theta^*_i} \mathrm{Cr}_i\{\theta_i\}$$

which is in contradiction with (C.23). Thus

$$\sup_{\boldsymbol{\theta} \neq \boldsymbol{\theta}^*} \mathrm{Cr}\{\boldsymbol{\theta}\} = \sup_{\theta_i \neq \theta^*_i} \mathrm{Cr}_i\{\theta_i\}. \tag{C.25}$$

It follows from (C.20), (C.22) and (C.25) that

$$\mathrm{Cr}\{\boldsymbol{\theta}^*\} + \sup_{\boldsymbol{\theta} \neq \boldsymbol{\theta}^*} \mathrm{Cr}\{\boldsymbol{\theta}\} = \mathrm{Cr}_i\{\theta^*_i\} + \sup_{\theta_i \neq \theta^*_i} \mathrm{Cr}_i\{\theta_i\} = 1.$$

Thus Cr satisfies the credibility extension condition. It follows from the credibility extension theorem that $\mathrm{Cr}\{A\}$ is just the unique extension of $\mathrm{Cr}\{\boldsymbol{\theta}\}$. The theorem is proved.

Definition C.3. *Let* $(\Theta_k, \mathcal{P}_k, \mathrm{Cr}_k), k = 1, 2, \cdots, n$ *be credibility spaces,* $\Theta = \Theta_1 \times \Theta_2 \times \cdots \times \Theta_n$ *and* $\mathrm{Cr} = \mathrm{Cr}_1 \wedge \mathrm{Cr}_2 \wedge \cdots \wedge \mathrm{Cr}_n$. *Then* $(\Theta, \mathcal{P}, \mathrm{Cr})$ *is called the product credibility space of* $(\Theta_k, \mathcal{P}_k, \mathrm{Cr}_k), k = 1, 2, \cdots, n$.

Theorem C.11. *Let* $(\Theta, \mathcal{P}, \mathrm{Cr})$ *be the product credibility space of* $(\Theta_k, \mathcal{P}_k, \mathrm{Cr}_k)$, $k = 1, 2, \cdots, n$. *Then for any* $A_k \in \mathcal{P}_k$, $k = 1, 2, \cdots, n$, *we have*

$$\mathrm{Cr}\{A_1 \times A_2 \times \cdots \times A_k\} = \mathrm{Cr}_1\{A_1\} \wedge \mathrm{Cr}_2\{A_2\} \wedge \cdots \wedge \mathrm{Cr}_n\{A_n\}.$$

Proof: We only prove the case of $n = 2$. If $\mathrm{Cr}_1\{A_1\} < 0.5$ or $\mathrm{Cr}_2\{A_2\} < 0.5$, then we have

$$\sup_{\theta_1 \in A_1} \mathrm{Cr}_1\{\theta_1\} < 0.5 \quad \text{or} \quad \sup_{\theta_2 \in A_2} \mathrm{Cr}_2\{\theta_2\} < 0.5.$$

It follows from

$$\sup_{(\theta_1, \theta_2) \in A_1 \times A_2} \mathrm{Cr}_1\{\theta_1\} \wedge \mathrm{Cr}_2\{\theta_2\} = \sup_{\theta_1 \in A_1} \mathrm{Cr}_1\{\theta_1\} \wedge \sup_{\theta_2 \in A_2} \mathrm{Cr}_2\{\theta_2\} < 0.5$$

that

$$\mathrm{Cr}\{A_1 \times A_2\} = \sup_{\theta_1 \in A_1} \mathrm{Cr}_1\{\theta_1\} \wedge \sup_{\theta_2 \in A_2} \mathrm{Cr}_2\{\theta_2\} = \mathrm{Cr}_1\{A_1\} \wedge \mathrm{Cr}_2\{A_2\}.$$

If $Cr_1\{A_1\} \geq 0.5$ and $Cr_2\{A_2\} \geq 0.5$, then we have

$$\sup_{\theta_1 \in A_1} Cr_1\{\theta_1\} \geq 0.5 \quad \text{and} \quad \sup_{\theta_2 \in A_2} Cr_2\{\theta_2\} \geq 0.5.$$

It follows from

$$\sup_{(\theta_1,\theta_2) \in A_1 \times A_2} Cr_1\{\theta_1\} \wedge Cr_2\{\theta_2\} = \sup_{\theta_1 \in A_1} Cr_1\{\theta_1\} \wedge \sup_{\theta_2 \in A_2} Cr_2\{\theta_2\} \geq 0.5$$

that

$$Cr\{A_1 \times A_2\} = 1 - \sup_{(\theta_1,\theta_2) \notin A_1 \times A_2} Cr_1\{\theta_1\} \wedge Cr_2\{\theta_2\}$$

$$= \left(1 - \sup_{\theta_1 \in A_1^c} Cr_1\{\theta_1\}\right) \wedge \left(1 - \sup_{\theta_2 \in A_2^c} Cr_2\{\theta_2\}\right)$$

$$= Cr_1\{A_1\} \wedge Cr_2\{A_2\}.$$

The theorem is proved.

C.2   Fuzzy Variable

Definition C.4. *A fuzzy variable is a (measurable) function from a credibility space* $(\Theta, \mathcal{P}, Cr)$ *to the set of real numbers.*

Example C.8: Take $(\Theta, \mathcal{P}, Cr)$ to be $\{\theta_1, \theta_2\}$ with $Cr\{\theta_1\} = Cr\{\theta_2\} = 0.5$. Then the function

$$\xi(\theta) = \begin{cases} 0, & \text{if } \theta = \theta_1 \\ 1, & \text{if } \theta = \theta_2 \end{cases}$$

is a fuzzy variable.

Example C.9: Take $(\Theta, \mathcal{P}, Cr)$ to be the interval $[0,1]$ with $Cr\{\theta\} = \theta/2$ for each $\theta \in [0,1]$. Then the identity function $\xi(\theta) = \theta$ is a fuzzy variable.

Example C.10: A crisp number c may be regarded as a special fuzzy variable. In fact, it is the constant function $\xi(\theta) \equiv c$ on the credibility space $(\Theta, \mathcal{P}, Cr)$.

Remark C.2: Since a fuzzy variable ξ is a function on a credibility space, for any set B of real numbers, the set

$$\{\xi \in B\} = \{\theta \in \Theta \mid \xi(\theta) \in B\} \tag{C.26}$$

is always an element in $\mathcal{P}$. In other words, the fuzzy variable ξ is always a measurable function and $\{\xi \in B\}$ is always an event.

Definition C.5. *Let ξ_1 and ξ_2 be fuzzy variables defined on the credibility space* $(\Theta, \mathcal{P}, Cr)$. *We say $\xi_1 = \xi_2$ if $\xi_1(\theta) = \xi_2(\theta)$ for almost all $\theta \in \Theta$.*

Fuzzy Vector

Definition C.6. *An n-dimensional fuzzy vector is defined as a function from a credibility space* $(\Theta, \mathcal{P}, \mathrm{Cr})$ *to the set of n-dimensional real vectors.*

Theorem C.12. *The vector* $(\xi_1, \xi_2, \cdots, \xi_n)$ *is a fuzzy vector if and only if* $\xi_1, \xi_2, \cdots, \xi_n$ *are fuzzy variables.*

Proof: Write $\boldsymbol{\xi} = (\xi_1, \xi_2, \cdots, \xi_n)$. Suppose that $\boldsymbol{\xi}$ is a fuzzy vector. Then $\xi_1, \xi_2, \cdots, \xi_n$ are functions from Θ to $\Re$. Thus $\xi_1, \xi_2, \cdots, \xi_n$ are fuzzy variables. Conversely, suppose that ξ_i are fuzzy variables defined on the credibility spaces $(\Theta_i, \mathcal{P}_i, \mathrm{Cr}_i)$, $i = 1, 2, \cdots, n$, respectively. It is clear that $(\xi_1, \xi_2, \cdots, \xi_n)$ is a function from the product credibility space $(\Theta, \mathcal{P}, \mathrm{Cr})$ to $\Re^n$, i.e.,

$$\boldsymbol{\xi}(\theta_1, \theta_2, \cdots, \theta_n) = (\xi_1(\theta_1), \xi_2(\theta_2), \cdots, \xi_n(\theta_n))$$

for all $(\theta_1, \theta_2, \cdots, \theta_n) \in \Theta$. Hence $\boldsymbol{\xi} = (\xi_1, \xi_2, \cdots, \xi_n)$ is a fuzzy vector.

Fuzzy Arithmetic

In this subsection, we will suppose that all fuzzy variables are defined on a common credibility space. Otherwise, we may embed them into the product credibility space.

Definition C.7. *Let* $f : \Re^n \to \Re$ *be a function, and* $\xi_1, \xi_2, \cdots, \xi_n$ *fuzzy variables on the credibility space* $(\Theta, \mathcal{P}, \mathrm{Cr})$. *Then* $\xi = f(\xi_1, \xi_2, \cdots, \xi_n)$ *is a fuzzy variable defined as*

$$\xi(\theta) = f(\xi_1(\theta), \xi_2(\theta), \cdots, \xi_n(\theta)) \tag{C.27}$$

for any $\theta \in \Theta$.

The reader may wonder whether $\xi(\theta_1, \theta_2, \cdots, \theta_n)$ defined by (C.27) is a fuzzy variable. The following theorem answers this question.

Theorem C.13. *Let* $\boldsymbol{\xi}$ *be an n-dimensional fuzzy vector, and* $f : \Re^n \to \Re$ *a function. Then* $f(\boldsymbol{\xi})$ *is a fuzzy variable.*

Proof: Since $f(\boldsymbol{\xi})$ is a function from a credibility space to the set of real numbers, it is a fuzzy variable.

C.3   Membership Function

Definition C.8. *Let* ξ *be a fuzzy variable defined on the credibility space* $(\Theta, \mathcal{P}, \mathrm{Cr})$. *Then its membership function is derived from the credibility measure by*

$$\mu(x) = (2\mathrm{Cr}\{\xi = x\}) \wedge 1, \quad x \in \Re. \tag{C.28}$$

Membership function represents the degree that the fuzzy variable ξ takes some prescribed value. How do we determine membership functions? There are several methods reported in the past literature. Anyway, the membership degree $\mu(x) = 0$ if x is an impossible point, and $\mu(x) = 1$ if x is the most possible point that ξ takes.

Example C.11: It is clear that a fuzzy variable has a unique membership function. However, a membership function may produce multiple fuzzy variables. For example, let $\Theta = \{\theta_1, \theta_2\}$ and $\mathrm{Cr}\{\theta_1\} = \mathrm{Cr}\{\theta_2\} = 0.5$. Then $(\Theta, \mathcal{P}, \mathrm{Cr})$ is a credibility space. We define

$$\xi_1(\theta) = \begin{cases} 0, & \text{if } \theta = \theta_1 \\ 1, & \text{if } \theta = \theta_2, \end{cases} \qquad \xi_2(\theta) = \begin{cases} 1, & \text{if } \theta = \theta_1 \\ 0, & \text{if } \theta = \theta_2. \end{cases}$$

It is clear that both of them are fuzzy variables and have the same membership function, $\mu(x) \equiv 1$ on $x = 0$ or 1.

Theorem C.14 *(Credibility Inversion Theorem). Let ξ be a fuzzy variable with membership function μ. Then for any set B of real numbers, we have*

$$\mathrm{Cr}\{\xi \in B\} = \frac{1}{2}\left(\sup_{x \in B} \mu(x) + 1 - \sup_{x \in B^c} \mu(x)\right). \qquad (C.29)$$

Proof: If $\mathrm{Cr}\{\xi \in B\} \leq 0.5$, then by Axiom 2, we have $\mathrm{Cr}\{\xi = x\} \leq 0.5$ for each $x \in B$. It follows from Axiom 4 that

$$\mathrm{Cr}\{\xi \in B\} = \frac{1}{2}\left(\sup_{x \in B}(2\mathrm{Cr}\{\xi = x\} \wedge 1)\right) = \frac{1}{2}\sup_{x \in B} \mu(x). \qquad (C.30)$$

The self-duality of credibility measure implies that $\mathrm{Cr}\{\xi \in B^c\} \geq 0.5$ and $\sup_{x \in B^c} \mathrm{Cr}\{\xi = x\} \geq 0.5$, i.e.,

$$\sup_{x \in B^c} \mu(x) = \sup_{x \in B^c}(2\mathrm{Cr}\{\xi = x\} \wedge 1) = 1. \qquad (C.31)$$

It follows from (C.30) and (C.31) that (C.29) holds.

If $\mathrm{Cr}\{\xi \in B\} \geq 0.5$, then $\mathrm{Cr}\{\xi \in B^c\} \leq 0.5$. It follows from the first case that

$$\mathrm{Cr}\{\xi \in B\} = 1 - \mathrm{Cr}\{\xi \in B^c\} = 1 - \frac{1}{2}\left(\sup_{x \in B^c} \mu(x) + 1 - \sup_{x \in B} \mu(x)\right)$$

$$= \frac{1}{2}\left(\sup_{x \in B} \mu(x) + 1 - \sup_{x \in B^c} \mu(x)\right).$$

The theorem is proved.

Example C.12: Let ξ be a fuzzy variable with membership function μ. Then the following equations follow immediately from Theorem C.14:

$$\mathrm{Cr}\{\xi = x\} = \frac{1}{2}\left(\mu(x) + 1 - \sup_{y \neq x} \mu(y)\right), \qquad \forall x \in \Re; \qquad (C.32)$$

$$\text{Cr}\{\xi \leq x\} = \frac{1}{2}\left(\sup_{y \leq x} \mu(y) + 1 - \sup_{y > x} \mu(y)\right), \quad \forall x \in \Re; \tag{C.33}$$

$$\text{Cr}\{\xi \geq x\} = \frac{1}{2}\left(\sup_{y \geq x} \mu(y) + 1 - \sup_{y < x} \mu(y)\right), \quad \forall x \in \Re. \tag{C.34}$$

Especially, if μ is a continuous function, then

$$\text{Cr}\{\xi = x\} = \frac{\mu(x)}{2}, \quad \forall x \in \Re. \tag{C.35}$$

Theorem C.15 *(Sufficient and Necessary Condition for Membership Function). A function* $\mu : \Re \rightarrow [0,1]$ *is a membership function if and only if* $\sup \mu(x) = 1$.

Proof: If μ is a membership function, then there exists a fuzzy variable ξ whose membership function is just μ, and

$$\sup_{x \in \Re} \mu(x) = \sup_{x \in \Re} (2\text{Cr}\{\xi = x\}) \wedge 1.$$

If there is some point $x \in \Re$ such that $\text{Cr}\{\xi = x\} \geq 0.5$, then $\sup \mu(x) = 1$. Otherwise, we have $\text{Cr}\{\xi = x\} < 0.5$ for each $x \in \Re$. It follows from Axiom 4 that

$$\sup_{x \in \Re} \mu(x) = \sup_{x \in \Re} (2\text{Cr}\{\xi = x\}) \wedge 1 = 2\sup_{x \in \Re} \text{Cr}\{\xi = x\} = 2\,(\text{Cr}\{\Theta\} \wedge 0.5) = 1.$$

Conversely, suppose that $\sup \mu(x) = 1$. For each $x \in \Re$, we define

$$\text{Cr}\{x\} = \frac{1}{2}\left(\mu(x) + 1 - \sup_{y \neq x} \mu(y)\right).$$

It is clear that

$$\sup_{x \in \Re} \text{Cr}\{x\} \geq \frac{1}{2}(1 + 1 - 1) = 0.5.$$

For any $x^* \in \Re$ with $\text{Cr}\{x^*\} \geq 0.5$, we have $\mu(x^*) = 1$ and

$$\text{Cr}\{x^*\} + \sup_{y \neq x^*} \text{Cr}\{y\}$$

$$= \frac{1}{2}\left(\mu(x^*) + 1 - \sup_{y \neq x^*} \mu(y)\right) + \sup_{y \neq x^*} \frac{1}{2}\left(\mu(y) + 1 - \sup_{z \neq y} \mu(z)\right)$$

$$= 1 - \frac{1}{2}\sup_{y \neq x^*} \mu(y) + \frac{1}{2}\sup_{y \neq x^*} \mu(y) = 1.$$

Thus $\text{Cr}\{x\}$ satisfies the credibility extension condition, and has a unique extension to credibility measure on $\mathcal{P}(\Re)$ by using the credibility extension theorem. Now we define a fuzzy variable ξ as an identity function from the credibility space $(\Re, \mathcal{P}(\Re), \mathrm{Cr})$ to $\Re$. Then the membership function of the fuzzy variable ξ is

$$(2\mathrm{Cr}\{\xi = x\}) \wedge 1 = \left(\mu(x) + 1 - \sup_{y \neq x} \mu(y) \right) \wedge 1 = \mu(x)$$

for each x. The theorem is proved.

Some Special Membership Functions

By an *equipossible fuzzy variable* we mean the fuzzy variable fully determined by the pair (a, b) of crisp numbers with $a < b$, whose membership function is given by

$$\mu_1(x) = 1, \quad a \leq x \leq b.$$

By a *triangular fuzzy variable* we mean the fuzzy variable fully determined by the triplet (a, b, c) of crisp numbers with $a < b < c$, whose membership function is given by

$$\mu_2(x) = \begin{cases} \dfrac{x - a}{b - a}, & \text{if } a \leq x \leq b \\[2mm] \dfrac{x - c}{b - c}, & \text{if } b \leq x \leq c. \end{cases}$$

By a *trapezoidal fuzzy variable* we mean the fuzzy variable fully determined by the quadruplet (a, b, c, d) of crisp numbers with $a < b < c < d$, whose membership function is given by

$$\mu_3(x) = \begin{cases} \dfrac{x - a}{b - a}, & \text{if } a \leq x \leq b \\[2mm] 1, & \text{if } b \leq x \leq c \\[2mm] \dfrac{x - d}{c - d}, & \text{if } c \leq x \leq d. \end{cases}$$

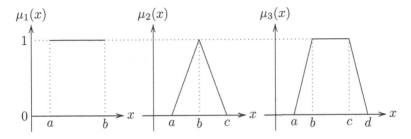

Figure C.1: Membership Functions μ_1, μ_2 and μ_3

C.4   Credibility Distribution

Definition C.9 *(Liu [112]). The credibility distribution* $\Phi : \Re \rightarrow [0, 1]$ *of a fuzzy variable* ξ *is defined by*

$$\Phi(x) = \mathrm{Cr} \left\{ \theta \in \Theta \mid \xi(\theta) \leq x \right\}. \tag{C.36}$$

That is, $\Phi(x)$ is the credibility that the fuzzy variable ξ takes a value less than or equal to x. Generally speaking, the credibility distribution Φ is neither left-continuous nor right-continuous.

Example C.13: The credibility distribution of an equipossible fuzzy variable (a, b) is

$$\Phi_1(x) = \begin{cases} 0, & \text{if } x < a \\ 1/2, & \text{if } a \leq x < b \\ 1, & \text{if } x \geq b. \end{cases}$$

Especially, if ξ is an equipossible fuzzy variable on $\Re$, then $\Phi_1(x) \equiv 1/2$.

Example C.14: The credibility distribution of a triangular fuzzy variable (a, b, c) is

$$\Phi_2(x) = \begin{cases} 0, & \text{if } x \leq a \\ \dfrac{x - a}{2(b - a)}, & \text{if } a \leq x \leq b \\ \dfrac{x + c - 2b}{2(c - b)}, & \text{if } b \leq x \leq c \\ 1, & \text{if } x \geq c. \end{cases}$$

Example C.15: The credibility distribution of a trapezoidal fuzzy variable (a, b, c, d) is

$$\Phi_3(x) = \begin{cases} 0, & \text{if } x \leq a \\ \dfrac{x - a}{2(b - a)}, & \text{if } a \leq x \leq b \\ \dfrac{1}{2}, & \text{if } b \leq x \leq c \\ \dfrac{x + d - 2c}{2(d - c)}, & \text{if } c \leq x \leq d \\ 1, & \text{if } x \geq d. \end{cases}$$

Theorem C.16. *Let* ξ *be a fuzzy variable with membership function* μ. *Then its credibility distribution is*

$$\Phi(x) = \frac{1}{2} \left(\sup_{y \leq x} \mu(y) + 1 - \sup_{y > x} \mu(y) \right), \ \forall x \in \Re. \tag{C.37}$$

Proof: It follows from the credibility inversion theorem immediately.

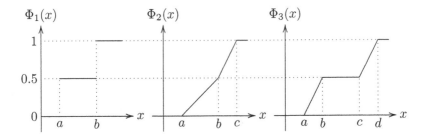

Figure C.2: Credibility Distributions Φ_1, Φ_2 and Φ_3

Theorem C.17 *(Liu [117], Sufficient and Necessary Condition for Credibility Distribution). A function* $\Phi : \Re \to [0,1]$ *is a credibility distribution if and only if it is an increasing function with*

$$\lim_{x \to -\infty} \Phi(x) \leq 0.5 \leq \lim_{x \to \infty} \Phi(x), \tag{C.38}$$

$$\lim_{y \downarrow x} \Phi(y) = \Phi(x) \text{ if } \lim_{y \downarrow x} \Phi(y) > 0.5 \text{ or } \Phi(x) \geq 0.5. \tag{C.39}$$

Proof: It is obvious that a credibility distribution Φ is an increasing function. The inequalities (C.38) follow from the credibility asymptotic theorem immediately. Assume that x is a point at which $\lim_{y \downarrow x} \Phi(y) > 0.5$. That is,

$$\lim_{y \downarrow x} \mathrm{Cr}\{\xi \leq y\} > 0.5.$$

Since $\{\xi \leq y\} \downarrow \{\xi \leq x\}$ as $y \downarrow x$, it follows from the credibility semicontinuity law that

$$\Phi(y) = \mathrm{Cr}\{\xi \leq y\} \downarrow \mathrm{Cr}\{\xi \leq x\} = \Phi(x)$$

as $y \downarrow x$. When x is a point at which $\Phi(x) \geq 0.5$, if $\lim_{y \downarrow x} \Phi(y) \neq \Phi(x)$, then we have

$$\lim_{y \downarrow x} \Phi(y) > \Phi(x) \geq 0.5.$$

For this case, we have proved that $\lim_{y \downarrow x} \Phi(y) = \Phi(x)$. Thus (C.38) and (C.39) are proved. Conversely, if $\Phi : \Re \to [0,1]$ is an increasing function satisfying (C.38) and (C.39), then

$$\mu(x) = \begin{cases} 2\Phi(x), & \text{if } \Phi(x) < 0.5 \\ 1, & \text{if } \lim_{y \uparrow x} \Phi(y) < 0.5 \leq \Phi(x) \\ 2 - 2\Phi(x), & \text{if } 0.5 \leq \lim_{y \uparrow x} \Phi(y) \end{cases} \tag{C.40}$$

takes values in $[0,1]$ and $\sup \mu(x) = 1$. It follows from Theorem C.15 that there is a fuzzy variable ξ whose membership function is just μ. Let us verify that Φ is the credibility distribution of ξ, i.e., $\mathrm{Cr}\{\xi \leq x\} = \Phi(x)$ for each x. The argument breaks down into two cases. (i) If $\Phi(x) < 0.5$, then we have $\sup_{y>x} \mu(y) = 1$, and $\mu(y) = 2\Phi(y)$ for each y with $y \leq x$. Thus

$$\mathrm{Cr}\{\xi \leq x\} = \frac{1}{2} \left(\sup_{y \leq x} \mu(y) + 1 - \sup_{y > x} \mu(y) \right) = \sup_{y \leq x} \Phi(y) = \Phi(x).$$

(ii) If $\Phi(x) \geq 0.5$, then we have $\sup_{y \leq x} \mu(y) = 1$ and $\Phi(y) \geq \Phi(x) \geq 0.5$ for each y with $y > x$. Thus $\mu(y) = 2 - 2\Phi(y)$ and

$$\begin{aligned}
\mathrm{Cr}\{\xi \leq x\} &= \frac{1}{2} \left(\sup_{y \leq x} \mu(y) + 1 - \sup_{y > x} \mu(y) \right) \\
&= \frac{1}{2} \left(1 + 1 - \sup_{y > x} (2 - 2\Phi(y)) \right) \\
&= \inf_{y > x} \Phi(y) = \lim_{y \downarrow x} \Phi(y) = \Phi(x).
\end{aligned}$$

The theorem is proved.

Example C.16: Let a and b be two numbers with $0 \leq a \leq 0.5 \leq b \leq 1$. We define a fuzzy variable by the following membership function,

$$\mu(x) = \begin{cases} 2a, & \text{if } x < 0 \\ 1, & \text{if } x = 0 \\ 2 - 2b, & \text{if } x > 0. \end{cases}$$

Then its credibility distribution is

$$\Phi(x) = \begin{cases} a, & \text{if } x < 0 \\ b, & \text{if } x \geq 0. \end{cases}$$

Thus we have

$$\lim_{x \to -\infty} \Phi(x) = a, \qquad \lim_{x \to +\infty} \Phi(x) = b.$$

C.5   Independence

The independence of fuzzy variables has been discussed by many authors from different angles. Here we use the following definition.

Definition C.10 (*Liu and Gao [142]*). *The fuzzy variables* $\xi_1, \xi_2, \cdots, \xi_m$ *are said to be independent if*

$$\mathrm{Cr}\left\{ \bigcap_{i=1}^{m} \{\xi_i \in B_i\} \right\} = \min_{1 \leq i \leq m} \mathrm{Cr}\{\xi_i \in B_i\} \qquad (\text{C.41})$$

for any sets $B_1, B_2, \cdots, B_m$ *of* $\Re$.

Theorem C.18. *The fuzzy variables $\xi_1, \xi_2, \cdots, \xi_m$ are independent if and only if*

$$\text{Cr}\left\{\bigcup_{i=1}^{m}\{\xi_i \in B_i\}\right\} = \max_{1 \leq i \leq m} \text{Cr}\{\xi_i \in B_i\} \qquad (C.42)$$

for any sets $B_1, B_2, \cdots, B_m$ of $\Re$.

Proof: It follows from the self-duality of credibility measure that $\xi_1, \xi_2, \cdots, \xi_m$ are independent if and only if

$$\text{Cr}\left\{\bigcup_{i=1}^{m}\{\xi_i \in B_i\}\right\} = 1 - \text{Cr}\left\{\bigcap_{i=1}^{m}\{\xi_i \in B_i^c\}\right\}$$

$$= 1 - \min_{1 \leq i \leq m} \text{Cr}\{\xi_i \in B_i^c\} = \max_{1 \leq i \leq m} \text{Cr}\{\xi_i \in B_i\}.$$

Thus (C.42) is verified. The proof is complete.

Theorem C.19. *The fuzzy variables $\xi_1, \xi_2, \cdots, \xi_m$ are independent if and only if*

$$\text{Cr}\left\{\bigcap_{i=1}^{m}\{\xi_i = x_i\}\right\} = \min_{1 \leq i \leq m} \text{Cr}\{\xi_i = x_i\} \qquad (C.43)$$

for any real numbers $x_1, x_2, \cdots, x_m$.

Proof: If $\xi_1, \xi_2, \cdots, \xi_m$ are independent, then we have (C.43) immediately by taking $B_i = \{x_i\}$ for each i. Conversely, if $\text{Cr}\{\cap_{i=1}^{m}(\xi_i \in B_i)\} \geq 0.5$, it follows from Theorem C.2 that (C.41) holds. Otherwise, we have $\text{Cr}\{\cap_{i=1}^{m}(\xi_i = x_i)\} < 0.5$ for any real numbers $x_i \in B_i$, $i = 1, 2, \cdots, m$, and

$$\text{Cr}\left\{\bigcap_{i=1}^{m}\{\xi_i \in B_i\}\right\} = \text{Cr}\left\{\bigcup_{x_i \in B_i, 1 \leq i \leq m} \bigcap_{i=1}^{m}\{\xi_i = x_i\}\right\}$$

$$= \sup_{x_i \in B_i, 1 \leq i \leq m} \text{Cr}\left\{\bigcap_{i=1}^{m}\{\xi_i = x_i\}\right\} = \sup_{x_i \in B_i, 1 \leq i \leq m} \min_{1 \leq i \leq m} \text{Cr}\{\xi_i = x_i\}$$

$$= \min_{1 \leq i \leq m} \sup_{x_i \in B_i} \text{Cr}\{\xi_i = x_i\} = \min_{1 \leq i \leq m} \text{Cr}\{\xi_i \in B_i\}.$$

Hence (C.41) is true, and $\xi_1, \xi_2, \cdots, \xi_m$ are independent. The theorem is thus proved.

Theorem C.20. *Let μ_i be membership functions of fuzzy variables ξ_i, $i = 1, 2, \cdots, m$, respectively, and μ the joint membership function of fuzzy vector $(\xi_1, \xi_2, \cdots, \xi_m)$. Then the fuzzy variables $\xi_1, \xi_2, \cdots, \xi_m$ are independent if and only if*

$$\mu(x_1, x_2, \cdots, x_m) = \min_{1 \leq i \leq m} \mu_i(x_i) \qquad (C.44)$$

for any real numbers $x_1, x_2, \cdots, x_m$.

Proof: Suppose that $\xi_1, \xi_2, \cdots, \xi_m$ are independent. It follows from Theorem C.19 that

$$
\begin{aligned}
\mu(x_1, x_2, \cdots, x_m) &= \left(2\mathrm{Cr}\left\{ \bigcap_{i=1}^{m} \{\xi_i = x_i\} \right\} \right) \wedge 1 \\
&= \left(2 \min_{1 \leq i \leq m} \mathrm{Cr}\{\xi_i = x_i\} \right) \wedge 1 \\
&= \min_{1 \leq i \leq m} \left(2\mathrm{Cr}\{\xi_i = x_i\} \right) \wedge 1 = \min_{1 \leq i \leq m} \mu_i(x_i).
\end{aligned}
$$

Conversely, for any real numbers $x_1, x_2, \cdots, x_m$ with $\mathrm{Cr}\{\cap_{i=1}^{m}\{\xi_i = x_i\}\} < 0.5$, we have

$$
\begin{aligned}
\mathrm{Cr}\left\{ \bigcap_{i=1}^{m} \{\xi_i = x_i\} \right\} &= \frac{1}{2} \left(2\mathrm{Cr}\left\{ \bigcap_{i=1}^{m} \{\xi_i = x_i\} \right\} \right) \wedge 1 \\
&= \frac{1}{2} \mu(x_1, x_2, \cdots, x_m) = \frac{1}{2} \min_{1 \leq i \leq m} \mu_i(x_i) \\
&= \frac{1}{2} \left(\min_{1 \leq i \leq m} \left(2\mathrm{Cr}\{\xi_i = x_i\} \right) \wedge 1 \right) \\
&= \min_{1 \leq i \leq m} \mathrm{Cr}\{\xi_i = x_i\}.
\end{aligned}
$$

It follows from Theorem C.19 that $\xi_1, \xi_2, \cdots, \xi_m$ are independent. The theorem is proved.

C.6   Extension Principle of Zadeh

Theorem C.21 *(Extension Principle of Zadeh). Let $\xi_1, \xi_2, \cdots, \xi_n$ be independent fuzzy variables with membership functions $\mu_1, \mu_2, \cdots, \mu_n$, respectively, and $f : \Re^n \to \Re$ a function. Then the membership function μ of $\xi = f(\xi_1, \xi_2, \cdots, \xi_n)$ is derived from the membership functions $\mu_1, \mu_2, \cdots, \mu_n$ by*

$$
\mu(x) = \sup_{x = f(x_1, x_2, \cdots, x_n)} \min_{1 \leq i \leq n} \mu_i(x_i) \tag{C.45}
$$

for any $x \in \Re$. Here we set $\mu(x) = 0$ if there are not real numbers $x_1, x_2, \cdots, x_n$ such that $x = f(x_1, x_2, \cdots, x_n)$.

Proof: It follows from Definition C.8 that the membership function of $\xi = f(\xi_1, \xi_2, \cdots, \xi_n)$ is

$$\mu(x) = (2\mathrm{Cr}\,\{f(\xi_1, \xi_2, \cdots, \xi_n) = x\}) \wedge 1$$

$$= \left(2 \sup_{x=f(x_1, x_2, \cdots, x_n)} \mathrm{Cr}\{\xi_1 = x_1, \xi_2 = x_2, \cdots, \xi_n = x_n\} \right) \wedge 1$$

$$= \left(2 \sup_{x=f(x_1, x_2, \cdots, x_n)} \min_{1 \leq k \leq n} \mathrm{Cr}\{\xi_i = x_i\} \right) \wedge 1 \quad \text{(by independence)}$$

$$= \sup_{x=f(x_1, x_2, \cdots, x_n)} \min_{1 \leq k \leq n} (2\mathrm{Cr}\{\xi_i = x_i\}) \wedge 1$$

$$= \sup_{x=f(x_1, x_2, \cdots, x_n)} \min_{1 \leq i \leq n} \mu_i(x_i).$$

The theorem is proved.

Remark C.3: The extension principle of Zadeh is only applicable to the operations on independent fuzzy variables. In the past literature, the extension principle is used as a postulate. However, it is treated as a theorem in credibility theory.

Example C.17: The sum of independent equipossible fuzzy variables $\xi = (a_1, a_2)$ and $\eta = (b_1, b_2)$ is also an equipossible fuzzy variable, and

$$\xi + \eta = (a_1 + b_1, a_2 + b_2).$$

Their product is also an equipossible fuzzy variable, and

$$\xi \cdot \eta = \left(\min_{a_1 \leq x \leq a_2, b_1 \leq y \leq b_2} xy, \max_{a_1 \leq x \leq a_2, b_1 \leq y \leq b_2} xy \right).$$

Example C.18: The sum of independent triangular fuzzy variables $\xi = (a_1, a_2, a_3)$ and $\eta = (b_1, b_2, b_3)$ is also a triangular fuzzy variable, and

$$\xi + \eta = (a_1 + b_1, a_2 + b_2, a_3 + b_3).$$

The product of a triangular fuzzy variable $\xi = (a_1, a_2, a_3)$ and a scalar number λ is

$$\lambda \cdot \xi = \begin{cases} (\lambda a_1, \lambda a_2, \lambda a_3), & \text{if } \lambda \geq 0 \\ (\lambda a_3, \lambda a_2, \lambda a_1), & \text{if } \lambda < 0. \end{cases}$$

That is, the product of a triangular fuzzy variable and a scalar number is also a triangular fuzzy variable. However, the product of two triangular fuzzy variables is not a triangular one.

Example C.19: The sum of independent trapezoidal fuzzy variables $\xi = (a_1, a_2, a_3, a_4)$ and $\eta = (b_1, b_2, b_3, b_4)$ is also a trapezoidal fuzzy variable, and

$$\xi + \eta = (a_1 + b_1, a_2 + b_2, a_3 + b_3, a_4 + b_4).$$

The product of a trapezoidal fuzzy variable $\xi = (a_1, a_2, a_3, a_4)$ and a scalar number λ is

$$\lambda \cdot \xi = \begin{cases} (\lambda a_1, \lambda a_2, \lambda a_3, \lambda a_4), & \text{if } \lambda \geq 0 \\ (\lambda a_4, \lambda a_3, \lambda a_2, \lambda a_1), & \text{if } \lambda < 0. \end{cases}$$

That is, the product of a trapezoidal fuzzy variable and a scalar number is also a trapezoidal fuzzy variable. However, the product of two trapezoidal fuzzy variables is not a trapezoidal one.

Example C.20: Let $\xi_1, \xi_2, \cdots, \xi_n$ be independent fuzzy variables with membership functions $\mu_1, \mu_2, \cdots, \mu_n$, respectively, and $f : \Re^n \to \Re$ a function. Then for any set B of real numbers, the credibility $\mathrm{Cr}\{f(\xi_1, \xi_2, \cdots, \xi_n) \in B\}$ is

$$\frac{1}{2} \left(\sup_{f(x_1, x_2, \cdots, x_n) \in B} \min_{1 \leq i \leq n} \mu_i(x_i) + 1 - \sup_{f(x_1, x_2, \cdots, x_n) \in B^c} \min_{1 \leq i \leq n} \mu_i(x_i) \right).$$

C.7 Expected Value

There are many ways to define an expected value operator for fuzzy variables. The most general definition of expected value operator of fuzzy variable was given by Liu and Liu [114]. This definition is applicable to not only continuous fuzzy variables but also discrete ones.

Definition C.11 (*Liu and Liu [114]*). *Let ξ be a fuzzy variable. Then the expected value of ξ is defined by*

$$E[\xi] = \int_0^{+\infty} \mathrm{Cr}\{\xi \geq r\} \mathrm{d}r - \int_{-\infty}^0 \mathrm{Cr}\{\xi \leq r\} \mathrm{d}r \qquad (C.46)$$

provided that at least one of the two integrals is finite.

Example C.21: Let ξ be the equipossible fuzzy variable (a, b). Then its expected value is $E[\xi] = (a + b)/2$.

Example C.22: The triangular fuzzy variable $\xi = (a, b, c)$ has an expected value $E[\xi] = (a + 2b + c)/4$.

Example C.23: The trapezoidal fuzzy variable $\xi = (a, b, c, d)$ has an expected value $E[\xi] = (a + b + c + d)/4$.

Example C.24: Let ξ be a continuous fuzzy variable with membership function μ. If its expected value exists, and there is a point x_0 such that $\mu(x)$ is increasing on $(-\infty, x_0)$ and decreasing on $(x_0, +\infty)$, then $\mathrm{Cr}\{\xi \geq x\} = \mu(x)/2$ for any $x > x_0$ and $\mathrm{Cr}\{\xi \leq x\} = \mu(x)/2$ for any $x < x_0$. Thus

$$E[\xi] = x_0 + \frac{1}{2} \int_{x_0}^{+\infty} \mu(x) \mathrm{d}x - \frac{1}{2} \int_{-\infty}^{x_0} \mu(x) \mathrm{d}x.$$

Example C.25: The definition of expected value operator is also applicable to discrete case. Assume that ξ is a simple fuzzy variable whose membership function is given by

$$\mu(x) = \begin{cases} \mu_1, & \text{if } x = x_1 \\ \mu_2, & \text{if } x = x_2 \\ \cdots \\ \mu_m, & \text{if } x = x_m \end{cases} \tag{C.47}$$

where $x_1, x_2, \cdots, x_m$ are distinct numbers. Note that $\mu_1 \vee \mu_2 \vee \cdots \vee \mu_m = 1$. Definition C.11 implies that the expected value of ξ is

$$E[\xi] = \sum_{i=1}^{m} w_i x_i \tag{C.48}$$

where the weights are given by

$$w_i = \frac{1}{2} \left(\max_{1 \leq j \leq m} \{\mu_j | x_j \leq x_i\} - \max_{1 \leq j \leq m} \{\mu_j | x_j < x_i\} \right.$$
$$\left. + \max_{1 \leq j \leq m} \{\mu_j | x_j \geq x_i\} - \max_{1 \leq j \leq m} \{\mu_j | x_j > x_i\} \right)$$

for $i = 1, 2, \cdots, m$. It is easy to verify that all $w_i \geq 0$ and the sum of all weights is just 1.

Example C.26: Consider the fuzzy variable ξ defined by (C.47). Suppose $x_1 < x_2 < \cdots < x_m$ and there exists an index k with $1 < k < m$ such that

$$\mu_1 \leq \mu_2 \leq \cdots \leq \mu_k \quad \text{and} \quad \mu_k \geq \mu_{k+1} \geq \cdots \geq \mu_m.$$

Note that $\mu_k \equiv 1$. Then the expected value is determined by (C.48) and the weights are given by

$$w_i = \begin{cases} \dfrac{\mu_1}{2}, & \text{if } i = 1 \\[2mm] \dfrac{\mu_i - \mu_{i-1}}{2}, & \text{if } i = 2, 3, \cdots, k-1 \\[2mm] 1 - \dfrac{\mu_{k-1} + \mu_{k+1}}{2}, & \text{if } i = k \\[2mm] \dfrac{\mu_i - \mu_{i+1}}{2}, & \text{if } i = k+1, k+2, \cdots, m-1 \\[2mm] \dfrac{\mu_m}{2}, & \text{if } i = m. \end{cases}$$

Linearity of Expected Value Operator

Theorem C.22 (*Liu and Liu [137]*). *Let ξ and η be independent fuzzy variables with finite expected values. Then for any numbers a and b, we have*

$$E[a\xi + b\eta] = aE[\xi] + bE[\eta]. \tag{C.49}$$

Proof: STEP 1: We first prove that $E[\xi + b] = E[\xi] + b$ for any real number b. If $b \geq 0$, we have

$$
\begin{aligned}
E[\xi + b] &= \int_0^\infty \mathrm{Cr}\{\xi + b \geq r\}\mathrm{d}r - \int_{-\infty}^0 \mathrm{Cr}\{\xi + b \leq r\}\mathrm{d}r \\
&= \int_0^\infty \mathrm{Cr}\{\xi \geq r - b\}\mathrm{d}r - \int_{-\infty}^0 \mathrm{Cr}\{\xi \leq r - b\}\mathrm{d}r \\
&= E[\xi] + \int_0^b \left(\mathrm{Cr}\{\xi \geq r - b\} + \mathrm{Cr}\{\xi < r - b\}\right)\mathrm{d}r \\
&= E[\xi] + b.
\end{aligned}
$$

If $b < 0$, then we have

$$
E[\xi + b] = E[\xi] - \int_b^0 \left(\mathrm{Cr}\{\xi \geq r - b\} + \mathrm{Cr}\{\xi < r - b\}\right)\mathrm{d}r = E[\xi] + b.
$$

STEP 2: We prove that $E[a\xi] = aE[\xi]$ for any real number a. If $a = 0$, then the equation $E[a\xi] = aE[\xi]$ holds trivially. If $a > 0$, we have

$$
\begin{aligned}
E[a\xi] &= \int_0^\infty \mathrm{Cr}\{a\xi \geq r\}\mathrm{d}r - \int_{-\infty}^0 \mathrm{Cr}\{a\xi \leq r\}\mathrm{d}r \\
&= \int_0^\infty \mathrm{Cr}\left\{\xi \geq \frac{r}{a}\right\}\mathrm{d}r - \int_{-\infty}^0 \mathrm{Cr}\left\{\xi \leq \frac{r}{a}\right\}\mathrm{d}r \\
&= a\int_0^\infty \mathrm{Cr}\left\{\xi \geq \frac{r}{a}\right\}\mathrm{d}\left(\frac{r}{a}\right) - a\int_{-\infty}^0 \mathrm{Cr}\left\{\xi \leq \frac{r}{a}\right\}\mathrm{d}\left(\frac{r}{a}\right) = aE[\xi].
\end{aligned}
$$

If $a < 0$, we have

$$
\begin{aligned}
E[a\xi] &= \int_0^\infty \mathrm{Cr}\{a\xi \geq r\}\mathrm{d}r - \int_{-\infty}^0 \mathrm{Cr}\{a\xi \leq r\}\mathrm{d}r \\
&= \int_0^\infty \mathrm{Cr}\left\{\xi \leq \frac{r}{a}\right\}\mathrm{d}r - \int_{-\infty}^0 \mathrm{Cr}\left\{\xi \geq \frac{r}{a}\right\}\mathrm{d}r \\
&= a\int_0^\infty \mathrm{Cr}\left\{\xi \geq \frac{r}{a}\right\}\mathrm{d}\left(\frac{r}{a}\right) - a\int_{-\infty}^0 \mathrm{Cr}\left\{\xi \leq \frac{r}{a}\right\}\mathrm{d}\left(\frac{r}{a}\right) = aE[\xi].
\end{aligned}
$$

STEP 3: We prove that $E[\xi + \eta] = E[\xi] + E[\eta]$ when both ξ and η are simple fuzzy variables with the following membership functions,

$$
\mu(x) = \begin{cases} \mu_1, & \text{if } x = a_1 \\ \mu_2, & \text{if } x = a_2 \\ \cdots \\ \mu_m, & \text{if } x = a_m, \end{cases}
\qquad
\nu(x) = \begin{cases} \nu_1, & \text{if } x = b_1 \\ \nu_2, & \text{if } x = b_2 \\ \cdots \\ \nu_n, & \text{if } x = b_n. \end{cases}
$$

Then $\xi + \eta$ is also a simple fuzzy variable taking values $a_i + b_j$ with membership degrees $\mu_i \wedge \nu_j$, $i = 1, 2, \cdots, m$, $j = 1, 2, \cdots, n$, respectively. Now we define

$$w_i' = \frac{1}{2} \left(\max_{1 \leq k \leq m} \{\mu_k | a_k \leq a_i\} - \max_{1 \leq k \leq m} \{\mu_k | a_k < a_i\} \right.$$
$$\left. + \max_{1 \leq k \leq m} \{\mu_k | a_k \geq a_i\} - \max_{1 \leq k \leq m} \{\mu_k | a_k > a_i\} \right),$$

$$w_j'' = \frac{1}{2} \left(\max_{1 \leq l \leq n} \{\nu_l | b_l \leq b_j\} - \max_{1 \leq l \leq n} \{\nu_l | b_l < b_j\} \right.$$
$$\left. + \max_{1 \leq l \leq n} \{\nu_l | b_l \geq b_j\} - \max_{1 \leq l \leq n} \{\nu_l | b_l > b_j\} \right),$$

$$w_{ij} = \frac{1}{2} \left(\max_{1 \leq k \leq m, 1 \leq l \leq n} \{\mu_k \wedge \nu_l | a_k + b_l \leq a_i + b_j\} \right.$$
$$- \max_{1 \leq k \leq m, 1 \leq l \leq n} \{\mu_k \wedge \nu_l | a_k + b_l < a_i + b_j\}$$
$$+ \max_{1 \leq k \leq m, 1 \leq l \leq n} \{\mu_k \wedge \nu_l | a_k + b_l \geq a_i + b_j\}$$
$$\left. - \max_{1 \leq k \leq m, 1 \leq l \leq n} \{\mu_k \wedge \nu_l | a_k + b_l > a_i + b_j\} \right)$$

for $i = 1, 2, \cdots, m$ and $j = 1, 2, \cdots, n$. It is also easy to verify that

$$w_i' = \sum_{j=1}^{n} w_{ij}, \qquad w_j'' = \sum_{i=1}^{m} w_{ij}$$

for $i = 1, 2, \cdots, m$ and $j = 1, 2, \cdots, n$. If $\{a_i\}$, $\{b_j\}$ and $\{a_i + b_j\}$ are sequences consisting of distinct elements, then

$$E[\xi] = \sum_{i=1}^{m} a_i w_i', \quad E[\eta] = \sum_{j=1}^{n} b_j w_j'', \quad E[\xi + \eta] = \sum_{i=1}^{m} \sum_{j=1}^{n} (a_i + b_j) w_{ij}.$$

Thus $E[\xi + \eta] = E[\xi] + E[\eta]$. If not, we may give them a small perturbation such that they are distinct, and prove the linearity by letting the perturbation tend to zero.

STEP 4: We prove that $E[\xi + \eta] = E[\xi] + E[\eta]$ when ξ and η are fuzzy variables such that

$$\lim_{y \uparrow 0} \mathrm{Cr}\{\xi \leq y\} \leq \frac{1}{2} \leq \mathrm{Cr}\{\xi \leq 0\},$$
$$\lim_{y \uparrow 0} \mathrm{Cr}\{\eta \leq y\} \leq \frac{1}{2} \leq \mathrm{Cr}\{\eta \leq 0\}. \tag{C.50}$$

We define simple fuzzy variables ξ_i via credibility distributions as follows,

$$\Phi_i(x) = \begin{cases} \dfrac{k-1}{2^i}, & \text{if } \dfrac{k-1}{2^i} \leq \text{Cr}\{\xi \leq x\} < \dfrac{k}{2^i}, \, k = 1, 2, \cdots, 2^{i-1} \\[2mm] \dfrac{k}{2^i}, & \text{if } \dfrac{k-1}{2^i} \leq \text{Cr}\{\xi \leq x\} < \dfrac{k}{2^i}, \, k = 2^{i-1}+1, \cdots, 2^i \\[2mm] 1, & \text{if } \text{Cr}\{\xi \leq x\} = 1 \end{cases}$$

for $i = 1, 2, \cdots$ Thus $\{\xi_i\}$ is a sequence of simple fuzzy variables satisfying

$$\text{Cr}\{\xi_i \leq r\} \uparrow \text{Cr}\{\xi \leq r\}, \quad \text{if } r \leq 0$$
$$\text{Cr}\{\xi_i \geq r\} \uparrow \text{Cr}\{\xi \geq r\}, \quad \text{if } r \geq 0$$

as $i \to \infty$. Similarly, we define simple fuzzy variables η_i via credibility distributions as follows,

$$\Psi_i(x) = \begin{cases} \dfrac{k-1}{2^i}, & \text{if } \dfrac{k-1}{2^i} \leq \text{Cr}\{\eta \leq x\} < \dfrac{k}{2^i}, \, k = 1, 2, \cdots, 2^{i-1} \\[2mm] \dfrac{k}{2^i}, & \text{if } \dfrac{k-1}{2^i} \leq \text{Cr}\{\eta \leq x\} < \dfrac{k}{2^i}, \, k = 2^{i-1}+1, \cdots, 2^i \\[2mm] 1, & \text{if } \text{Cr}\{\eta \leq x\} = 1 \end{cases}$$

for $i = 1, 2, \cdots$ Thus $\{\eta_i\}$ is a sequence of simple fuzzy variables satisfying

$$\text{Cr}\{\eta_i \leq r\} \uparrow \text{Cr}\{\eta \leq r\}, \quad \text{if } r \leq 0$$
$$\text{Cr}\{\eta_i \geq r\} \uparrow \text{Cr}\{\eta \geq r\}, \quad \text{if } r \geq 0$$

as $i \to \infty$. It is also clear that $\{\xi_i + \eta_i\}$ is a sequence of simple fuzzy variables. Furthermore, when $r \leq 0$, it follows from (C.50) that

$$\begin{aligned} \lim_{i \to \infty} \text{Cr}\{\xi_i + \eta_i \leq r\} &= \lim_{i \to \infty} \sup_{x \leq 0, y \leq 0, x+y \leq r} \text{Cr}\{\xi_i \leq x\} \wedge \text{Cr}\{\eta_i \leq y\} \\ &= \sup_{x \leq 0, y \leq 0, x+y \leq r} \lim_{i \to \infty} \text{Cr}\{\xi_i \leq x\} \wedge \text{Cr}\{\eta_i \leq y\} \\ &= \sup_{x \leq 0, y \leq 0, x+y \leq r} \text{Cr}\{\xi \leq x\} \wedge \text{Cr}\{\eta \leq y\} \\ &= \text{Cr}\{\xi + \eta \leq r\}. \end{aligned}$$

That is,

$$\text{Cr}\{\xi_i + \eta_i \leq r\} \uparrow \text{Cr}\{\xi + \eta \leq r\}, \quad \text{if } r \leq 0.$$

A similar way may prove that

$$\text{Cr}\{\xi_i + \eta_i \geq r\} \uparrow \text{Cr}\{\xi + \eta \geq r\}, \quad \text{if } r \geq 0.$$

Since the expected values $E[\xi]$ and $E[\eta]$ exist, we have

$$
\begin{aligned}
E[\xi_i] &= \int_0^{+\infty} \mathrm{Cr}\{\xi_i \geq r\}\mathrm{d}r - \int_{-\infty}^0 \mathrm{Cr}\{\xi_i \leq r\}\mathrm{d}r \\
&\to \int_0^{+\infty} \mathrm{Cr}\{\xi \geq r\}\mathrm{d}r - \int_{-\infty}^0 \mathrm{Cr}\{\xi \leq r\}\mathrm{d}r = E[\xi],
\end{aligned}
$$

$$
\begin{aligned}
E[\eta_i] &= \int_0^{+\infty} \mathrm{Cr}\{\eta_i \geq r\}\mathrm{d}r - \int_{-\infty}^0 \mathrm{Cr}\{\eta_i \leq r\}\mathrm{d}r \\
&\to \int_0^{+\infty} \mathrm{Cr}\{\eta \geq r\}\mathrm{d}r - \int_{-\infty}^0 \mathrm{Cr}\{\eta \leq r\}\mathrm{d}r = E[\eta],
\end{aligned}
$$

$$
\begin{aligned}
E[\xi_i + \eta_i] &= \int_0^{+\infty} \mathrm{Cr}\{\xi_i + \eta_i \geq r\}\mathrm{d}r - \int_{-\infty}^0 \mathrm{Cr}\{\xi_i + \eta_i \leq r\}\mathrm{d}r \\
&\to \int_0^{+\infty} \mathrm{Cr}\{\xi + \eta \geq r\}\mathrm{d}r - \int_{-\infty}^0 \mathrm{Cr}\{\xi + \eta \leq r\}\mathrm{d}r = E[\xi + \eta]
\end{aligned}
$$

as $i \to \infty$. It follows from Step 3 that $E[\xi + \eta] = E[\xi] + E[\eta]$.

STEP 5: We prove that $E[\xi + \eta] = E[\xi] + E[\eta]$ when ξ and η are arbitrary fuzzy variables. Since they have finite expected values, there exist two numbers c and d such that

$$
\lim_{y \uparrow 0} \mathrm{Cr}\{\xi + c \leq y\} \leq \frac{1}{2} \leq \mathrm{Cr}\{\xi + c \leq 0\},
$$

$$
\lim_{y \uparrow 0} \mathrm{Cr}\{\eta + d \leq y\} \leq \frac{1}{2} \leq \mathrm{Cr}\{\eta + d \leq 0\}.
$$

It follows from Steps 1 and 4 that

$$
\begin{aligned}
E[\xi + \eta] &= E[(\xi + c) + (\eta + d) - c - d] \\
&= E[(\xi + c) + (\eta + d)] - c - d \\
&= E[\xi + c] + E[\eta + d] - c - d \\
&= E[\xi] + c + E[\eta] + d - c - d \\
&= E[\xi] + E[\eta].
\end{aligned}
$$

STEP 6: We prove that $E[a\xi + b\eta] = aE[\xi] + bE[\eta]$ for any real numbers a and b. In fact, the equation follows immediately from Steps 2 and 5. The theorem is proved.

C.8 Variance

Definition C.12 *(Liu and Liu [114]). Let ξ be a fuzzy variable with finite expected value e. Then the variance of ξ is defined by $V[\xi] = E[(\xi - e)^2]$.*

The variance of a fuzzy variable provides a measure of the spread of the distribution around its expected value.

Example C.27: A fuzzy variable ξ is called normally distributed if it has a normal membership function

$$\mu(x) = 2 \left(1 + \exp\left(\frac{\pi|x - e|}{\sqrt{6}\sigma} \right) \right)^{-1}, \quad x \in \Re, \, \sigma > 0. \tag{C.51}$$

The expected value is e and variance is σ^2. Let ξ_1 and ξ_2 be independently and normally distributed fuzzy variables with expected values e_1 and e_2, variances σ_1^2 and σ_2^2, respectively. Then for any real numbers a_1 and a_2, the fuzzy variable $a_1\xi_1 + a_2\xi_2$ is also normally distributed with expected value $a_1e_1 + a_2e_2$ and variance $(|a_1|\sigma_1 + |a_2|\sigma_2)^2$.

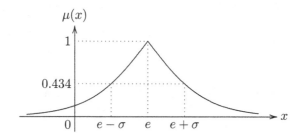

Figure C.3: Normal Membership Function

Theorem C.23. *If ξ is a fuzzy variable whose variance exists, a and b are real numbers, then $V[a\xi + b] = a^2 V[\xi]$.*

Proof: It follows from the definition of variance that

$$V[a\xi + b] = E\left[(a\xi + b - aE[\xi] - b)^2\right] = a^2 E[(\xi - E[\xi])^2] = a^2 V[\xi].$$

Theorem C.24. *Let ξ be a fuzzy variable with expected value e. Then $V[\xi] = 0$ if and only if $\mathrm{Cr}\{\xi = e\} = 1$.*

Proof: If $V[\xi] = 0$, then $E[(\xi - e)^2] = 0$. Note that

$$E[(\xi - e)^2] = \int_0^{+\infty} \mathrm{Cr}\{(\xi - e)^2 \geq r\} dr$$

which implies $\mathrm{Cr}\{(\xi - e)^2 \geq r\} = 0$ for any $r > 0$. Hence we have $\mathrm{Cr}\{(\xi - e)^2 = 0\} = 1$, i.e., $\mathrm{Cr}\{\xi = e\} = 1$. Conversely, if $\mathrm{Cr}\{\xi = e\} = 1$, then we have $\mathrm{Cr}\{(\xi - e)^2 = 0\} = 1$ and $\mathrm{Cr}\{(\xi - e)^2 \geq r\} = 0$ for any $r > 0$. Thus

$$V[\xi] = \int_0^{+\infty} \mathrm{Cr}\{(\xi - e)^2 \geq r\} dr = 0.$$

C.9   Moments

Definition C.13 *(Liu [116]). Let ξ be a fuzzy variable, and k a positive number. Then*
(a) the expected value $E[\xi^k]$ is called the kth moment;
(b) the expected value $E[|\xi|^k]$ is called the kth absolute moment;
(c) the expected value $E[(\xi - E[\xi])^k]$ is called the kth central moment;
(d) the expected value $E[|\xi - E[\xi]|^k]$ is called the kth absolute central moment.

Note that the first central moment is always 0, the first moment is just the expected value, and the second central moment is just the variance.

Example C.28: A fuzzy variable ξ is called exponentially distributed if it has an exponential membership function

$$\mu(x) = 2 \left(1 + \exp\left(\frac{\pi x}{\sqrt{6}m}\right)\right)^{-1}, \quad x \geq 0, \, m > 0. \quad (C.52)$$

The expected value is $(\sqrt{6}m \ln 2)/\pi$ and the second moment is m^2. Let ξ_1 and ξ_2 be independently and exponentially distributed fuzzy variables with second moments m_1^2 and m_2^2, respectively. Then for any positive real numbers a_1 and a_2, the fuzzy variable $a_1\xi_1 + a_2\xi_2$ is also exponentially distributed with second moment $(a_1m_1 + a_2m_2)^2$.

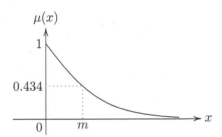

Figure C.4: Exponential Membership Function

C.10   Critical Values

In order to rank fuzzy variables, we may use two critical values: optimistic value and pessimistic value.

Definition C.14 *(Liu [112]). Let ξ be a fuzzy variable, and $\alpha \in (0,1]$. Then*

$$\xi_{\sup}(\alpha) = \sup\left\{r \mid \text{Cr}\left\{\xi \geq r\right\} \geq \alpha\right\} \quad (C.53)$$

is called the α-optimistic value to ξ, and

$$\xi_{\inf}(\alpha) = \inf\left\{r \mid \text{Cr}\left\{\xi \leq r\right\} \geq \alpha\right\} \quad (C.54)$$

is called the α-pessimistic value to ξ.

Example C.29: Let $\xi = (a, b, c)$ be a triangular fuzzy variable. Then its α-optimistic and α-pessimistic values are

$$\xi_{\sup}(\alpha) = \begin{cases} 2\alpha b + (1 - 2\alpha)c, & \text{if } \alpha \leq 0.5 \\ (2\alpha - 1)a + (2 - 2\alpha)b, & \text{if } \alpha > 0.5, \end{cases}$$

$$\xi_{\inf}(\alpha) = \begin{cases} (1 - 2\alpha)a + 2\alpha b, & \text{if } \alpha \leq 0.5 \\ (2 - 2\alpha)b + (2\alpha - 1)c, & \text{if } \alpha > 0.5. \end{cases}$$

Theorem C.25. *Let ξ be a fuzzy variable, and $\alpha \in (0, 1]$. If $\alpha > 0.5$, then we have*

$$\text{Cr}\{\xi \leq \xi_{\inf}(\alpha)\} \geq \alpha, \quad \text{Cr}\{\xi \geq \xi_{\sup}(\alpha)\} \geq \alpha. \tag{C.55}$$

Proof: It follows from the definition of α-pessimistic value that there exists a decreasing sequence $\{x_i\}$ such that $\text{Cr}\{\xi \leq x_i\} \geq \alpha$ and $x_i \downarrow \xi_{\inf}(\alpha)$ as $i \to \infty$. Since $\{\xi \leq x_i\} \downarrow \{\xi \leq \xi_{\inf}(\alpha)\}$ and $\lim_{i \to \infty} \text{Cr}\{\xi \leq x_i\} \geq \alpha > 0.5$, it follows from the credibility semicontinuity law that

$$\text{Cr}\{\xi \leq \xi_{\inf}(\alpha)\} = \lim_{i \to \infty} \text{Cr}\{\xi \leq x_i\} \geq \alpha.$$

Similarly, there exists an increasing sequence $\{x_i\}$ such that $\text{Cr}\{\xi \geq x_i\} \geq \alpha$ and $x_i \uparrow \xi_{\sup}(\alpha)$ as $i \to \infty$. Since $\{\xi \geq x_i\} \downarrow \{\xi \geq \xi_{\sup}(\alpha)\}$ and $\lim_{i \to \infty} \text{Cr}\{\xi \geq x_i\} \geq \alpha > 0.5$, it follows from the credibility semicontinuity law that

$$\text{Cr}\{\xi \geq \xi_{\sup}(\alpha)\} = \lim_{i \to \infty} \text{Cr}\{\xi \geq x_i\} \geq \alpha.$$

The theorem is proved.

Theorem C.26. *Let ξ be a fuzzy variable, and $\alpha \in (0, 1]$. Then we have*
(a) $\xi_{\inf}(\alpha)$ is an increasing and left-continuous function of α;
(b) $\xi_{\sup}(\alpha)$ is a decreasing and left-continuous function of α.

Proof: (a) Let α_1 and α_2 be two numbers with $0 < \alpha_1 < \alpha_2 \leq 1$. Then for any number $r < \xi_{\sup}(\alpha_2)$, we have

$$\text{Cr}\{\xi \geq r\} \geq \alpha_2 > \alpha_1.$$

Thus, by the definition of optimistic value, we obtain $\xi_{\sup}(\alpha_1) \geq \xi_{\sup}(\alpha_2)$. That is, the value $\xi_{\sup}(\alpha)$ is a decreasing function of α. Next, we prove the left-continuity of $\xi_{\inf}(\alpha)$ with respect to α. Let $\{\alpha_i\}$ be an arbitrary sequence of positive numbers such that $\alpha_i \uparrow \alpha$. Then $\{\xi_{\inf}(\alpha_i)\}$ is an increasing sequence. If the limitation is equal to $\xi_{\inf}(\alpha)$, then the left-continuity is proved. Otherwise, there exists a number z^* such that

$$\lim_{i \to \infty} \xi_{\inf}(\alpha_i) < z^* < \xi_{\inf}(\alpha).$$

Thus $\text{Cr}\{\xi \leq z^*\} \geq \alpha_i$ for each i. Letting $i \to \infty$, we get $\text{Cr}\{\xi \leq z^*\} \geq \alpha$. Hence $z^* \geq \xi_{\inf}(\alpha)$. A contradiction proves the left-continuity of $\xi_{\inf}(\alpha)$ with respect to α. The part (b) may be proved similarly.

C.11 Entropy

Fuzzy entropy is a measure of uncertainty and has been studied by many researchers such as De Luca and Termini [27], Kaufmann [68], Yager [210], Kosko [78], Pal and Pal [165], Bhandari and Pal [7], and Pal and Bezdek [167]. Those definitions of entropy characterize the uncertainty resulting primarily from the linguistic vagueness rather than resulting from information deficiency, and vanishes when the fuzzy variable is an equipossible one.

Liu [119] suggested that an entropy of fuzzy variables should meet at least the following three basic requirements: (i) *minimum:* the entropy of a crisp number is minimum, i.e., 0; (ii) *maximum:* the entropy of an equipossible fuzzy variable is maximum; (iii) *universality:* the entropy is applicable not only to finite and infinite cases but also to discrete and continuous cases.

In order to meet those requirements, Li and Liu [89] provided a new definition of fuzzy entropy to characterize the uncertainty resulting from information deficiency which is caused by the impossibility to predict the specified value that a fuzzy variable takes.

Entropy of Discrete Fuzzy Variables

Definition C.15 *(Li and Liu [89]). Let ξ be a discrete fuzzy variable taking values in $\{x_1, x_2, \cdots\}$. Then its entropy is defined by*

$$H[\xi] = \sum_{i=1}^{\infty} S(\mathrm{Cr}\{\xi = x_i\}) \qquad (C.56)$$

where $S(t) = -t \ln t - (1 - t) \ln(1 - t)$.

Example C.30: Suppose that ξ is a discrete fuzzy variable taking values in $\{x_1, x_2, \cdots\}$. If there exists some index k such that the membership function $\mu(x_k) = 1$, and 0 otherwise, then its entropy $H[\xi] = 0$.

Example C.31: Suppose that ξ is a simple fuzzy variable taking values in $\{x_1, x_2, \cdots, x_n\}$. If its membership function $\mu(x) \equiv 1$, then its entropy $H[\xi] = n \ln 2$.

Theorem C.27. *Suppose that ξ is a discrete fuzzy variable taking values in $\{x_1, x_2, \cdots\}$. Then*

$$H[\xi] \geq 0 \qquad (C.57)$$

and equality holds if and only if ξ is essentially a crisp number.

Proof: The nonnegativity is clear. In addition, $H[\xi] = 0$ if and only if $\mathrm{Cr}\{\xi = x_i\} = 0$ or 1 for each i. That is, there exists one and only one index k such that $\mathrm{Cr}\{\xi = x_k\} = 1$, i.e., ξ is essentially a crisp number.

This theorem states that the entropy of a fuzzy variable reaches its minimum 0 when the fuzzy variable degenerates to a crisp number. In this case, there is no uncertainty.

Theorem C.28. *Suppose that ξ is a simple fuzzy variable taking values in* $\{x_1, x_2, \cdots, x_n\}$. *Then*

$$H[\xi] \leq n \ln 2 \tag{C.58}$$

and equality holds if and only if ξ is an equipossible fuzzy variable.

Proof: Since the function $S(t)$ reaches its maximum $\ln 2$ at $t = 0.5$, we have

$$H[\xi] = \sum_{i=1}^{n} S(\mathrm{Cr}\{\xi = x_i\}) \leq n \ln 2$$

and equality holds if and only if $\mathrm{Cr}\{\xi = x_i\} = 0.5$, i.e., $\mu(x_i) \equiv 1$ for all $i = 1, 2, \cdots, n$.

This theorem states that the entropy of a fuzzy variable reaches its maximum when the fuzzy variable is an equipossible one. In this case, there is no preference among all the values that the fuzzy variable will take.

Entropy of Continuous Fuzzy Variables

Definition C.16 *(Li and Liu [89]). Let ξ be a continuous fuzzy variable. Then its entropy is defined by*

$$H[\xi] = \int_{-\infty}^{+\infty} S(\mathrm{Cr}\{\xi = x\}) \mathrm{d}x \tag{C.59}$$

where $S(t) = -t \ln t - (1-t) \ln(1-t)$.

For any continuous fuzzy variable ξ with membership function μ, we have $\mathrm{Cr}\{\xi = x\} = \mu(x)/2$ for each $x \in \Re$. Thus

$$H[\xi] = -\int_{-\infty}^{+\infty} \left(\frac{\mu(x)}{2} \ln \frac{\mu(x)}{2} + \left(1 - \frac{\mu(x)}{2} \right) \ln \left(1 - \frac{\mu(x)}{2} \right) \right) \mathrm{d}x. \tag{C.60}$$

Example C.32: Let ξ be an equipossible fuzzy variable (a, b). Then $\mu(x) = 1$ if $a \leq x \leq b$, and 0 otherwise. Thus its entropy is

$$H[\xi] = -\int_{a}^{b} \left(\frac{1}{2} \ln \frac{1}{2} + \left(1 - \frac{1}{2} \right) \ln \left(1 - \frac{1}{2} \right) \right) \mathrm{d}x = (b-a) \ln 2.$$

Example C.33: Let ξ be a triangular fuzzy variable (a, b, c). Then its entropy is $H[\xi] = (c-a)/2$.

Example C.34: Let ξ be a trapezoidal fuzzy variable (a, b, c, d). Then its entropy is $H[\xi] = (d-a)/2 + (\ln 2 - 0.5)(c-b)$.

Example C.35: Let ξ be an exponentially distributed fuzzy variable with second moment m^2. Then its entropy is $H[\xi] = \pi m / \sqrt{6}$.

Example C.36: Let ξ be a normally distributed fuzzy variable with expected value e and variance σ^2. Then its entropy is $H[\xi] = \sqrt{6}\pi\sigma/3$.

Theorem C.29. *Let ξ be a continuous fuzzy variable taking values on the interval $[a, b]$. Then*

$$H[\xi] \le (b - a)\ln 2 \tag{C.61}$$

and equality holds if and only if ξ is an equipossible fuzzy variable (a, b).

Proof: The theorem follows from the fact that the function $S(t)$ reaches its maximum $\ln 2$ at $t = 0.5$.

Maximum Entropy Principle

Given some constraints, for example, expected value and variance, there are usually multiple compatible membership functions. Which membership function shall we take? The *maximum entropy principle* attempts to select the membership function that maximizes the value of entropy and satisfies the prescribed constraints.

Theorem C.30 *(Li and Liu [92]). Let ξ be a continuous nonnegative fuzzy variable with finite second moment m^2. Then*

$$H[\xi] \le \frac{\pi m}{\sqrt{6}} \tag{C.62}$$

and the equality holds if ξ is an exponentially distributed fuzzy variable with second moment m^2.

Proof: Let μ be the membership function of ξ. Note that μ is a continuous function. The proof is based on the following two steps.

STEP 1: Suppose that μ is a decreasing function on $[0, +\infty)$. For this case, we have $\mathrm{Cr}\{\xi \ge x\} = \mu(x)/2$ for any $x > 0$. Thus the second moment

$$E[\xi^2] = \int_0^{+\infty} \mathrm{Cr}\{\xi^2 \ge x\}\mathrm{d}x = \int_0^{+\infty} 2x\mathrm{Cr}\{\xi \ge x\}\mathrm{d}x = \int_0^{+\infty} x\mu(x)\mathrm{d}x.$$

The maximum entropy membership function μ should maximize the entropy

$$-\int_0^{+\infty} \left(\frac{\mu(x)}{2} \ln \frac{\mu(x)}{2} + \left(1 - \frac{\mu(x)}{2}\right) \ln \left(1 - \frac{\mu(x)}{2}\right) \right) \mathrm{d}x$$

subject to the moment constraint

$$\int_0^{+\infty} x\mu(x)\mathrm{d}x = m^2.$$

The Lagrangian is

$$L = -\int_0^{+\infty} \left(\frac{\mu(x)}{2} \ln \frac{\mu(x)}{2} + \left(1 - \frac{\mu(x)}{2}\right) \ln \left(1 - \frac{\mu(x)}{2}\right) \right) \mathrm{d}x$$

$$-\lambda \left(\int_0^{+\infty} x\mu(x)\mathrm{d}x - m^2 \right).$$

The maximum entropy membership function meets Euler-Lagrange equation

$$\frac{1}{2} \ln \frac{\mu(x)}{2} - \frac{1}{2} \ln \left(1 - \frac{\mu(x)}{2}\right) + \lambda x = 0$$

and has the form $\mu(x) = 2\left(1 + \exp(2\lambda x)\right)^{-1}$. Substituting it into the moment constraint, we get

$$\mu^*(x) = 2\left(1 + \exp\left(\frac{\pi x}{\sqrt{6}m}\right)\right)^{-1}, \quad x \geq 0$$

which is just the exponential membership function with second moment m^2, and the maximum entropy is $H[\xi^*] = \pi m/\sqrt{6}$.

STEP 2: Let ξ be a general fuzzy variable with second moment m^2. Now we define a fuzzy variable $\widehat{\xi}$ via membership function

$$\widehat{\mu}(x) = \sup_{y \geq x} \mu(y), \quad x \geq 0.$$

Then $\widehat{\mu}$ is a decreasing function on $[0, +\infty)$, and

$$\mathrm{Cr}\{\widehat{\xi}^2 \geq x\} = \frac{1}{2} \sup_{y \geq \sqrt{x}} \widehat{\mu}(y) = \frac{1}{2} \sup_{y \geq \sqrt{x}} \sup_{z \geq y} \mu(z) = \frac{1}{2} \sup_{z \geq \sqrt{x}} \mu(z) \leq \mathrm{Cr}\{\xi^2 \geq x\}$$

for any $x > 0$. Thus we have

$$E[\widehat{\xi}^2] = \int_0^{+\infty} \mathrm{Cr}\{\widehat{\xi}^2 \geq x\}\mathrm{d}x \leq \int_0^{+\infty} \mathrm{Cr}\{\xi^2 \geq x\}\mathrm{d}x = E[\xi^2] = m^2.$$

It follows from $\mu(x) \leq \widehat{\mu}(x)$ and Step 1 that

$$H[\xi] \leq H[\widehat{\xi}] \leq \frac{\pi\sqrt{E[\widehat{\xi}^2]}}{\sqrt{6}} \leq \frac{\pi m}{\sqrt{6}}.$$

The theorem is thus proved.

Theorem C.31 (*Li and Liu [92]*). *Let ξ be a continuous fuzzy variable with finite expected value e and variance σ^2. Then*

$$H[\xi] \leq \frac{\sqrt{6}\pi\sigma}{3} \tag{C.63}$$

and the equality holds if ξ is a normally distributed fuzzy variable with expected value e and variance σ^2.

Proof: Let μ be the continuous membership function of ξ. The proof is based on the following two steps.

STEP 1: Let $\mu(x)$ be a unimodal and symmetric function about $x = e$. For this case, the variance is

$$V[\xi] = \int_0^{+\infty} \mathrm{Cr}\{(\xi - e)^2 \geq x\}\mathrm{d}x = \int_0^{+\infty} \mathrm{Cr}\{\xi - e \geq \sqrt{x}\}\mathrm{d}x$$

$$= \int_e^{+\infty} 2(x - e)\mathrm{Cr}\{\xi \geq x\}\mathrm{d}x = \int_e^{+\infty} (x - e)\mu(x)\mathrm{d}x$$

and the entropy is

$$H[\xi] = -2\int_e^{+\infty} \left(\frac{\mu(x)}{2}\ln\frac{\mu(x)}{2} + \left(1 - \frac{\mu(x)}{2}\right)\ln\left(1 - \frac{\mu(x)}{2}\right)\right)\mathrm{d}x.$$

The maximum entropy membership function μ should maximize the entropy subject to the variance constraint. The Lagrangian is

$$L = -2\int_e^{+\infty} \left(\frac{\mu(x)}{2}\ln\frac{\mu(x)}{2} + \left(1 - \frac{\mu(x)}{2}\right)\ln\left(1 - \frac{\mu(x)}{2}\right)\right)\mathrm{d}x$$

$$-\lambda\left(\int_e^{+\infty} (x - e)\mu(x)\mathrm{d}x - \sigma^2\right).$$

The maximum entropy membership function meets Euler-Lagrange equation

$$\ln\frac{\mu(x)}{2} - \ln\left(1 - \frac{\mu(x)}{2}\right) + \lambda(x - e) = 0$$

and has the form $\mu(x) = 2\left(1 + \exp\left(\lambda(x - e)\right)\right)^{-1}$. Substituting it into the variance constraint, we get

$$\mu^*(x) = 2\left(1 + \exp\left(\frac{\pi|x - e|}{\sqrt{6}\sigma}\right)\right)^{-1}, \quad x \in \Re$$

which is just the normal membership function with expected value e and variance σ^2, and the maximum entropy is $H[\xi^*] = \sqrt{6}\pi\sigma/3$.

STEP 2: Let ξ be a general fuzzy variable with expected value e and variance σ^2. We define a fuzzy variable $\widehat{\xi}$ by the membership function

$$\widehat{\mu}(x) = \begin{cases} \sup\limits_{y \leq x}(\mu(y) \vee \mu(2e - y)), & \text{if } x \leq e \\[2mm] \sup\limits_{y \geq x}(\mu(y) \vee \mu(2e - y)), & \text{if } x > e. \end{cases}$$

It is easy to verify that $\widehat{\mu}(x)$ is a unimodal and symmetric function about $x = e$. Furthermore,

$$\mathrm{Cr}\left\{(\widehat{\xi} - e)^2 \geq r\right\} = \frac{1}{2} \sup_{x \geq e + \sqrt{r}} \widehat{\mu}(x) = \frac{1}{2} \sup_{x \geq e + \sqrt{r}} \sup_{y \geq x}(\mu(y) \vee \mu(2e - y))$$

$$= \frac{1}{2} \sup_{y \geq e + \sqrt{r}} (\mu(y) \vee \mu(2e - y)) = \frac{1}{2} \sup_{(y-e)^2 \geq r} \mu(y)$$

$$\leq \mathrm{Cr}\left\{(\xi - e)^2 \geq r\right\}$$

for any $r > 0$. Thus

$$V[\widehat{\xi}] = \int_0^{+\infty} \mathrm{Cr}\{(\widehat{\xi} - e)^2 \geq r\}dr \leq \int_0^{+\infty} \mathrm{Cr}\{(\xi - e)^2 \geq r\}dr = \sigma^2.$$

It follows from $\mu(x) \leq \widehat{\mu}(x)$ and Step 1 that

$$H[\xi] \leq H[\widehat{\xi}] \leq \frac{\sqrt{6}\pi\sqrt{V[\widehat{\xi}]}}{3} \leq \frac{\sqrt{6}\pi\sigma}{3}.$$

The proof is complete.

C.12   Conditional Credibility

We now consider the credibility of an event A after it has been learned that some other event B has occurred. This new credibility of A is called the conditional credibility of A given B.

In order to define a conditional credibility measure $\mathrm{Cr}\{A|B\}$, at first we have to enlarge $\mathrm{Cr}\{A \cap B\}$ because $\mathrm{Cr}\{A \cap B\} < 1$ for all events whenever $\mathrm{Cr}\{B\} < 1$. It seems that we have no alternative but to divide $\mathrm{Cr}\{A \cap B\}$ by $\mathrm{Cr}\{B\}$. Unfortunately, $\mathrm{Cr}\{A \cap B\}/\mathrm{Cr}\{B\}$ is not always a credibility measure. However, the value $\mathrm{Cr}\{A|B\}$ should not be greater than $\mathrm{Cr}\{A \cap B\}/\mathrm{Cr}\{B\}$ (otherwise the normality will be lost), i.e.,

$$\mathrm{Cr}\{A|B\} \leq \frac{\mathrm{Cr}\{A \cap B\}}{\mathrm{Cr}\{B\}}. \tag{C.64}$$

On the other hand, in order to preserve the self-duality, we should have

$$\mathrm{Cr}\{A|B\} = 1 - \mathrm{Cr}\{A^c|B\} \geq 1 - \frac{\mathrm{Cr}\{A^c \cap B\}}{\mathrm{Cr}\{B\}}. \tag{C.65}$$

Furthermore, since $(A \cap B) \cup (A^c \cap B) = B$, we have $\mathrm{Cr}\{B\} \leq \mathrm{Cr}\{A \cap B\} + \mathrm{Cr}\{A^c \cap B\}$ by using the credibility subadditivity theorem. Thus

$$0 \leq 1 - \frac{\mathrm{Cr}\{A^c \cap B\}}{\mathrm{Cr}\{B\}} \leq \frac{\mathrm{Cr}\{A \cap B\}}{\mathrm{Cr}\{B\}} \leq 1. \tag{C.66}$$

Hence any numbers between $1 - \mathrm{Cr}\{A^c \cap B\}/\mathrm{Cr}\{B\}$ and $\mathrm{Cr}\{A \cap B\}/\mathrm{Cr}\{B\}$ are reasonable values that the conditional credibility may take. Based on the maximum uncertainty principle, we have the following conditional credibility measure.

Definition C.17 *(Liu [120]). Let $(\Theta, \mathcal{P}, \mathrm{Cr})$ be a credibility space, and $A, B \in \mathcal{P}$. Then the conditional credibility measure of A given B is defined by*

$$
\mathrm{Cr}\{A|B\} =
\begin{cases}
\dfrac{\mathrm{Cr}\{A \cap B\}}{\mathrm{Cr}\{B\}}, & \text{if } \dfrac{\mathrm{Cr}\{A \cap B\}}{\mathrm{Cr}\{B\}} < 0.5 \\[2ex]
1 - \dfrac{\mathrm{Cr}\{A^c \cap B\}}{\mathrm{Cr}\{B\}}, & \text{if } \dfrac{\mathrm{Cr}\{A^c \cap B\}}{\mathrm{Cr}\{B\}} < 0.5 \\[2ex]
0.5, & \text{otherwise}
\end{cases}
\tag{C.67}
$$

provided that $\mathrm{Cr}\{B\} > 0$.

It follows immediately from the definition of conditional credibility that

$$
1 - \frac{\mathrm{Cr}\{A^c \cap B\}}{\mathrm{Cr}\{B\}} \leq \mathrm{Cr}\{A|B\} \leq \frac{\mathrm{Cr}\{A \cap B\}}{\mathrm{Cr}\{B\}}.
\tag{C.68}
$$

Furthermore, the value of $\mathrm{Cr}\{A|B\}$ takes values as close to 0.5 as possible in the interval. In other words, it accords with the maximum uncertainty principle.

Theorem C.32. *Let $(\Theta, \mathcal{P}, \mathrm{Cr})$ be a credibility space, and B an event with $\mathrm{Cr}\{B\} > 0$. Then $\mathrm{Cr}\{\cdot|B\}$ defined by (C.67) is a credibility measure, and $(\Theta, \mathcal{P}, \mathrm{Cr}\{\cdot|B\})$ is a credibility space.*

Proof: It is sufficient to prove that $\mathrm{Cr}\{\cdot|B\}$ satisfies the normality, monotonicity, self-duality and maximality axioms. At first, it satisfies the normality axiom, i.e.,

$$
\mathrm{Cr}\{\Theta|B\} = 1 - \frac{\mathrm{Cr}\{\Theta^c \cap B\}}{\mathrm{Cr}\{B\}} = 1 - \frac{\mathrm{Cr}\{\emptyset\}}{\mathrm{Cr}\{B\}} = 1.
$$

For any events A_1 and A_2 with $A_1 \subset A_2$, if

$$
\frac{\mathrm{Cr}\{A_1 \cap B\}}{\mathrm{Cr}\{B\}} \leq \frac{\mathrm{Cr}\{A_2 \cap B\}}{\mathrm{Cr}\{B\}} < 0.5,
$$

then

$$
\mathrm{Cr}\{A_1|B\} = \frac{\mathrm{Cr}\{A_1 \cap B\}}{\mathrm{Cr}\{B\}} \leq \frac{\mathrm{Cr}\{A_2 \cap B\}}{\mathrm{Cr}\{B\}} = \mathrm{Cr}\{A_2|B\}.
$$

If

$$
\frac{\mathrm{Cr}\{A_1 \cap B\}}{\mathrm{Cr}\{B\}} \leq 0.5 \leq \frac{\mathrm{Cr}\{A_2 \cap B\}}{\mathrm{Cr}\{B\}},
$$

then $\mathrm{Cr}\{A_1|B\} \leq 0.5 \leq \mathrm{Cr}\{A_2|B\}$. If

$$0.5 < \frac{\text{Cr}\{A_1 \cap B\}}{\text{Cr}\{B\}} \leq \frac{\text{Cr}\{A_2 \cap B\}}{\text{Cr}\{B\}},$$

then we have

$$\text{Cr}\{A_1|B\} = \left(1 - \frac{\text{Cr}\{A_1^c \cap B\}}{\text{Cr}\{B\}}\right) \vee 0.5 \leq \left(1 - \frac{\text{Cr}\{A_2^c \cap B\}}{\text{Cr}\{B\}}\right) \vee 0.5 = \text{Cr}\{A_2|B\}.$$

This means that $\text{Cr}\{\cdot|B\}$ satisfies the monotonicity axiom. For any event A, if

$$\frac{\text{Cr}\{A \cap B\}}{\text{Cr}\{B\}} \geq 0.5, \quad \frac{\text{Cr}\{A^c \cap B\}}{\text{Cr}\{B\}} \geq 0.5,$$

then we have $\text{Cr}\{A|B\} + \text{Cr}\{A^c|B\} = 0.5 + 0.5 = 1$ immediately. Otherwise, without loss of generality, suppose

$$\frac{\text{Cr}\{A \cap B\}}{\text{Cr}\{B\}} < 0.5 < \frac{\text{Cr}\{A^c \cap B\}}{\text{Cr}\{B\}},$$

then we have

$$\text{Cr}\{A|B\} + \text{Cr}\{A^c|B\} = \frac{\text{Cr}\{A \cap B\}}{\text{Cr}\{B\}} + \left(1 - \frac{\text{Cr}\{A \cap B\}}{\text{Cr}\{B\}}\right) = 1.$$

That is, $\text{Cr}\{\cdot|B\}$ satisfies the self-duality axiom. Finally, for any events $\{A_i\}$ with $\sup_i \text{Cr}\{A_i|B\} < 0.5$, we have $\sup_i \text{Cr}\{A_i \cap B\} < 0.5$ and

$$\sup_i \text{Cr}\{A_i|B\} = \frac{\sup_i \text{Cr}\{A_i \cap B\}}{\text{Cr}\{B\}} = \frac{\text{Cr}\{\cup_i A_i \cap B\}}{\text{Cr}\{B\}} = \text{Cr}\{\cup_i A_i|B\}.$$

Thus $\text{Cr}\{\cdot|B\}$ satisfies the maximality axiom. Hence $\text{Cr}\{\cdot|B\}$ is a credibility measure. Furthermore, $(\Theta, \mathcal{P}, \text{Cr}\{\cdot|B\})$ is a credibility space.

Definition C.18 (Liu [120]). *The conditional membership function of a fuzzy variable ξ given B is defined by*

$$\mu(x|B) = (2\text{Cr}\{\xi = x|B\}) \wedge 1, \quad x \in \Re \tag{C.69}$$

provided that $\text{Cr}\{B\} > 0$.

Example C.37: Let ξ be a fuzzy variable with membership function $\mu(x)$, and X a set of real numbers such that $\mu(x) > 0$ for some $x \in X$. Then the conditional membership function of ξ given $\xi \in X$ is

$$\mu(x|X) = \begin{cases} \dfrac{2\mu(x)}{\sup\limits_{x \in X} \mu(x)} \wedge 1, & \text{if } \sup\limits_{x \in X} \mu(x) < 1 \\[4mm] \dfrac{2\mu(x)}{2 - \sup\limits_{x \in X^c} \mu(x)} \wedge 1, & \text{if } \sup\limits_{x \in X} \mu(x) = 1 \end{cases} \tag{C.70}$$

for $x \in X$. Please mention that $\mu(x|X) \equiv 0$ if $x \notin X$.

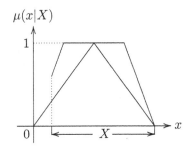 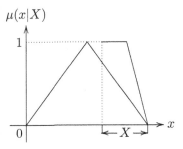

Figure C.5: Conditional Membership Function $\mu(x|X)$

Example C.38: Let ξ and η be two fuzzy variables with joint membership function $\mu(x,y)$, and Y a set of real numbers. Then the conditional membership function of ξ given $\eta \in Y$ is

$$
\mu(x|Y) = \begin{cases}
\dfrac{2\sup\limits_{y\in Y} \mu(x,y)}{\sup\limits_{x\in\Re,y\in Y} \mu(x,y)} \wedge 1, & \text{if } \sup\limits_{x\in\Re,y\in Y} \mu(x,y) < 1 \\[4ex]
\dfrac{2\sup\limits_{y\in Y} \mu(x,y)}{2 - \sup\limits_{x\in\Re,y\in Y^c} \mu(x,y)} \wedge 1, & \text{if } \sup\limits_{x\in\Re,y\in Y} \mu(x,y) = 1
\end{cases}
\tag{C.71}
$$

provided that $\mu(x,y) > 0$ for some $x \in \Re$ and $y \in Y$. Especially, the conditional membership function of ξ given $\eta = y$ is

$$
\mu(x|y) = \begin{cases}
\dfrac{2\mu(x,y)}{\sup\limits_{x\in\Re} \mu(x,y)} \wedge 1, & \text{if } \sup\limits_{x\in\Re} \mu(x,y) < 1 \\[4ex]
\dfrac{2\mu(x,y)}{2 - \sup\limits_{x\in\Re,z\neq y} \mu(x,z)} \wedge 1, & \text{if } \sup\limits_{x\in\Re} \mu(x,y) = 1
\end{cases}
$$

provided that $\mu(x,y) > 0$ for some $x \in \Re$.

Definition C.19 (Liu [120]). *The conditional credibility distribution* Φ: $\Re \to [0,1]$ *of a fuzzy variable* ξ *given* B *is defined by*

$$
\Phi(x|B) = \operatorname{Cr}\{\xi \leq x|B\}
\tag{C.72}
$$

provided that $\operatorname{Cr}\{B\} > 0$.

Definition C.20 (Liu [120]). *Let* ξ *be a fuzzy variable. Then the conditional expected value of* ξ *given* B *is defined by*

$$
E[\xi|B] = \int_0^{+\infty} \operatorname{Cr}\{\xi \geq r|B\}\mathrm{d}r - \int_{-\infty}^0 \operatorname{Cr}\{\xi \leq r|B\}\mathrm{d}r
\tag{C.73}
$$

provided that at least one of the two integrals is finite.

C.13   Fuzzy Set

Zadeh [222] introduced the concept of fuzzy set via membership function in 1965: *A fuzzy set is defined by its membership function μ which assigns to each element x a real number $\mu(x)$ in the interval $[0,1]$, where the value of $\mu(x)$ represents the grade of membership of x in the fuzzy set.* Liu [120] redefined a fuzzy set as a function from a credibility space to a collection of sets.

Definition C.21 *(Liu [120]). A fuzzy set is a function ξ from a credibility space $(\Theta, \mathcal{P}, \mathrm{Cr})$ to a collection of sets of real numbers.*

Warning: The fuzzy set in the sense of Definition C.21 is not equivalent to Zadeh's definition of fuzzy set. Thus the fuzzy set theory based on Definition C.21 is inconsistent with Zadeh's fuzzy set theory. For example, the extension principle of Zadeh is no longer valid.

Let ξ and η be two nonempty fuzzy sets. Then the strong membership degree of η to ξ is defined as the credibility measure that η is strongly included in ξ, i.e., $\mathrm{Cr}\{\eta \subset \xi\}$. The weak membership degree of η to ξ is defined as the credibility measure that η is weakly included in ξ, i.e., $\mathrm{Cr}\{\eta \not\subset \xi^c\}$.

Definition C.22 . *Let ξ and η be two nonempty fuzzy sets. Then the membership degree of η to ξ is defined as the average of strong and weak membership degrees, i.e.,*

$$\mathrm{Cr}\{\eta \rhd \xi\} = \frac{1}{2}\left(\mathrm{Cr}\{\eta \subset \xi\} + \mathrm{Cr}\{\eta \not\subset \xi^c\}\right). \tag{C.74}$$

The membership degree is understood as the credibility measure that η is imaginarily included in ξ.

Note that if η degenerates to a single point a, then the strong inclusion is identical with the weak inclusion, and $\mathrm{Cr}\{a \rhd \xi\} = \mathrm{Cr}\{a \in \xi\} = \mathrm{Cr}\{a \not\in \xi^c\}$.

Definition C.23. *Let ξ be a nonempty fuzzy set. Then the function*

$$\Phi(x) = \mathrm{Cr}\{\xi \rhd (-\infty, x]\}, \quad \forall x \in \Re \tag{C.75}$$

is called the credibility distribution of ξ.

The concept of membership function is only applicable to a special class of fuzzy sets. In other words, it is not true that every fuzzy set has its own membership function.

Definition C.24. *A fuzzy set ξ is said to have a membership function μ if the range of ξ is just the total class of μ, and*

$$\mathrm{Cr}\{\xi \in W_\alpha\} = \alpha, \quad \forall \alpha \in [0,1] \tag{C.76}$$

where W_α is the α-class of μ.

A representation theorem states that, if ξ is a fuzzy set with membership function μ, then ξ may be represented by

$$\xi = \bigcup_{0 \leq \alpha \leq 1} \alpha \cdot \mu_\alpha \qquad (C.77)$$

where μ_α is the α-cut of membership function μ.

Warning: The complement ξ^c, union $\xi \cup \eta$, intersection $\xi \cap \eta$, sum $\xi + \eta$ and product $\xi \times \eta$ of fuzzy sets have no membership functions even though the original fuzzy sets have their own membership functions.

Definition C.25. *Let ξ be a nonempty fuzzy set. Then the expected value of ξ is defined by*

$$E[\xi] = \int_0^{+\infty} \mathrm{Cr}\{\xi \rhd [r, +\infty)\}\mathrm{d}r - \int_{-\infty}^0 \mathrm{Cr}\{\xi \rhd (-\infty, r]\}\mathrm{d}r \qquad (C.78)$$

provided that at least one of the two integrals is finite.

Let ξ be a nonempty fuzzy set with credibility distribution Φ. If ξ has a finite expected value, then

$$E[\xi] = \int_0^{+\infty} (1 - \Phi(x))\mathrm{d}x - \int_{-\infty}^0 \Phi(x)\mathrm{d}x. \qquad (C.79)$$

Definition C.26. *Let ξ be a fuzzy set, and $\alpha \in (0, 1]$. Then*

$$\xi_{\sup}(\alpha) = \sup\left\{r \mid \mathrm{Cr}\left\{\xi \rhd [r, +\infty)\right\} \geq \alpha\right\} \qquad (C.80)$$

is called the α-optimistic value to ξ, and

$$\xi_{\inf}(\alpha) = \inf\left\{r \mid \mathrm{Cr}\left\{\xi \rhd (-\infty, r]\right\} \geq \alpha\right\} \qquad (C.81)$$

is called the α-pessimistic value to ξ.

Let ξ be a fuzzy set with credibility distribution Φ. Then its α-optimistic value and α-pessimistic value are

$$\xi_{\sup}(\alpha) = \Phi^{-1}(1 - \alpha), \quad \xi_{\inf}(\alpha) = \Phi^{-1}(\alpha) \qquad (C.82)$$

for any α with $0 < \alpha < 1$.

Appendix D

Chance Theory

Fuzziness and randomness are two basic types of uncertainty. In many cases, fuzziness and randomness simultaneously appear in a system. In order to describe this phenomena, a fuzzy random variable was introduced by Kwakernaak [80] as a random element taking "fuzzy variable" values. In addition, a random fuzzy variable was proposed by Liu [112] as a fuzzy element taking "random variable" values. For example, it might be known that the lifetime of a modern engine is an exponentially distributed random variable with an unknown parameter. If the parameter is provided as a fuzzy variable, then the lifetime is a random fuzzy variable.

More generally, a hybrid variable was introduced by Liu [118] in 2006 as a tool to describe the quantities with fuzziness and randomness. Fuzzy random variable and random fuzzy variable are instances of hybrid variable. In order to measure hybrid events, a concept of chance measure was introduced by Li and Liu [94] in 2009. Chance theory is a hybrid of probability theory and credibility theory. Perhaps the reader would like to know what axioms we should assume for chance theory. In fact, chance theory will be based on the three axioms of probability and five axioms of credibility.

The emphasis in this appendix is mainly on chance space, hybrid variable, chance measure, chance distribution, expected value, variance, critical values and conditional chance.

D.1 Chance Space

Chance theory begins with the concept of chance space that inherits the mathematical foundations of both probability theory and credibility theory.

Definition D.1 *(Liu [118]). Suppose that* $(\Theta, \mathcal{P}, \mathrm{Cr})$ *is a credibility space and* $(\Omega, \mathcal{A}, \mathrm{Pr})$ *is a probability space. The product* $(\Theta, \mathcal{P}, \mathrm{Cr}) \times (\Omega, \mathcal{A}, \mathrm{Pr})$ *is called a chance space.*

The universal set $\Theta \times \Omega$ is clearly the set of all ordered pairs of the form (θ, ω), where $\theta \in \Theta$ and $\omega \in \Omega$. What is the product σ-algebra $\mathcal{P} \times \mathcal{A}$? What is the product measure $\mathrm{Cr} \times \mathrm{Pr}$? Let us discuss these two basic problems.

What is the product σ-algebra $\mathcal{P} \times \mathcal{A}$?

Generally speaking, it is not true that all subsets of $\Theta \times \Omega$ are measurable. Let Λ be a subset of $\Theta \times \Omega$. Write

$$\Lambda(\theta) = \{\omega \in \Omega \mid (\theta, \omega) \in \Lambda\}. \tag{D.1}$$

It is clear that $\Lambda(\theta)$ is a subset of Ω. If $\Lambda(\theta) \in \mathcal{A}$ holds for each $\theta \in \Theta$, then Λ may be regarded as a *measurable set*.

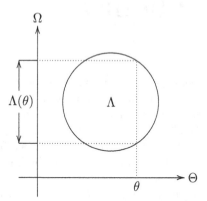

Figure D.1: $\Theta \times \Omega$, Λ and $\Lambda(\theta)$

Definition D.2 *(Liu [120]). Let $(\Theta, \mathcal{P}, \mathrm{Cr}) \times (\Omega, \mathcal{A}, \mathrm{Pr})$ be a chance space. A subset $\Lambda \subset \Theta \times \Omega$ is called an event if $\Lambda(\theta) \in \mathcal{A}$ for each $\theta \in \Theta$.*

Example D.1: Empty set $\emptyset$ and universal set $\Theta \times \Omega$ are clearly events.

Example D.2: Let $X \in \mathcal{P}$ and $Y \in \mathcal{A}$. Then $X \times Y$ is a subset of $\Theta \times \Omega$. Since the set

$$(X \times Y)(\theta) = \begin{cases} Y, & \text{if } \theta \in X \\ \emptyset, & \text{if } \theta \in X^c \end{cases}$$

is in the σ-algebra $\mathcal{A}$ for each $\theta \in \Theta$, the rectangle $X \times Y$ is an event.

Theorem D.1 *(Liu [120]). Let $(\Theta, \mathcal{P}, \mathrm{Cr}) \times (\Omega, \mathcal{A}, \mathrm{Pr})$ be a chance space. The class of all events is a σ-algebra over $\Theta \times \Omega$, and denoted by $\mathcal{P} \times \mathcal{A}$.*

Proof: At first, it is obvious that $\Theta \times \Omega \in \mathcal{P} \times \mathcal{A}$. For any event Λ, we always have

$$\Lambda(\theta) \in \mathcal{A}, \quad \forall \theta \in \Theta.$$

Thus for each $\theta \in \Theta$, the set

$$\Lambda^c(\theta) = \{\omega \in \Omega \mid (\theta, \omega) \in \Lambda^c\} = (\Lambda(\theta))^c \in \mathcal{A}$$

which implies that $\Lambda^c \in \mathcal{P} \times \mathcal{A}$. Finally, let $\Lambda_1, \Lambda_2, \cdots$ be events. Then for each $\theta \in \Theta$, we have

$$\left(\bigcup_{i=1}^{\infty} \Lambda_i\right)(\theta) = \left\{\omega \in \Omega \mid (\theta, \omega) \in \bigcup_{i=1}^{\infty} \Lambda_i\right\} = \bigcup_{i=1}^{\infty} \{\omega \in \Omega \mid (\theta, \omega) \in \Lambda_i\} \in \mathcal{A}.$$

That is, the countable union $\cup_i \Lambda_i \in \mathcal{P} \times \mathcal{A}$. Hence $\mathcal{P} \times \mathcal{A}$ is a σ-algebra.

What is the product measure $\text{Cr} \times \text{Pr}$**?**

Product probability is a probability measure, and product credibility is a credibility measure. What is the product measure $\text{Cr} \times \text{Pr}$? We will call it *chance measure* and define it as follows.

Definition D.3 *(Li and Liu [94]). Let* $(\Theta, \mathcal{P}, \text{Cr}) \times (\Omega, \mathcal{A}, \text{Pr})$ *be a chance space. Then a chance measure of an event* Λ *is defined as*

$$
\text{Ch}\{\Lambda\} =
\begin{cases}
\sup_{\theta \in \Theta}(\text{Cr}\{\theta\} \wedge \text{Pr}\{\Lambda(\theta)\}), \\
\quad \textit{if} \ \sup_{\theta \in \Theta}(\text{Cr}\{\theta\} \wedge \text{Pr}\{\Lambda(\theta)\}) < 0.5 \\
1 - \sup_{\theta \in \Theta}(\text{Cr}\{\theta\} \wedge \text{Pr}\{\Lambda^c(\theta)\}), \\
\quad \textit{if} \ \sup_{\theta \in \Theta}(\text{Cr}\{\theta\} \wedge \text{Pr}\{\Lambda(\theta)\}) \geq 0.5.
\end{cases}
\tag{D.2}
$$

Example D.3: Take a credibility space $(\Theta, \mathcal{P}, \text{Cr})$ to be $\{\theta_1, \theta_2\}$ with $\text{Cr}\{\theta_1\} = 0.6$ and $\text{Cr}\{\theta_2\} = 0.4$, and take a probability space $(\Omega, \mathcal{A}, \text{Pr})$ to be $\{\omega_1, \omega_2\}$ with $\text{Pr}\{\omega_1\} = 0.7$ and $\text{Pr}\{\omega_2\} = 0.3$. Then

$$\text{Ch}\{(\theta_1, \omega_1)\} = 0.6, \quad \text{Ch}\{(\theta_2, \omega_2)\} = 0.3.$$

Theorem D.2. *Let* $(\Theta, \mathcal{P}, \text{Cr}) \times (\Omega, \mathcal{A}, \text{Pr})$ *be a chance space and* Ch *a chance measure. Then we have*

$$\text{Ch}\{\emptyset\} = 0, \tag{D.3}$$

$$\text{Ch}\{\Theta \times \Omega\} = 1, \tag{D.4}$$

$$0 \leq \text{Ch}\{\Lambda\} \leq 1 \tag{D.5}$$

for any event Λ.

Proof: It follows from the definition immediately.

Theorem D.3. *Let* $(\Theta, \mathcal{P}, \text{Cr}) \times (\Omega, \mathcal{A}, \text{Pr})$ *be a chance space and* Ch *a chance measure. Then for any event* Λ, *we have*

$$\sup_{\theta \in \Theta}(\text{Cr}\{\theta\} \wedge \text{Pr}\{\Lambda(\theta)\}) \vee \sup_{\theta \in \Theta}(\text{Cr}\{\theta\} \wedge \text{Pr}\{\Lambda^c(\theta)\}) \geq 0.5, \tag{D.6}$$

$$\sup_{\theta \in \Theta}(\text{Cr}\{\theta\} \wedge \text{Pr}\{\Lambda(\theta)\}) + \sup_{\theta \in \Theta}(\text{Cr}\{\theta\} \wedge \text{Pr}\{\Lambda^c(\theta)\}) \leq 1, \tag{D.7}$$

$$\sup_{\theta \in \Theta}(\text{Cr}\{\theta\} \wedge \text{Pr}\{\Lambda(\theta)\}) \leq \text{Ch}\{\Lambda\} \leq 1 - \sup_{\theta \in \Theta}(\text{Cr}\{\theta\} \wedge \text{Pr}\{\Lambda^c(\theta)\}). \tag{D.8}$$

Proof: It follows from the basic properties of probability and credibility that

$$\sup_{\theta\in\Theta}(\mathrm{Cr}\{\theta\}\wedge\mathrm{Pr}\{\Lambda(\theta)\})\vee\sup_{\theta\in\Theta}(\mathrm{Cr}\{\theta\}\wedge\mathrm{Pr}\{\Lambda^c(\theta)\})$$

$$\geq\sup_{\theta\in\Theta}(\mathrm{Cr}\{\theta\}\wedge(\mathrm{Pr}\{\Lambda(\theta)\}\vee\mathrm{Pr}\{\Lambda^c(\theta)\}))$$

$$\geq\sup_{\theta\in\Theta}\mathrm{Cr}\{\theta\}\wedge0.5=0.5$$

and

$$\sup_{\theta\in\Theta}(\mathrm{Cr}\{\theta\}\wedge\mathrm{Pr}\{\Lambda(\theta)\})+\sup_{\theta\in\Theta}(\mathrm{Cr}\{\theta\}\wedge\mathrm{Pr}\{\Lambda^c(\theta)\})$$

$$=\sup_{\theta_1,\theta_2\in\Theta}(\mathrm{Cr}\{\theta_1\}\wedge\mathrm{Pr}\{\Lambda(\theta_1)\}+\mathrm{Cr}\{\theta_2\}\wedge\mathrm{Pr}\{\Lambda^c(\theta_2)\})$$

$$\leq\sup_{\theta_1\neq\theta_2}(\mathrm{Cr}\{\theta_1\}+\mathrm{Cr}\{\theta_2\})\vee\sup_{\theta\in\Theta}(\mathrm{Pr}\{\Lambda(\theta)\}+\mathrm{Pr}\{\Lambda^c(\theta)\})$$

$$\leq1\vee1=1.$$

The inequalities (D.8) follows immediately from the above inequalities and the definition of chance measure.

Theorem D.4 *(Li and Liu [94]). The chance measure is increasing. That is,*

$$\mathrm{Ch}\{\Lambda_1\}\leq\mathrm{Ch}\{\Lambda_2\} \tag{D.9}$$

for any events Λ_1 and Λ_2 with $\Lambda_1\subset\Lambda_2$.

Proof: Since $\Lambda_1(\theta)\subset\Lambda_2(\theta)$ and $\Lambda_2^c(\theta)\subset\Lambda_1^c(\theta)$ for each $\theta\in\Theta$, we have

$$\sup_{\theta\in\Theta}(\mathrm{Cr}\{\theta\}\wedge\mathrm{Pr}\{\Lambda_1(\theta)\})\leq\sup_{\theta\in\Theta}(\mathrm{Cr}\{\theta\}\wedge\mathrm{Pr}\{\Lambda_2(\theta)\}),$$

$$\sup_{\theta\in\Theta}(\mathrm{Cr}\{\theta\}\wedge\mathrm{Pr}\{\Lambda_2^c(\theta)\})\leq\sup_{\theta\in\Theta}(\mathrm{Cr}\{\theta\}\wedge\mathrm{Pr}\{\Lambda_1^c(\theta)\}).$$

The argument breaks down into three cases.

Case 1: $\sup_{\theta\in\Theta}(\mathrm{Cr}\{\theta\}\wedge\mathrm{Pr}\{\Lambda_2(\theta)\})<0.5$. For this case, we have

$$\sup_{\theta\in\Theta}(\mathrm{Cr}\{\theta\}\wedge\mathrm{Pr}\{\Lambda_1(\theta)\})<0.5,$$

$$\mathrm{Ch}\{\Lambda_2\}=\sup_{\theta\in\Theta}(\mathrm{Cr}\{\theta\}\wedge\mathrm{Pr}\{\Lambda_2(\theta)\})\geq\sup_{\theta\in\Theta}(\mathrm{Cr}\{\theta\}\wedge\mathrm{Pr}\{\Lambda_1(\theta)\}=\mathrm{Ch}\{\Lambda_1\}.$$

Case 2: $\sup_{\theta\in\Theta}(\mathrm{Cr}\{\theta\}\wedge\mathrm{Pr}\{\Lambda_2(\theta)\})\geq0.5$ and $\sup_{\theta\in\Theta}(\mathrm{Cr}\{\theta\}\wedge\mathrm{Pr}\{\Lambda_1(\theta)\})<0.5$. It follows from Theorem D.3 that

$$\mathrm{Ch}\{\Lambda_2\}\geq\sup_{\theta\in\Theta}(\mathrm{Cr}\{\theta\}\wedge\mathrm{Pr}\{\Lambda_2(\theta)\})\geq0.5>\mathrm{Ch}\{\Lambda_1\}.$$

Case 3: $\sup_{\theta\in\Theta}(\mathrm{Cr}\{\theta\}\wedge\mathrm{Pr}\{\Lambda_2(\theta)\})\geq0.5$ and $\sup_{\theta\in\Theta}(\mathrm{Cr}\{\theta\}\wedge\mathrm{Pr}\{\Lambda_1(\theta)\})\geq0.5$. For this case, we have

$$\mathrm{Ch}\{\Lambda_2\}=1-\sup_{\theta\in\Theta}(\mathrm{Cr}\{\theta\}\wedge\mathrm{Pr}\{\Lambda_2^c(\theta)\})\geq1-\sup_{\theta\in\Theta}(\mathrm{Cr}\{\theta\}\wedge\mathrm{Pr}\{\Lambda_1^c(\theta)\})=\mathrm{Ch}\{\Lambda_1\}.$$

Thus Ch is an increasing measure.

Theorem D.5 *(Li and Liu [94]). The chance measure is self-dual. That is,*

$$\text{Ch}\{\Lambda\} + \text{Ch}\{\Lambda^c\} = 1 \tag{D.10}$$

for any event Λ.

Proof: For any event Λ, please note that

$$\text{Ch}\{\Lambda^c\} = \begin{cases} \sup_{\theta\in\Theta}(\text{Cr}\{\theta\} \wedge \text{Pr}\{\Lambda^c(\theta)\}), & \text{if } \sup_{\theta\in\Theta}(\text{Cr}\{\theta\} \wedge \text{Pr}\{\Lambda^c(\theta)\}) < 0.5 \\ 1 - \sup_{\theta\in\Theta}(\text{Cr}\{\theta\} \wedge \text{Pr}\{\Lambda(\theta)\}), & \text{if } \sup_{\theta\in\Theta}(\text{Cr}\{\theta\} \wedge \text{Pr}\{\Lambda^c(\theta)\}) \geq 0.5. \end{cases}$$

The argument breaks down into three cases.

Case 1: $\sup_{\theta\in\Theta}(\text{Cr}\{\theta\} \wedge \text{Pr}\{\Lambda(\theta)\}) < 0.5$. For this case, we have

$$\sup_{\theta\in\Theta}(\text{Cr}\{\theta\} \wedge \text{Pr}\{\Lambda^c(\theta)\}) \geq 0.5,$$

$$\text{Ch}\{\Lambda\} + \text{Ch}\{\Lambda^c\} = \sup_{\theta\in\Theta}(\text{Cr}\{\theta\}\wedge\text{Pr}\{\Lambda(\theta)\}) + 1 - \sup_{\theta\in\Theta}(\text{Cr}\{\theta\}\wedge\text{Pr}\{\Lambda(\theta)\}) = 1.$$

Case 2: $\sup_{\theta\in\Theta}(\text{Cr}\{\theta\} \wedge \text{Pr}\{\Lambda(\theta)\}) \geq 0.5$ and $\sup_{\theta\in\Theta}(\text{Cr}\{\theta\} \wedge \text{Pr}\{\Lambda^c(\theta)\}) < 0.5$. For this case, we have

$$\text{Ch}\{\Lambda\} + \text{Ch}\{\Lambda^c\} = 1 - \sup_{\theta\in\Theta}(\text{Cr}\{\theta\}\wedge\text{Pr}\{\Lambda^c(\theta)\}) + \sup_{\theta\in\Theta}(\text{Cr}\{\theta\}\wedge\text{Pr}\{\Lambda^c(\theta)\}) = 1.$$

Case 3: $\sup_{\theta\in\Theta}(\text{Cr}\{\theta\} \wedge \text{Pr}\{\Lambda(\theta)\}) \geq 0.5$ and $\sup_{\theta\in\Theta}(\text{Cr}\{\theta\} \wedge \text{Pr}\{\Lambda^c(\theta)\}) \geq 0.5$. For this case, it follows from Theorem D.3 that

$$\sup_{\theta\in\Theta}(\text{Cr}\{\theta\} \wedge \text{Pr}\{\Lambda(\theta)\}) = \sup_{\theta\in\Theta}(\text{Cr}\{\theta\} \wedge \text{Pr}\{\Lambda^c(\theta)\}) = 0.5.$$

Hence $\text{Ch}\{\Lambda\} + \text{Ch}\{\Lambda^c\} = 0.5 + 0.5 = 1$. The theorem is proved.

Theorem D.6 *(Li and Liu [94]). For any event $X \times Y$, we have*

$$\text{Ch}\{X \times Y\} = \text{Cr}\{X\} \wedge \text{Pr}\{Y\}. \tag{D.11}$$

Proof: The argument breaks down into three cases.

Case 1: $\text{Cr}\{X\} < 0.5$. For this case, we have

$$\sup_{\theta\in X} \text{Cr}\{\theta\} \wedge \text{Pr}\{Y\} = \text{Cr}\{X\} \wedge \text{Cr}\{Y\} < 0.5,$$

$$\text{Ch}\{X \times Y\} = \sup_{\theta\in X} \text{Cr}\{\theta\} \wedge \text{Pr}\{Y\} = \text{Cr}\{X\} \wedge \text{Pr}\{Y\}.$$

Case 2: $\text{Cr}\{X\} \geq 0.5$ and $\text{Pr}\{Y\} < 0.5$. Then we have

$$\sup_{\theta \in X} \mathrm{Cr}\{\theta\} \geq 0.5,$$

$$\sup_{\theta \in X} \mathrm{Cr}\{\theta\} \wedge \mathrm{Pr}\{Y\} = \mathrm{Pr}\{Y\} < 0.5,$$

$$\mathrm{Ch}\{X \times Y\} = \sup_{\theta \in X} \mathrm{Cr}\{\theta\} \wedge \mathrm{Pr}\{Y\} = \mathrm{Pr}\{Y\} = \mathrm{Cr}\{X\} \wedge \mathrm{Pr}\{Y\}.$$

Case 3: $\mathrm{Cr}\{X\} \geq 0.5$ and $\mathrm{Pr}\{Y\} \geq 0.5$. Then we have

$$\sup_{\theta \in \Theta} (\mathrm{Cr}\{\theta\} \wedge \mathrm{Pr}\{(X \times Y)(\theta)\}) \geq \sup_{\theta \in X} \mathrm{Cr}\{\theta\} \wedge \mathrm{Pr}\{Y\} \geq 0.5,$$

$$\mathrm{Ch}\{X \times Y\} = 1 - \sup_{\theta \in \Theta} (\mathrm{Cr}\{\theta\} \wedge \mathrm{Pr}\{(X \times Y)^c(\theta)\}) = \mathrm{Cr}\{X\} \wedge \mathrm{Pr}\{Y\}.$$

The theorem is proved.

Example D.4: It follows from Theorem D.6 that for any events $X \times \Omega$ and $\Theta \times Y$, we have

$$\mathrm{Ch}\{X \times \Omega\} = \mathrm{Cr}\{X\}, \quad \mathrm{Ch}\{\Theta \times Y\} = \mathrm{Pr}\{Y\}. \tag{D.12}$$

Theorem D.7 *(Li and Liu [94], Chance Subadditivity Theorem).* *The chance measure is subadditive. That is,*

$$\mathrm{Ch}\{\Lambda_1 \cup \Lambda_2\} \leq \mathrm{Ch}\{\Lambda_1\} + \mathrm{Ch}\{\Lambda_2\} \tag{D.13}$$

for any events Λ_1 and Λ_2. In fact, chance measure is not only finitely sub-additive but also countably subadditive.

Proof: The proof breaks down into three cases.

Case 1: $\mathrm{Ch}\{\Lambda_1 \cup \Lambda_2\} < 0.5$. Then $\mathrm{Ch}\{\Lambda_1\} < 0.5$, $\mathrm{Ch}\{\Lambda_2\} < 0.5$ and

$$
\begin{aligned}
\mathrm{Ch}\{\Lambda_1 \cup \Lambda_2\} &= \sup_{\theta \in \Theta}(\mathrm{Cr}\{\theta\} \wedge \mathrm{Pr}\{(\Lambda_1 \cup \Lambda_2)(\theta)\}) \\
&\leq \sup_{\theta \in \Theta}(\mathrm{Cr}\{\theta\} \wedge (\mathrm{Pr}\{\Lambda_1(\theta)\} + \mathrm{Pr}\{\Lambda_2(\theta)\})) \\
&\leq \sup_{\theta \in \Theta}(\mathrm{Cr}\{\theta\} \wedge \mathrm{Pr}\{\Lambda_1(\theta)\} + \mathrm{Cr}\{\theta\} \wedge \mathrm{Pr}\{\Lambda_2(\theta)\}) \\
&\leq \sup_{\theta \in \Theta}(\mathrm{Cr}\{\theta\} \wedge \mathrm{Pr}\{\Lambda_1(\theta)\}) + \sup_{\theta \in \Theta}(\mathrm{Cr}\{\theta\} \wedge \mathrm{Pr}\{\Lambda_2(\theta)\}) \\
&= \mathrm{Ch}\{\Lambda_1\} + \mathrm{Ch}\{\Lambda_2\}.
\end{aligned}
$$

Case 2: $\mathrm{Ch}\{\Lambda_1 \cup \Lambda_2\} \geq 0.5$ and $\mathrm{Ch}\{\Lambda_1\} \vee \mathrm{Ch}\{\Lambda_2\} < 0.5$. We first have

$$\sup_{\theta \in \Theta}(\mathrm{Cr}\{\theta\} \wedge \mathrm{Pr}\{(\Lambda_1 \cup \Lambda_2)(\theta)\}) \geq 0.5.$$

For any sufficiently small number $\varepsilon > 0$, there exists a point θ such that

$$\mathrm{Cr}\{\theta\} \wedge \mathrm{Pr}\{(\Lambda_1 \cup \Lambda_2)(\theta)\} > 0.5 - \varepsilon > \mathrm{Ch}\{\Lambda_1\} \vee \mathrm{Ch}\{\Lambda_2\},$$

$$\mathrm{Cr}\{\theta\} > 0.5 - \varepsilon > \mathrm{Pr}\{\Lambda_1(\theta)\},$$
$$\mathrm{Cr}\{\theta\} > 0.5 - \varepsilon > \mathrm{Pr}\{\Lambda_2(\theta)\}.$$

Thus we have

$$\mathrm{Cr}\{\theta\} \wedge \mathrm{Pr}\{(\Lambda_1 \cup \Lambda_2)^c(\theta)\} + \mathrm{Cr}\{\theta\} \wedge \mathrm{Pr}\{\Lambda_1(\theta)\} + \mathrm{Cr}\{\theta\} \wedge \mathrm{Pr}\{\Lambda_2(\theta)\}$$
$$= \mathrm{Cr}\{\theta\} \wedge \mathrm{Pr}\{(\Lambda_1 \cup \Lambda_2)^c(\theta)\} + \mathrm{Pr}\{\Lambda_1(\theta)\} + \mathrm{Pr}\{\Lambda_2(\theta)\}$$
$$\geq \mathrm{Cr}\{\theta\} \wedge \mathrm{Pr}\{(\Lambda_1 \cup \Lambda_2)^c(\theta)\} + \mathrm{Pr}\{(\Lambda_1 \cup \Lambda_2)(\theta)\} \geq 1 - 2\varepsilon$$

because if $\mathrm{Cr}\{\theta\} \geq \mathrm{Pr}\{(\Lambda_1 \cup \Lambda_2)^c(\theta)\}$, then

$$\mathrm{Cr}\{\theta\} \wedge \mathrm{Pr}\{(\Lambda_1 \cup \Lambda_2)^c(\theta)\} + \mathrm{Pr}\{(\Lambda_1 \cup \Lambda_2)(\theta)\}$$
$$= \mathrm{Pr}\{(\Lambda_1 \cup \Lambda_2)^c(\theta)\} + \mathrm{Pr}\{(\Lambda_1 \cup \Lambda_2)(\theta)\}$$
$$= 1 \geq 1 - 2\varepsilon$$

and if $\mathrm{Cr}\{\theta\} < \mathrm{Pr}\{(\Lambda_1 \cup \Lambda_2)^c(\theta)\}$, then

$$\mathrm{Cr}\{\theta\} \wedge \mathrm{Pr}\{(\Lambda_1 \cup \Lambda_2)^c(\theta)\} + \mathrm{Pr}\{(\Lambda_1 \cup \Lambda_2)(\theta)\}$$
$$= \mathrm{Cr}\{\theta\} + \mathrm{Pr}\{(\Lambda_1 \cup \Lambda_2)(\theta)\}$$
$$\geq (0.5 - \varepsilon) + (0.5 - \varepsilon) = 1 - 2\varepsilon.$$

Taking supremum on both sides and letting $\varepsilon \to 0$, we obtain

$$\begin{aligned}
\mathrm{Ch}\{\Lambda_1 \cup \Lambda_2\} &= 1 - \sup_{\theta \in \Theta}(\mathrm{Cr}\{\theta\} \wedge \mathrm{Pr}\{(\Lambda_1 \cup \Lambda_2)^c(\theta)\}) \\
&\leq \sup_{\theta \in \Theta}(\mathrm{Cr}\{\theta\} \wedge \mathrm{Pr}\{\Lambda_1(\theta)\}) + \sup_{\theta \in \Theta}(\mathrm{Cr}\{\theta\} \wedge \mathrm{Pr}\{\Lambda_2(\theta)\}) \\
&= \mathrm{Ch}\{\Lambda_1\} + \mathrm{Ch}\{\Lambda_2\}.
\end{aligned}$$

Case 3: $\mathrm{Ch}\{\Lambda_1 \cup \Lambda_2\} \geq 0.5$ and $\mathrm{Ch}\{\Lambda_1\} \vee \mathrm{Ch}\{\Lambda_2\} \geq 0.5$. Without loss of generality, suppose $\mathrm{Ch}\{\Lambda_1\} \geq 0.5$. For each θ, we first have

$$\begin{aligned}
\mathrm{Cr}\{\theta\} \wedge \mathrm{Pr}\{\Lambda_1^c(\theta)\} &= \mathrm{Cr}\{\theta\} \wedge \mathrm{Pr}\{(\Lambda_1^c(\theta) \cap \Lambda_2^c(\theta)) \cup (\Lambda_1^c(\theta) \cap \Lambda_2(\theta))\} \\
&\leq \mathrm{Cr}\{\theta\} \wedge (\mathrm{Pr}\{(\Lambda_1 \cup \Lambda_2)^c(\theta)\} + \mathrm{Pr}\{\Lambda_2(\theta)\}) \\
&\leq \mathrm{Cr}\{\theta\} \wedge \mathrm{Pr}\{(\Lambda_1 \cup \Lambda_2)^c(\theta)\} + \mathrm{Cr}\{\theta\} \wedge \mathrm{Pr}\{\Lambda_2(\theta)\},
\end{aligned}$$

i.e., $\mathrm{Cr}\{\theta\} \wedge \mathrm{Pr}\{(\Lambda_1 \cup \Lambda_2)^c(\theta)\} \geq \mathrm{Cr}\{\theta\} \wedge \mathrm{Pr}\{\Lambda_1^c(\theta)\} - \mathrm{Cr}\{\theta\} \wedge \mathrm{Pr}\{\Lambda_2(\theta)\}$. It follows from Theorem D.3 that

$$\begin{aligned}
\mathrm{Ch}\{\Lambda_1 \cup \Lambda_2\} &= 1 - \sup_{\theta \in \Theta}(\mathrm{Cr}\{\theta\} \wedge \mathrm{Pr}\{(\Lambda_1 \cup \Lambda_2)^c(\theta)\}) \\
&\leq 1 - \sup_{\theta \in \Theta}(\mathrm{Cr}\{\theta\} \wedge \mathrm{Pr}\{\Lambda_1^c(\theta)\}) + \sup_{\theta \in \Theta}(\mathrm{Cr}\{\theta\} \wedge \mathrm{Pr}\{\Lambda_2(\theta)\}) \\
&\leq \mathrm{Ch}\{\Lambda_1\} + \mathrm{Ch}\{\Lambda_2\}.
\end{aligned}$$

The theorem is proved.

Remark D.1: For any events Λ_1 and Λ_2, it follows from the chance subadditivity theorem that the chance measure is null-additive, i.e., $\text{Ch}\{\Lambda_1 \cup \Lambda_2\} = \text{Ch}\{\Lambda_1\} + \text{Ch}\{\Lambda_2\}$ if either $\text{Ch}\{\Lambda_1\} = 0$ or $\text{Ch}\{\Lambda_2\} = 0$.

Theorem D.8. *Let $\{\Lambda_i\}$ be a decreasing sequence of events with $\text{Ch}\{\Lambda_i\} \to 0$ as $i \to \infty$. Then for any event Λ, we have*

$$\lim_{i \to \infty} \text{Ch}\{\Lambda \cup \Lambda_i\} = \lim_{i \to \infty} \text{Ch}\{\Lambda \backslash \Lambda_i\} = \text{Ch}\{\Lambda\}. \qquad (D.14)$$

Proof: Since chance measure is increasing and subadditive, we immediately have

$$\text{Ch}\{\Lambda\} \le \text{Ch}\{\Lambda \cup \Lambda_i\} \le \text{Ch}\{\Lambda\} + \text{Ch}\{\Lambda_i\}$$

for each i. Thus we get $\text{Ch}\{\Lambda \cup \Lambda_i\} \to \text{Ch}\{\Lambda\}$ by using $\text{Ch}\{\Lambda_i\} \to 0$. Since $(\Lambda \backslash \Lambda_i) \subset \Lambda \subset ((\Lambda \backslash \Lambda_i) \cup \Lambda_i)$, we have

$$\text{Ch}\{\Lambda \backslash \Lambda_i\} \le \text{Ch}\{\Lambda\} \le \text{Ch}\{\Lambda \backslash \Lambda_i\} + \text{Ch}\{\Lambda_i\}.$$

Hence $\text{Ch}\{\Lambda \backslash \Lambda_i\} \to \text{Ch}\{\Lambda\}$ by using $\text{Ch}\{\Lambda_i\} \to 0$.

Theorem D.9 *(Li and Liu [94], Chance Semicontinuity Law). For events $\Lambda_1, \Lambda_2, \cdots$, we have*

$$\lim_{i \to \infty} \text{Ch}\{\Lambda_i\} = \text{Ch}\left\{\lim_{i \to \infty} \Lambda_i\right\} \qquad (D.15)$$

if one of the following conditions is satisfied:
(a) $\text{Ch}\{\Lambda\} \le 0.5$ and $\Lambda_i \uparrow \Lambda$;   (b) $\lim_{i \to \infty} \text{Ch}\{\Lambda_i\} < 0.5$ and $\Lambda_i \uparrow \Lambda$;
(c) $\text{Ch}\{\Lambda\} \ge 0.5$ and $\Lambda_i \downarrow \Lambda$;   (d) $\lim_{i \to \infty} \text{Ch}\{\Lambda_i\} > 0.5$ and $\Lambda_i \downarrow \Lambda$.

Proof: (a) Assume $\text{Ch}\{\Lambda\} \le 0.5$ and $\Lambda_i \uparrow \Lambda$. We first have

$$\text{Ch}\{\Lambda\} = \sup_{\theta \in \Theta}(\text{Cr}\{\theta\} \wedge \text{Pr}\{\Lambda(\theta)\}), \quad \text{Ch}\{\Lambda_i\} = \sup_{\theta \in \Theta}(\text{Cr}\{\theta\} \wedge \text{Pr}\{\Lambda_i(\theta)\})$$

for $i = 1, 2, \cdots$ For each $\theta \in \Theta$, since $\Lambda_i(\theta) \uparrow \Lambda(\theta)$, it follows from the probability continuity theorem that

$$\lim_{i \to \infty} \text{Cr}\{\theta\} \wedge \text{Pr}\{\Lambda_i(\theta)\} = \text{Cr}\{\theta\} \wedge \text{Pr}\{\Lambda(\theta)\}.$$

Taking supremum on both sides, we obtain

$$\lim_{i \to \infty} \sup_{\theta \in \Theta}(\text{Cr}\{\theta\} \wedge \text{Pr}\{\Lambda_i(\theta)\}) = \sup_{\theta \in \Theta}(\text{Cr}\{\theta\} \wedge \text{Pr}\{\Lambda(\theta)\}).$$

The part (a) is verified.

(b) Assume $\lim_{i \to \infty} \text{Ch}\{\Lambda_i\} < 0.5$ and $\Lambda_i \uparrow \Lambda$. For each $\theta \in \Theta$, since

$$\text{Cr}\{\theta\} \wedge \text{Pr}\{\Lambda(\theta)\} = \lim_{i \to \infty} \text{Cr}\{\theta\} \wedge \text{Pr}\{\Lambda_i(\theta)\},$$

we have

$$\sup_{\theta \in \Theta}(\mathrm{Cr}\{\theta\} \wedge \mathrm{Pr}\{\Lambda(\theta)\}) \leq \lim_{i \to \infty} \sup_{\theta \in \Theta}(\mathrm{Cr}\{\theta\} \wedge \mathrm{Pr}\{\Lambda_i(\theta)\}) < 0.5.$$

It follows that $\mathrm{Ch}\{\Lambda\} < 0.5$ and the part (b) holds by using (a).

(c) Assume $\mathrm{Ch}\{\Lambda\} \geq 0.5$ and $\Lambda_i \downarrow \Lambda$. We have $\mathrm{Ch}\{\Lambda^c\} \leq 0.5$ and $\Lambda_i^c \uparrow \Lambda^c$. It follows from (a) that

$$\lim_{i \to \infty} \mathrm{Ch}\{\Lambda_i\} = 1 - \lim_{i \to \infty} \mathrm{Ch}\{\Lambda_i^c\} = 1 - \mathrm{Ch}\{\Lambda^c\} = \mathrm{Ch}\{\Lambda\}.$$

(d) Assume $\lim_{i \to \infty} \mathrm{Ch}\{\Lambda_i\} > 0.5$ and $\Lambda_i \downarrow \Lambda$. We have $\lim_{i \to \infty} \mathrm{Ch}\{\Lambda_i^c\} < 0.5$ and $\Lambda_i^c \uparrow \Lambda^c$. It follows from (b) that

$$\lim_{i \to \infty} \mathrm{Ch}\{\Lambda_i\} = 1 - \lim_{i \to \infty} \mathrm{Ch}\{\Lambda_i^c\} = 1 - \mathrm{Ch}\{\Lambda^c\} = \mathrm{Ch}\{\Lambda\}.$$

The theorem is proved.

Theorem D.10 *(Chance Asymptotic Theorem). For any events $\Lambda_1, \Lambda_2, \cdots$, we have*

$$\lim_{i \to \infty} \mathrm{Ch}\{\Lambda_i\} \geq 0.5, \quad \textit{if } \Lambda_i \uparrow \Theta \times \Omega, \tag{D.16}$$

$$\lim_{i \to \infty} \mathrm{Ch}\{\Lambda_i\} \leq 0.5, \quad \textit{if } \Lambda_i \downarrow \emptyset. \tag{D.17}$$

Proof: Assume $\Lambda_i \uparrow \Theta \times \Omega$. If $\lim_{i \to \infty} \mathrm{Ch}\{\Lambda_i\} < 0.5$, it follows from the chance semicontinuity law that

$$\mathrm{Ch}\{\Theta \times \Omega\} = \lim_{i \to \infty} \mathrm{Ch}\{\Lambda_i\} < 0.5$$

which is in contradiction with $\mathrm{Cr}\{\Theta \times \Omega\} = 1$. The first inequality is proved. The second one may be verified similarly.

D.2 Hybrid Variable

Recall that a random variable is a measurable function from a probability space to the set of real numbers, and a fuzzy variable is a function from a credibility space to the set of real numbers. In order to describe a quantity with both fuzziness and randomness, we introduce a concept of hybrid variable as follows.

Definition D.4 *(Liu [118]). A hybrid variable is a measurable function from a chance space $(\Theta, \mathcal{P}, \mathrm{Cr}) \times (\Omega, \mathcal{A}, \mathrm{Pr})$ to the set of real numbers, i.e., for any Borel set B of real numbers, the set*

$$\{\xi \in B\} = \{(\theta, \omega) \in \Theta \times \Omega \mid \xi(\theta, \omega) \in B\} \tag{D.18}$$

is an event.

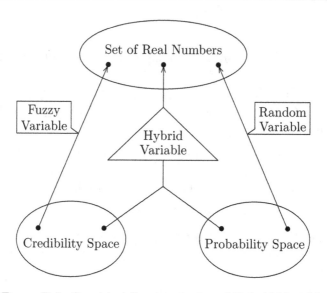

Figure D.2: Graphical Representation of Hybrid Variable

Remark D.2: A hybrid variable degenerates to a fuzzy variable if the value of $\xi(\theta, \omega)$ does not vary with ω. For example,

$$\xi(\theta, \omega) = \theta, \quad \xi(\theta, \omega) = \theta^2 + 1, \quad \xi(\theta, \omega) = \sin\theta.$$

Remark D.3: A hybrid variable degenerates to a random variable if the value of $\xi(\theta, \omega)$ does not vary with θ. For example,

$$\xi(\theta, \omega) = \omega, \quad \xi(\theta, \omega) = \omega^2 + 1, \quad \xi(\theta, \omega) = \sin\omega.$$

Remark D.4: For each fixed $\theta \in \Theta$, it is clear that the hybrid variable $\xi(\theta, \omega)$ is a measurable function from the probability space $(\Omega, \mathcal{A}, \Pr)$ to the set of real numbers. Thus it is a random variable and we will denote it by $\xi(\theta, \cdot)$. Then a hybrid variable $\xi(\theta, \omega)$ may also be regarded as a function from a credibility space $(\Theta, \mathcal{P}, \mathrm{Cr})$ to the set $\{\xi(\theta, \cdot)|\theta \in \Theta\}$ of random variables. Thus ξ is a *random fuzzy variable* defined by Liu [112].

Remark D.5: For each fixed $\omega \in \Omega$, it is clear that the hybrid variable $\xi(\theta, \omega)$ is a function from the credibility space $(\Theta, \mathcal{P}, \mathrm{Cr})$ to the set of real numbers. Thus it is a fuzzy variable and we will denote it by $\xi(\cdot, \omega)$. Then a hybrid variable $\xi(\theta, \omega)$ may be regarded as a function from a probability space $(\Omega, \mathcal{A}, \Pr)$ to the set $\{\xi(\cdot, \omega)|\omega \in \Omega\}$ of fuzzy variables. If $\mathrm{Cr}\{\xi(\cdot, \omega) \in B\}$ is a measurable function of ω for any Borel set B of real numbers, then ξ is a *fuzzy random variable* in the sense of Liu and Liu [136].

Model I

If $\tilde{a}$ is a fuzzy variable and η is a random variable, then the sum $\xi = \tilde{a} + \eta$ is a hybrid variable. The product $\xi = \tilde{a} \cdot \eta$ is also a hybrid variable. Generally speaking, if $f : \Re^2 \to \Re$ is a measurable function, then

$$\xi = f(\tilde{a}, \eta) \qquad (D.19)$$

is a hybrid variable. Suppose that $\tilde{a}$ has a membership function μ, and η has a probability density function ϕ. Then for any Borel set B of real numbers, we have

$$\mathrm{Ch}\{f(\tilde{a}, \eta) \in B\} = \begin{cases} \sup_x \left(\dfrac{\mu(x)}{2} \wedge \displaystyle\int_{f(x,y)\in B} \phi(y)\mathrm{d}y \right), \\[2ex] \qquad \text{if } \sup_x \left(\dfrac{\mu(x)}{2} \wedge \displaystyle\int_{f(x,y)\in B} \phi(y)\mathrm{d}y \right) < 0.5 \\[3ex] 1 - \sup_x \left(\dfrac{\mu(x)}{2} \wedge \displaystyle\int_{f(x,y)\in B^c} \phi(y)\mathrm{d}y \right), \\[2ex] \qquad \text{if } \sup_x \left(\dfrac{\mu(x)}{2} \wedge \displaystyle\int_{f(x,y)\in B} \phi(y)\mathrm{d}y \right) \geq 0.5. \end{cases}$$

More generally, let $\tilde{a}_1, \tilde{a}_2, \cdots, \tilde{a}_m$ be fuzzy variables, and let $\eta_1, \eta_2, \cdots, \eta_n$ be random variables. If $f : \Re^{m+n} \to \Re$ is a measurable function, then

$$\xi = f(\tilde{a}_1, \tilde{a}_2, \cdots, \tilde{a}_m; \eta_1, \eta_2, \cdots, \eta_n) \qquad (D.20)$$

is a hybrid variable. The chance $\mathrm{Ch}\{f(\tilde{a}_1, \tilde{a}_2, \cdots, \tilde{a}_m; \eta_1, \eta_2, \cdots, \eta_n) \in B\}$ may be calculated in a similar way provided that μ is the joint membership function and ϕ is the joint probability density function.

Model II

Let $\tilde{a}_1, \tilde{a}_2, \cdots, \tilde{a}_m$ be fuzzy variables, and let $p_1, p_2, \cdots, p_m$ be nonnegative numbers with $p_1 + p_2 + \cdots + p_m = 1$. Then

$$\xi = \begin{cases} \tilde{a}_1 \text{ with probability } p_1 \\ \tilde{a}_2 \text{ with probability } p_2 \\ \cdots \\ \tilde{a}_m \text{ with probability } p_m \end{cases} \qquad (D.21)$$

is clearly a hybrid variable. If $\tilde{a}_1, \tilde{a}_2, \cdots, \tilde{a}_m$ have membership functions $\mu_1, \mu_2, \cdots, \mu_m$, respectively, then for any set B of real numbers, we have

$$
\mathrm{Ch}\{\xi \in B\} =
\begin{cases}
\displaystyle \sup_{x_1,x_2\cdots,x_m} \left(\left(\min_{1\le i\le m} \frac{\mu_i(x_i)}{2} \right) \wedge \sum_{i=1}^{m}\{p_i \mid x_i \in B\} \right), \\[2mm]
\quad \text{if } \displaystyle \sup_{x_1,x_2\cdots,x_m} \left(\left(\min_{1\le i\le m} \frac{\mu_i(x_i)}{2} \right) \wedge \sum_{i=1}^{m}\{p_i \mid x_i \in B\} \right) < 0.5 \\[4mm]
\displaystyle 1 - \sup_{x_1,x_2\cdots,x_m} \left(\left(\min_{1\le i\le m} \frac{\mu_i(x_i)}{2} \right) \wedge \sum_{i=1}^{m}\{p_i \mid x_i \in B^c\} \right), \\[2mm]
\quad \text{if } \displaystyle \sup_{x_1,x_2\cdots,x_m} \left(\left(\min_{1\le i\le m} \frac{\mu_i(x_i)}{2} \right) \wedge \sum_{i=1}^{m}\{p_i \mid x_i \in B\} \right) \ge 0.5.
\end{cases}
$$

Model III

Let $\eta_1, \eta_2, \cdots, \eta_m$ be random variables, and let $u_1, u_2, \cdots, u_m$ be nonnegative numbers with $u_1 \vee u_2 \vee \cdots \vee u_m = 1$. Then

$$
\xi =
\begin{cases}
\eta_1 \text{ with membership degree } u_1 \\
\eta_2 \text{ with membership degree } u_2 \\
\quad \cdots \\
\eta_m \text{ with membership degree } u_m
\end{cases}
\tag{D.22}
$$

is clearly a hybrid variable. If $\eta_1, \eta_2, \cdots, \eta_m$ have probability density functions $\phi_1, \phi_2, \cdots, \phi_m$, respectively, then for any Borel set B of real numbers, we have

$$
\mathrm{Ch}\{\xi \in B\} =
\begin{cases}
\displaystyle \max_{1\le i\le m} \left(\frac{u_i}{2} \wedge \int_B \phi_i(x)\mathrm{d}x \right), \\[2mm]
\quad \text{if } \displaystyle \max_{1\le i\le m} \left(\frac{u_i}{2} \wedge \int_B \phi_i(x)\mathrm{d}x \right) < 0.5 \\[4mm]
\displaystyle 1 - \max_{1\le i\le m} \left(\frac{u_i}{2} \wedge \int_{B^c} \phi_i(x)\mathrm{d}x \right), \\[2mm]
\quad \text{if } \displaystyle \max_{1\le i\le m} \left(\frac{u_i}{2} \wedge \int_B \phi_i(x)\mathrm{d}x \right) \ge 0.5.
\end{cases}
$$

Model IV

In many statistics problems, the probability density function is completely known except for the values of one or more parameters. For example, it might be known that the lifetime ξ of a modern engine is an exponentially distributed random variable with an unknown expected value β. Usually, there is some relevant information in practice. It is thus possible to specify an interval in which the value of β is likely to lie, or to give an approximate estimate of the value of β. It is typically not possible to determine the value of β exactly. If the value of β is provided as a fuzzy variable, then ξ is a hybrid variable. More generally, suppose that ξ has a probability density function

$$\phi(x; \tilde{a}_1, \tilde{a}_2, \cdots, \tilde{a}_m), \quad x \in \Re \tag{D.23}$$

in which the parameters $\tilde{a}_1, \tilde{a}_2, \cdots, \tilde{a}_m$ are fuzzy variables rather than crisp numbers. Then ξ is a hybrid variable provided that $\phi(x; y_1, y_2, \cdots, y_m)$ is a probability density function for any $(y_1, y_2, \cdots, y_m)$ that $(\tilde{a}_1, \tilde{a}_2, \cdots, \tilde{a}_m)$ may take. If $\tilde{a}_1, \tilde{a}_2, \cdots, \tilde{a}_m$ have membership functions $\mu_1, \mu_2, \cdots, \mu_m$, respectively, then for any Borel set B of real numbers, the chance $\mathrm{Ch}\{\xi \in B\}$ is

$$
\begin{cases}
\sup\limits_{y_1, y_2 \cdots, y_m} \left(\left(\min\limits_{1 \le i \le m} \dfrac{\mu_i(y_i)}{2} \right) \wedge \displaystyle\int_B \phi(x; y_1, y_2, \cdots, y_m) \mathrm{d}x \right), \\[2mm]
\quad \text{if} \sup\limits_{y_1, y_2, \cdots, y_m} \left(\left(\min\limits_{1 \le i \le m} \dfrac{\mu_i(y_i)}{2} \right) \wedge \displaystyle\int_B \phi(x; y_1, y_2, \cdots, y_m) \mathrm{d}x \right) < 0.5 \\[2mm]
1 - \sup\limits_{y_1, y_2 \cdots, y_m} \left(\left(\min\limits_{1 \le i \le m} \dfrac{\mu_i(y_i)}{2} \right) \wedge \displaystyle\int_{B^c} \phi(x; y_1, y_2, \cdots, y_m) \mathrm{d}x \right), \\[2mm]
\quad \text{if} \sup\limits_{y_1, y_2, \cdots, y_m} \left(\left(\min\limits_{1 \le i \le m} \dfrac{\mu_i(y_i)}{2} \right) \wedge \displaystyle\int_B \phi(x; y_1, y_2, \cdots, y_m) \mathrm{d}x \right) \ge 0.5.
\end{cases}
$$

When are two hybrid variables equal to each other?

Definition D.5. *Let ξ_1 and ξ_2 be hybrid variables defined on the chance space $(\Theta, \mathcal{P}, \mathrm{Cr}) \times (\Omega, \mathcal{A}, \mathrm{Pr})$. We say $\xi_1 = \xi_2$ if $\xi_1(\theta, \omega) = \xi_2(\theta, \omega)$ for almost all $(\theta, \omega) \in \Theta \times \Omega$.*

Hybrid Vectors

Definition D.6. *An n-dimensional hybrid vector is a measurable function from a chance space $(\Theta, \mathcal{P}, \mathrm{Cr}) \times (\Omega, \mathcal{A}, \mathrm{Pr})$ to the set of n-dimensional real vectors, i.e., for any Borel set B of $\Re^n$, the set*

$$\{\boldsymbol{\xi} \in B\} = \left\{ (\theta, \omega) \in \Theta \times \Omega \mid \boldsymbol{\xi}(\theta, \omega) \in B \right\} \tag{D.24}$$

is an event.

Theorem D.11. *The vector $(\xi_1, \xi_2, \cdots, \xi_n)$ is a hybrid vector if and only if $\xi_1, \xi_2, \cdots, \xi_n$ are hybrid variables.*

Proof: Write $\boldsymbol{\xi} = (\xi_1, \xi_2, \cdots, \xi_n)$. Suppose that $\boldsymbol{\xi}$ is a hybrid vector on the chance space $(\Theta, \mathcal{P}, \mathrm{Cr}) \times (\Omega, \mathcal{A}, \mathrm{Pr})$. For any Borel set B of $\Re$, the set $B \times \Re^{n-1}$ is a Borel set of $\Re^n$. Thus the set

$$\{\xi_1 \in B\} = \{\xi_1 \in B, \xi_2 \in \Re, \cdots, \xi_n \in \Re\} = \{\boldsymbol{\xi} \in B \times \Re^{n-1}\}$$

is an event. Hence ξ_1 is a hybrid variable. A similar process may prove that $\xi_2, \xi_3, \cdots, \xi_n$ are hybrid variables. Conversely, suppose that all $\xi_1, \xi_2, \cdots, \xi_n$ are hybrid variables on the chance space $(\Theta, \mathcal{P}, \mathrm{Cr}) \times (\Omega, \mathcal{A}, \mathrm{Pr})$. We define

$$\mathcal{B} = \left\{ B \subset \mathfrak{R}^n \mid \{\boldsymbol{\xi} \in B\} \text{ is an event} \right\}.$$

The vector $\boldsymbol{\xi} = (\xi_1, \xi_2, \cdots, \xi_n)$ is proved to be a hybrid vector if we can prove that $\mathcal{B}$ contains all Borel sets of $\mathfrak{R}^n$. First, the class $\mathcal{B}$ contains all open intervals of $\mathfrak{R}^n$ because

$$\left\{ \boldsymbol{\xi} \in \prod_{i=1}^n (a_i, b_i) \right\} = \bigcap_{i=1}^n \{\xi_i \in (a_i, b_i)\}$$

is an event. Next, the class $\mathcal{B}$ is a σ-algebra of $\mathfrak{R}^n$ because (i) we have $\mathfrak{R}^n \in \mathcal{B}$ since $\{\boldsymbol{\xi} \in \mathfrak{R}^n\} = \Theta \times \Omega$; (ii) if $B \in \mathcal{B}$, then $\{\boldsymbol{\xi} \in B\}$ is an event, and

$$\{\boldsymbol{\xi} \in B^c\} = \{\boldsymbol{\xi} \in B\}^c$$

is an event. This means that $B^c \in \mathcal{B}$; (iii) if $B_i \in \mathcal{B}$ for $i = 1, 2, \cdots$, then $\{\boldsymbol{\xi} \in B_i\}$ are events and

$$\left\{ \boldsymbol{\xi} \in \bigcup_{i=1}^\infty B_i \right\} = \bigcup_{i=1}^\infty \{\boldsymbol{\xi} \in B_i\}$$

is an event. This means that $\cup_i B_i \in \mathcal{B}$. Since the smallest σ-algebra containing all open intervals of $\mathfrak{R}^n$ is just the Borel algebra of $\mathfrak{R}^n$, the class $\mathcal{B}$ contains all Borel sets of $\mathfrak{R}^n$. The theorem is proved.

Hybrid Arithmetic

Definition D.7. *Let $f : \mathfrak{R}^n \to \mathfrak{R}$ be a measurable function, and $\xi_1, \xi_2, \cdots, \xi_n$ hybrid variables on the chance space $(\Theta, \mathcal{P}, \mathrm{Cr}) \times (\Omega, \mathcal{A}, \mathrm{Pr})$. Then $\xi = f(\xi_1, \xi_2, \cdots, \xi_n)$ is a hybrid variable defined as*

$$\xi(\theta, \omega) = f(\xi_1(\theta, \omega), \xi_2(\theta, \omega), \cdots, \xi_n(\theta, \omega)), \quad \forall (\theta, \omega) \in \Theta \times \Omega. \quad (\mathrm{D}.25)$$

Example D.5: Let ξ_1 and ξ_2 be two hybrid variables. Then the sum $\xi = \xi_1 + \xi_2$ is a hybrid variable defined by

$$\xi(\theta, \omega) = \xi_1(\theta, \omega) + \xi_2(\theta, \omega), \quad \forall (\theta, \omega) \in \Theta \times \Omega.$$

The product $\xi = \xi_1 \xi_2$ is also a hybrid variable defined by

$$\xi(\theta, \omega) = \xi_1(\theta, \omega) \cdot \xi_2(\theta, \omega), \quad \forall (\theta, \omega) \in \Theta \times \Omega.$$

Theorem D.12. *Let $\boldsymbol{\xi}$ be an n-dimensional hybrid vector, and $f : \mathfrak{R}^n \to \mathfrak{R}$ a measurable function. Then $f(\boldsymbol{\xi})$ is a hybrid variable.*

Proof: Assume that $\boldsymbol{\xi}$ is a hybrid vector on the chance space $(\Theta, \mathcal{P}, \mathrm{Cr}) \times (\Omega, \mathcal{A}, \mathrm{Pr})$. For any Borel set B of $\Re$, since f is a measurable function, the $f^{-1}(B)$ is a Borel set of $\Re^n$. Thus the set

$$\{f(\boldsymbol{\xi}) \in B\} = \{\boldsymbol{\xi} \in f^{-1}(B)\}$$

is an event for any Borel set B. Hence $f(\boldsymbol{\xi})$ is a hybrid variable.

D.3   Chance Distribution

Chance distribution has been defined in several ways. Here we accept the following definition of chance distribution of hybrid variables.

Definition D.8 *(Li and Liu [94]). The chance distribution $\Phi \colon \Re \to [0,1]$ of a hybrid variable ξ is defined by*

$$\Phi(x) = \mathrm{Ch}\left\{\xi \leq x\right\}. \tag{D.26}$$

Example D.6: Let η be a random variable on a probability space $(\Omega, \mathcal{A}, \mathrm{Pr})$. It is clear that η may be regarded as a hybrid variable on the chance space $(\Theta, \mathcal{P}, \mathrm{Cr}) \times (\Omega, \mathcal{A}, \mathrm{Pr})$ as follows,

$$\xi(\theta, \omega) = \eta(\omega), \quad \forall(\theta, \omega) \in \Theta \times \Omega.$$

Thus its chance distribution is

$$\Phi(x) = \mathrm{Ch}\{\xi \leq x\} = \mathrm{Ch}\{\Theta \times \{\eta \leq x\}\} = \mathrm{Cr}\{\Theta\} \wedge \mathrm{Pr}\{\eta \leq x\} = \mathrm{Pr}\{\eta \leq x\}$$

which is just the probability distribution of the random variable η.

Example D.7: Let $\tilde{a}$ be a fuzzy variable on a credibility space $(\Theta, \mathcal{P}, \mathrm{Cr})$. It is clear that $\tilde{a}$ may be regarded as a hybrid variable on the chance space $(\Theta, \mathcal{P}, \mathrm{Cr}) \times (\Omega, \mathcal{A}, \mathrm{Pr})$ as follows,

$$\xi(\theta, \omega) = \tilde{a}(\theta), \quad \forall(\theta, \omega) \in \Theta \times \Omega.$$

Thus its chance distribution is

$$\Phi(x) = \mathrm{Ch}\{\xi \leq x\} = \mathrm{Ch}\{\{\tilde{a} \leq x\} \times \Omega\} = \mathrm{Cr}\{\tilde{a} \leq x\} \wedge \mathrm{Pr}\{\Omega\} = \mathrm{Cr}\{\tilde{a} \leq x\}$$

which is just the credibility distribution of the fuzzy variable $\tilde{a}$.

Theorem D.13 *(Sufficient and Necessary Condition for Chance Distribution). A function $\Phi : \Re \to [0,1]$ is a chance distribution if and only if it is an increasing function with*

$$\lim_{x \to -\infty} \Phi(x) \leq 0.5 \leq \lim_{x \to +\infty} \Phi(x), \tag{D.27}$$

$$\lim_{y \downarrow x} \Phi(y) = \Phi(x) \ \textit{if} \ \lim_{y \downarrow x} \Phi(y) > 0.5 \ \textit{or} \ \Phi(x) \geq 0.5. \tag{D.28}$$

Proof: It is obvious that a chance distribution Φ is an increasing function. The inequalities (D.27) follow from the chance asymptotic theorem immediately. Assume that x is a point at which $\lim_{y\downarrow x}\Phi(y) > 0.5$. That is,

$$\lim_{y\downarrow x}\mathrm{Ch}\{\xi \leq y\} > 0.5.$$

Since $\{\xi \leq y\} \downarrow \{\xi \leq x\}$ as $y \downarrow x$, it follows from the chance semicontinuity law that

$$\Phi(y) = \mathrm{Ch}\{\xi \leq y\} \downarrow \mathrm{Ch}\{\xi \leq x\} = \Phi(x)$$

as $y \downarrow x$. When x is a point at which $\Phi(x) \geq 0.5$, if $\lim_{y\downarrow x}\Phi(y) \neq \Phi(x)$, then we have

$$\lim_{y\downarrow x}\Phi(y) > \Phi(x) \geq 0.5.$$

For this case, we have proved that $\lim_{y\downarrow x}\Phi(y) = \Phi(x)$. Thus (D.28) is proved.

Conversely, suppose $\Phi : \Re \rightarrow [0,1]$ is an increasing function satisfying (D.27) and (D.28). Theorem C.17 states that there is a fuzzy variable whose credibility distribution is just $\Phi(x)$. Since a fuzzy variable is a special hybrid variable, the theorem is proved.

D.4   Expected Value

Expected value has been defined in several ways. This book uses the following definition of expected value operator of hybrid variables.

Definition D.9 *(Li and Liu [94]). Let ξ be a hybrid variable. Then the expected value of ξ is defined by*

$$E[\xi] = \int_0^{+\infty} \mathrm{Ch}\{\xi \geq r\}\mathrm{d}r - \int_{-\infty}^0 \mathrm{Ch}\{\xi \leq r\}\mathrm{d}r \qquad (D.29)$$

provided that at least one of the two integrals is finite.

Example D.8: If a hybrid variable ξ degenerates to a random variable η, then

$$\mathrm{Ch}\{\xi \leq x\} = \mathrm{Pr}\{\eta \leq x\}, \quad \mathrm{Ch}\{\xi \geq x\} = \mathrm{Pr}\{\eta \geq x\}, \quad \forall x \in \Re.$$

It follows from (D.29) that $E[\xi] = E[\eta]$. In other words, the expected value operator of hybrid variable coincides with that of random variable.

Example D.9: If a hybrid variable ξ degenerates to a fuzzy variable $\tilde{a}$, then

$$\mathrm{Ch}\{\xi \leq x\} = \mathrm{Cr}\{\tilde{a} \leq x\}, \quad \mathrm{Ch}\{\xi \geq x\} = \mathrm{Cr}\{\tilde{a} \geq x\}, \quad \forall x \in \Re.$$

It follows from (D.29) that $E[\xi] = E[\tilde{a}]$. In other words, the expected value operator of hybrid variable coincides with that of fuzzy variable.

Example D.10: Let $\tilde{a}$ be a fuzzy variable and η a random variable with finite expected values. Then the hybrid variable $\xi = \tilde{a} + \eta$ has expected value $E[\xi] = E[\tilde{a}] + E[\eta]$.

Theorem D.14. *Let ξ be a hybrid variable with finite expected values. Then for any real numbers a and b, we have*

$$E[a\xi + b] = aE[\xi] + b. \tag{D.30}$$

Proof: STEP 1: We first prove that $E[\xi + b] = E[\xi] + b$ for any real number b. If $b \geq 0$, we have

$$
\begin{aligned}
E[\xi + b] &= \int_0^{+\infty} \mathrm{Ch}\{\xi + b \geq r\}\mathrm{d}r - \int_{-\infty}^0 \mathrm{Ch}\{\xi + b \leq r\}\mathrm{d}r \\
&= \int_0^{+\infty} \mathrm{Ch}\{\xi \geq r - b\}\mathrm{d}r - \int_{-\infty}^0 \mathrm{Ch}\{\xi \leq r - b\}\mathrm{d}r \\
&= E[\xi] + \int_0^b (\mathrm{Ch}\{\xi \geq r - b\} + \mathrm{Ch}\{\xi < r - b\})\mathrm{d}r \\
&= E[\xi] + b.
\end{aligned}
$$

If $b < 0$, then we have

$$E[a\xi + b] = E[\xi] - \int_b^0 (\mathrm{Ch}\{\xi \geq r - b\} + \mathrm{Ch}\{\xi < r - b\})\mathrm{d}r = E[\xi] + b.$$

STEP 2: We prove $E[a\xi] = aE[\xi]$. If $a = 0$, then the equation $E[a\xi] = aE[\xi]$ holds trivially. If $a > 0$, we have

$$
\begin{aligned}
E[a\xi] &= \int_0^{+\infty} \mathrm{Ch}\{a\xi \geq r\}\mathrm{d}r - \int_{-\infty}^0 \mathrm{Ch}\{a\xi \leq r\}\mathrm{d}r \\
&= \int_0^{+\infty} \mathrm{Ch}\{\xi \geq r/a\}\mathrm{d}r - \int_{-\infty}^0 \mathrm{Ch}\{\xi \leq r/a\}\mathrm{d}r \\
&= a\int_0^{+\infty} \mathrm{Ch}\{\xi \geq t\}\mathrm{d}t - a\int_{-\infty}^0 \mathrm{Ch}\{\xi \leq t\}\mathrm{d}t \\
&= aE[\xi].
\end{aligned}
$$

If $a < 0$, we have

$$
\begin{aligned}
E[a\xi] &= \int_0^{+\infty} \mathrm{Ch}\{a\xi \geq r\}\mathrm{d}r - \int_{-\infty}^0 \mathrm{Ch}\{a\xi \leq r\}\mathrm{d}r \\
&= \int_0^{+\infty} \mathrm{Ch}\{\xi \leq r/a\}\mathrm{d}r - \int_{-\infty}^0 \mathrm{Ch}\{\xi \geq r/a\}\mathrm{d}r \\
&= a\int_0^{+\infty} \mathrm{Ch}\{\xi \geq t\}\mathrm{d}t - a\int_{-\infty}^0 \mathrm{Ch}\{\xi \leq t\}\mathrm{d}t \\
&= aE[\xi].
\end{aligned}
$$

STEP 3: For any real numbers a and b, it follows from Steps 1 and 2 that

$$E[a\xi + b] = E[a\xi] + b = aE[\xi] + b.$$

The theorem is proved.

D.5   Variance

Definition D.10 *(Li and Liu [94]). Let ξ be a hybrid variable with finite expected value e. Then the variance of ξ is defined by $V[\xi] = E[(\xi - e)^2]$.*

Theorem D.15. *If ξ is a hybrid variable with finite expected value, a and b are real numbers, then $V[a\xi + b] = a^2 V[\xi]$.*

Proof: It follows from the definition of variance that

$$V[a\xi + b] = E\left[(a\xi + b - aE[\xi] - b)^2\right] = a^2 E[(\xi - E[\xi])^2] = a^2 V[\xi].$$

Theorem D.16. *Let ξ be a hybrid variable with expected value e. Then $V[\xi] = 0$ if and only if $\mathrm{Ch}\{\xi = e\} = 1$.*

Proof: If $V[\xi] = 0$, then $E[(\xi - e)^2] = 0$. Note that

$$E[(\xi - e)^2] = \int_0^{+\infty} \mathrm{Ch}\{(\xi - e)^2 \geq r\}\mathrm{d}r$$

which implies $\mathrm{Ch}\{(\xi - e)^2 \geq r\} = 0$ for any $r > 0$. Hence we have

$$\mathrm{Ch}\{(\xi - e)^2 = 0\} = 1.$$

That is, $\mathrm{Ch}\{\xi = e\} = 1$. Conversely, if $\mathrm{Ch}\{\xi = e\} = 1$, then we have $\mathrm{Ch}\{(\xi - e)^2 = 0\} = 1$ and $\mathrm{Ch}\{(\xi - e)^2 \geq r\} = 0$ for any $r > 0$. Thus

$$V[\xi] = \int_0^{+\infty} \mathrm{Ch}\{(\xi - e)^2 \geq r\}\mathrm{d}r = 0.$$

The theorem is proved.

D.6   Critical Values

In order to rank hybrid variables, we introduce the following definition of critical values of hybrid variables.

Definition D.11 *(Li and Liu [94]). Let ξ be a hybrid variable, and $\alpha \in (0, 1]$. Then*

$$\xi_{\sup}(\alpha) = \sup\left\{r \mid \mathrm{Ch}\left\{\xi \geq r\right\} \geq \alpha\right\} \tag{D.31}$$

is called the α-optimistic value to ξ, and

$$\xi_{\inf}(\alpha) = \inf\left\{r \mid \mathrm{Ch}\left\{\xi \leq r\right\} \geq \alpha\right\} \tag{D.32}$$

is called the α-pessimistic value to ξ.

The hybrid variable ξ reaches upwards of the α-optimistic value $\xi_{\sup}(\alpha)$, and is below the α-pessimistic value $\xi_{\inf}(\alpha)$ with chance α.

Example D.11: If a hybrid variable ξ degenerates to a random variable η, then

$$\mathrm{Ch}\{\xi \leq x\} = \mathrm{Pr}\{\eta \leq x\}, \quad \mathrm{Ch}\{\xi \geq x\} = \mathrm{Pr}\{\eta \geq x\}, \quad \forall x \in \Re.$$

It follows from the definition of critical values that

$$\xi_{\sup}(\alpha) = \eta_{\sup}(\alpha), \quad \xi_{\inf}(\alpha) = \eta_{\inf}(\alpha), \quad \forall \alpha \in (0,1].$$

In other words, the critical values of hybrid variable coincide with that of random variable.

Example D.12: If a hybrid variable ξ degenerates to a fuzzy variable $\tilde{a}$, then

$$\mathrm{Ch}\{\xi \leq x\} = \mathrm{Cr}\{\tilde{a} \leq x\}, \quad \mathrm{Ch}\{\xi \geq x\} = \mathrm{Cr}\{\tilde{a} \geq x\}, \quad \forall x \in \Re.$$

It follows from the definition of critical values that

$$\xi_{\sup}(\alpha) = \tilde{a}_{\sup}(\alpha), \quad \xi_{\inf}(\alpha) = \tilde{a}_{\inf}(\alpha), \quad \forall \alpha \in (0,1].$$

In other words, the critical values of hybrid variable coincide with that of fuzzy variable.

Theorem D.17. *Let ξ be a hybrid variable, and $\alpha \in (0,1]$. If $\alpha > 0.5$, then we have*

$$\mathrm{Ch}\{\xi \leq \xi_{\inf}(\alpha)\} \geq \alpha, \quad \mathrm{Ch}\{\xi \geq \xi_{\sup}(\alpha)\} \geq \alpha. \tag{D.33}$$

Proof: It follows from the definition of α-pessimistic value that there exists a decreasing sequence $\{x_i\}$ such that $\mathrm{Ch}\{\xi \leq x_i\} \geq \alpha$ and $x_i \downarrow \xi_{\inf}(\alpha)$ as $i \to \infty$. Since $\{\xi \leq x_i\} \downarrow \{\xi \leq \xi_{\inf}(\alpha)\}$ and $\lim_{i \to \infty} \mathrm{Ch}\{\xi \leq x_i\} \geq \alpha > 0.5$, it follows from the chance semicontinuity theorem that

$$\mathrm{Ch}\{\xi \leq \xi_{\inf}(\alpha)\} = \lim_{i \to \infty} \mathrm{Ch}\{\xi \leq x_i\} \geq \alpha.$$

Similarly, there exists an increasing sequence $\{x_i\}$ such that $\mathrm{Ch}\{\xi \geq x_i\} \geq \alpha$ and $x_i \uparrow \xi_{\sup}(\alpha)$ as $i \to \infty$. Since $\{\xi \geq x_i\} \downarrow \{\xi \geq \xi_{\sup}(\alpha)\}$ and $\lim_{i \to \infty} \mathrm{Ch}\{\xi \geq x_i\} \geq \alpha > 0.5$, it follows from the chance semicontinuity theorem that

$$\mathrm{Ch}\{\xi \geq \xi_{\sup}(\alpha)\} = \lim_{i \to \infty} \mathrm{Ch}\{\xi \geq x_i\} \geq \alpha.$$

The theorem is proved.

Theorem D.18. *Let ξ be a hybrid variable, and $\alpha \in (0,1]$. Then we have*
(a) $\xi_{\sup}(\alpha)$ is a decreasing and left-continuous function of α;
(b) $\xi_{\inf}(\alpha)$ is an increasing and left-continuous function of α.

Proof: (a) Let α_1 and α_2 be two numbers with $0 < \alpha_1 < \alpha_2 \leq 1$. Then for any number $r < \xi_{\sup}(\alpha_2)$, we have

$$\mathrm{Ch}\{\xi \geq r\} \geq \alpha_2 > \alpha_1.$$

Thus, by the definition of optimistic value, we obtain $\xi_{\sup}(\alpha_1) \geq \xi_{\sup}(\alpha_2)$. That is, the value $\xi_{\sup}(\alpha)$ is a decreasing function of α. Next, we prove the left-continuity of $\xi_{\inf}(\alpha)$ with respect to α. Let $\{\alpha_i\}$ be an arbitrary sequence of positive numbers such that $\alpha_i \uparrow \alpha$. Then $\{\xi_{\inf}(\alpha_i)\}$ is an increasing sequence. If the limitation is equal to $\xi_{\inf}(\alpha)$, then the left-continuity is proved. Otherwise, there exists a number z^* such that

$$\lim_{i \to \infty} \xi_{\inf}(\alpha_i) < z^* < \xi_{\inf}(\alpha).$$

Thus $\mathrm{Ch}\{\xi \leq z^*\} \geq \alpha_i$ for each i. Letting $i \to \infty$, we get $\mathrm{Ch}\{\xi \leq z^*\} \geq \alpha$. Hence $z^* \geq \xi_{\inf}(\alpha)$. A contradiction proves the left-continuity of $\xi_{\inf}(\alpha)$ with respect to α. The part (b) may be proved similarly.

D.7   Conditional Chance

We consider the chance measure of an event A after it has been learned that some other event B has occurred. This new chance measure of A is called the conditional chance measure of A given B.

In order to define a conditional chance measure $\mathrm{Ch}\{A|B\}$, at first we have to enlarge $\mathrm{Ch}\{A \cap B\}$ because $\mathrm{Ch}\{A \cap B\} < 1$ for all events whenever $\mathrm{Ch}\{B\} < 1$. It seems that we have no alternative but to divide $\mathrm{Ch}\{A \cap B\}$ by $\mathrm{Ch}\{B\}$. Unfortunately, $\mathrm{Ch}\{A \cap B\}/\mathrm{Ch}\{B\}$ is not always a chance measure. However, the value $\mathrm{Ch}\{A|B\}$ should not be greater than $\mathrm{Ch}\{A \cap B\}/\mathrm{Ch}\{B\}$ (otherwise the normality will be lost), i.e.,

$$\mathrm{Ch}\{A|B\} \leq \frac{\mathrm{Ch}\{A \cap B\}}{\mathrm{Ch}\{B\}}. \tag{D.34}$$

On the other hand, in order to preserve the self-duality, we should have

$$\mathrm{Ch}\{A|B\} = 1 - \mathrm{Ch}\{A^c|B\} \geq 1 - \frac{\mathrm{Ch}\{A^c \cap B\}}{\mathrm{Ch}\{B\}}. \tag{D.35}$$

Furthermore, since $(A \cap B) \cup (A^c \cap B) = B$, we have $\mathrm{Ch}\{B\} \leq \mathrm{Ch}\{A \cap B\} + \mathrm{Ch}\{A^c \cap B\}$ by using the chance subadditivity theorem. Thus

$$0 \leq 1 - \frac{\mathrm{Ch}\{A^c \cap B\}}{\mathrm{Ch}\{B\}} \leq \frac{\mathrm{Ch}\{A \cap B\}}{\mathrm{Ch}\{B\}} \leq 1. \tag{D.36}$$

Hence any numbers between $1 - \mathrm{Ch}\{A^c \cap B\}/\mathrm{Ch}\{B\}$ and $\mathrm{Ch}\{A \cap B\}/\mathrm{Ch}\{B\}$ are reasonable values that the conditional chance may take. Based on the maximum uncertainty principle, we have the following conditional chance measure.

Definition D.12 *(Li and Liu [97]). Let* $(\Theta, \mathcal{P}, \mathrm{Cr}) \times (\Omega, \mathcal{A}, \mathrm{Pr})$ *be a chance space and* A, B *two events. Then the conditional chance measure of* A *given* B *is defined by*

$$
\mathrm{Ch}\{A|B\} = \begin{cases} \dfrac{\mathrm{Ch}\{A \cap B\}}{\mathrm{Ch}\{B\}}, & \text{if } \dfrac{\mathrm{Ch}\{A \cap B\}}{\mathrm{Ch}\{B\}} < 0.5 \\[2mm] 1 - \dfrac{\mathrm{Ch}\{A^c \cap B\}}{\mathrm{Ch}\{B\}}, & \text{if } \dfrac{\mathrm{Ch}\{A^c \cap B\}}{\mathrm{Ch}\{B\}} < 0.5 \\[2mm] 0.5, & \text{otherwise} \end{cases} \qquad (D.37)
$$

provided that $\mathrm{Ch}\{B\} > 0$.

Remark D.6: It follows immediately from the definition of conditional chance that

$$
1 - \frac{\mathrm{Ch}\{A^c \cap B\}}{\mathrm{Ch}\{B\}} \leq \mathrm{Ch}\{A|B\} \leq \frac{\mathrm{Ch}\{A \cap B\}}{\mathrm{Ch}\{B\}}. \qquad (D.38)
$$

Furthermore, it is clear that the conditional chance measure obeys the maximum uncertainty principle.

Remark D.7: Let X and Y be events in the credibility space. Then the conditional chance measure of $X \times \Omega$ given $Y \times \Omega$ is

$$
\mathrm{Ch}\{X \times \Omega | Y \times \Omega\} = \begin{cases} \dfrac{\mathrm{Cr}\{X \cap Y\}}{\mathrm{Cr}\{Y\}}, & \text{if } \dfrac{\mathrm{Cr}\{X \cap Y\}}{\mathrm{Cr}\{Y\}} < 0.5 \\[2mm] 1 - \dfrac{\mathrm{Cr}\{X^c \cap Y\}}{\mathrm{Cr}\{Y\}}, & \text{if } \dfrac{\mathrm{Cr}\{X^c \cap Y\}}{\mathrm{Cr}\{Y\}} < 0.5 \\[2mm] 0.5, & \text{otherwise} \end{cases}
$$

which is just the conditional credibility of X given Y.

Remark D.8: Let X and Y be events in the probability space. Then the conditional chance measure of $\Theta \times X$ given $\Theta \times Y$ is

$$
\mathrm{Ch}\{\Theta \times X | \Theta \times Y\} = \frac{\mathrm{Pr}\{X \cap Y\}}{\mathrm{Pr}\{Y\}}
$$

which is just the conditional probability of X given Y.

Theorem D.19 *(Li and Liu [97]). Conditional chance measure is normal, increasing, self-dual and countably subadditive.*

Proof: At first, the conditional chance measure $\mathrm{Ch}\{\cdot|B\}$ is normal, i.e.,

$$
\mathrm{Ch}\{\Theta \times \Omega | B\} = 1 - \frac{\mathrm{Ch}\{\emptyset\}}{\mathrm{Ch}\{B\}} = 1.
$$

For any events A_1 and A_2 with $A_1 \subset A_2$, if

$$\frac{\text{Ch}\{A_1 \cap B\}}{\text{Ch}\{B\}} \leq \frac{\text{Ch}\{A_2 \cap B\}}{\text{Ch}\{B\}} < 0.5,$$

then

$$\text{Ch}\{A_1|B\} = \frac{\text{Ch}\{A_1 \cap B\}}{\text{Ch}\{B\}} \leq \frac{\text{Ch}\{A_2 \cap B\}}{\text{Ch}\{B\}} = \text{Ch}\{A_2|B\}.$$

If

$$\frac{\text{Ch}\{A_1 \cap B\}}{\text{Ch}\{B\}} \leq 0.5 \leq \frac{\text{Ch}\{A_2 \cap B\}}{\text{Ch}\{B\}},$$

then $\text{Ch}\{A_1|B\} \leq 0.5 \leq \text{Ch}\{A_2|B\}$. If

$$0.5 < \frac{\text{Ch}\{A_1 \cap B\}}{\text{Ch}\{B\}} \leq \frac{\text{Ch}\{A_2 \cap B\}}{\text{Ch}\{B\}},$$

then we have

$$\text{Ch}\{A_1|B\} = \left(1 - \frac{\text{Ch}\{A_1^c \cap B\}}{\text{Ch}\{B\}}\right) \vee 0.5 \leq \left(1 - \frac{\text{Ch}\{A_2^c \cap B\}}{\text{Ch}\{B\}}\right) \vee 0.5 = \text{Ch}\{A_2|B\}.$$

This means that $\text{Ch}\{\cdot|B\}$ is increasing. For any event A, if

$$\frac{\text{Ch}\{A \cap B\}}{\text{Ch}\{B\}} \geq 0.5, \quad \frac{\text{Ch}\{A^c \cap B\}}{\text{Ch}\{B\}} \geq 0.5,$$

then we have $\text{Ch}\{A|B\} + \text{Ch}\{A^c|B\} = 0.5 + 0.5 = 1$ immediately. Otherwise, without loss of generality, suppose

$$\frac{\text{Ch}\{A \cap B\}}{\text{Ch}\{B\}} < 0.5 < \frac{\text{Ch}\{A^c \cap B\}}{\text{Ch}\{B\}},$$

then we have

$$\text{Ch}\{A|B\} + \text{Ch}\{A^c|B\} = \frac{\text{Ch}\{A \cap B\}}{\text{Ch}\{B\}} + \left(1 - \frac{\text{Ch}\{A \cap B\}}{\text{Ch}\{B\}}\right) = 1.$$

That is, $\text{Ch}\{\cdot|B\}$ is self-dual. Finally, for any countable sequence $\{A_i\}$ of events, if $\text{Ch}\{A_i|B\} < 0.5$ for all i, it follows from the countable subadditivity of chance measure that

$$\text{Ch}\left\{\bigcup_{i=1}^{\infty} A_i \cap B\right\} \leq \frac{\text{Ch}\left\{\bigcup_{i=1}^{\infty} A_i \cap B\right\}}{\text{Ch}\{B\}} \leq \frac{\sum_{i=1}^{\infty}\text{Ch}\{A_i \cap B\}}{\text{Ch}\{B\}} = \sum_{i=1}^{\infty}\text{Ch}\{A_i|B\}.$$

Suppose there is one term greater than 0.5, say

$$\text{Ch}\{A_1|B\} \geq 0.5, \quad \text{Ch}\{A_i|B\} < 0.5, \quad i = 2, 3, \cdots$$

If $\mathrm{Ch}\{\cup_i A_i | B\} = 0.5$, then we immediately have

$$\mathrm{Ch}\left\{\bigcup_{i=1}^{\infty} A_i \cap B\right\} \le \sum_{i=1}^{\infty} \mathrm{Ch}\{A_i | B\}.$$

If $\mathrm{Ch}\{\cup_i A_i | B\} > 0.5$, we may prove the above inequality by the following facts:

$$A_1^c \cap B \subset \bigcup_{i=2}^{\infty} (A_i \cap B) \cup \left(\bigcap_{i=1}^{\infty} A_i^c \cap B\right),$$

$$\mathrm{Ch}\{A_1^c \cap B\} \le \sum_{i=2}^{\infty} \mathrm{Ch}\{A_i \cap B\} + \mathrm{Ch}\left\{\bigcap_{i=1}^{\infty} A_i^c \cap B\right\},$$

$$\mathrm{Ch}\left\{\bigcup_{i=1}^{\infty} A_i | B\right\} = 1 - \frac{\mathrm{Ch}\left\{\bigcap_{i=1}^{\infty} A_i^c \cap B\right\}}{\mathrm{Ch}\{B\}},$$

$$\sum_{i=1}^{\infty} \mathrm{Ch}\{A_i | B\} \ge 1 - \frac{\mathrm{Ch}\{A_1^c \cap B\}}{\mathrm{Ch}\{B\}} + \frac{\sum_{i=2}^{\infty} \mathrm{Ch}\{A_i \cap B\}}{\mathrm{Ch}\{B\}}.$$

If there are at least two terms greater than 0.5, then the countable subadditivity is clearly true. Thus $\mathrm{Ch}\{\cdot | B\}$ is countably subadditive. Hence the theorem is verified.

Definition D.13 *(Li and Liu [97]). The conditional chance distribution Φ: $\Re \to [0,1]$ of a hybrid variable ξ given B is defined by*

$$\Phi(x|B) = \mathrm{Ch}\{\xi \le x | B\} \tag{D.39}$$

provided that $\mathrm{Ch}\{B\} > 0$.

Definition D.14 *(Li and Liu [97]). Let ξ be a hybrid variable. Then the conditional expected value of ξ given B is defined by*

$$E[\xi|B] = \int_0^{+\infty} \mathrm{Ch}\{\xi \ge r | B\}\mathrm{d}r - \int_{-\infty}^{0} \mathrm{Ch}\{\xi \le r | B\}\mathrm{d}r \tag{D.40}$$

provided that at least one of the two integrals is finite.

Bibliography

[1] Alefeld, G., Herzberger, J.: Introduction to Interval Computations. Academic Press, New York (1983)

[2] Atanassov, K.T.: Intuitionistic Fuzzy Sets: Theory and Applications. Physica-Verlag, Heidelberg (1999)

[3] Bamber, D., Goodman, I.R., Nguyen, H.T.: Extension of the concept of propositional deduction from classical logic to probability: An overview of probability-selection approaches. Information Sciences 131, 195–250 (2001)

[4] Bandemer, H., Nather, W.: Fuzzy Data Analysis. Kluwer, Dordrecht (1992)

[5] Bedford, T., Cooke, M.R.: Probabilistic Risk Analysis. Cambridge University Press, Cambridge (2001)

[6] Bellman, R.E., Zadeh, L.A.: Decision making in a fuzzy environment. Management Science 17, 141–164 (1970)

[7] Bhandari, D., Pal, N.R.: Some new information measures of fuzzy sets. Information Sciences 67, 209–228 (1993)

[8] Black, F., Scholes, M.: The pricing of option and corporate liabilities. Journal of Political Economy 81, 637–654 (1973)

[9] Bouchon-Meunier, B., Mesiar, R., Ralescu, D.A.: Linear non-additive set-functions. International Journal of General Systems 33(1), 89–98 (2004)

[10] Buckley, J.J.: Possibility and necessity in optimization. Fuzzy Sets and Systems 25, 1–13 (1988)

[11] Buckley, J.J.: Stochastic versus possibilistic programming. Fuzzy Sets and Systems 34, 173–177 (1990)

[12] Cadenas, J.M., Verdegay, J.L.: Using fuzzy numbers in linear programming. IEEE Transactions on Systems, Man and Cybernetics–Part B 27(6), 1016–1022 (1997)

[13] Campos, L., González, A.: A subjective approach for ranking fuzzy numbers. Fuzzy Sets and Systems 29, 145–153 (1989)

[14] Campos, L., Verdegay, J.L.: Linear programming problems and ranking of fuzzy numbers. Fuzzy Sets and Systems 32, 1–11 (1989)

[15] Campos, F.A., Villar, J., Jimenez, M.: Robust solutions using fuzzy chance constraints. Engineering Optimization 38(6), 627–645 (2006)

[16] Chen, S.J., Hwang, C.L.: Fuzzy Multiple Attribute Decision Making: Methods and Applications. Springer, Berlin (1992)

[17] Chen, X.W., Liu, B.: Existence and uniqueness theorem for uncertain differential equations. Fuzzy Optimization and Decision Making 9(1), 69–81 (2010)

[18] Chen, X.W., Ralescu, D.A.: A note on truth value in uncertain logic. In: Proceedings of the Eighth International Conference on Information and Management Sciences, Kunming, China, July 20-28, pp. 739–741 (2009), http://orsc.edu.cn/online/090211.pdf

[19] Chen, X.W., Dai, W.: Maximum entropy principle for uncertain variables, http://orsc.edu.cn/online/090618.pdf

[20] Chen, X.W.: Stability analysis of linear uncertain differential equations, http://orsc.edu.cn/online/091001.pdf

[21] Chen, Y., Fung, R.Y.K., Yang, J.: Fuzzy expected value modelling approach for determining target values of engineering characteristics in QFD. International Journal of Production Research 43(17), 3583–3604 (2005)

[22] Chen, Y., Fung, R.Y.K., Tang, J.F.: Rating technical attributes in fuzzy QFD by integrating fuzzy weighted average method and fuzzy expected value operator. European Journal of Operational Research 174(3), 1553–1566 (2006)

[23] Choquet, G.: Theory of capacities. Annals de l'Institute Fourier 5, 131–295 (1954)

[24] Dai, W., Chen, X.W.: Entropy of function of uncertain variables, http://orsc.edu.cn/online/090805.pdf

[25] Das, B., Maity, K., Maiti, A.: A two warehouse supply-chain model under possibility/necessity/credibility measures. Mathematical and Computer Modelling 46(3-4), 398–409 (2007)

[26] De Cooman, G.: Possibility theory I-III. International Journal of General Systems 25, 291–371 (1997)

[27] De Luca, A., Termini, S.: A definition of nonprobabilistic entropy in the setting of fuzzy sets theory. Information and Control 20, 301–312 (1972)

[28] Dempster, A.P.: Upper and lower probabilities induced by a multivalued mapping. Ann. Math. Stat. 38(2), 325–339 (1967)

[29] Dubois, D., Prade, H.: Fuzzy logics and generalized modus ponens revisited. Cybernetics and Systems 15, 293–331 (1984)

[30] Dubois, D., Prade, H.: The mean value of a fuzzy number. Fuzzy Sets and Systems 24, 279–300 (1987)

[31] Dubois, D., Prade, H.: Possibility Theory: An Approach to Computerized Processing of Uncertainty. Plenum, New York (1988)

[32] Elkan, C.: The paradoxical success of fuzzy logic. IEEE Expert 9(4), 3–8 (1994)

[33] Elkan, C.: The paradoxical controversy over fuzzy logic. IEEE Expert 9(4), 47–49 (1994)

[34] Esogbue, A.O., Liu, B.: Reservoir operations optimization via fuzzy criterion decision processes. Fuzzy Optimization and Decision Making 5(3), 289–305 (2006)

[35] Feng, Y., Yang, L.X.: A two-objective fuzzy k-cardinality assignment problem. Journal of Computational and Applied Mathematics 197(1), 233–244 (2006)

[36] Fung, R.Y.K., Chen, Y.Z., Chen, L.: A fuzzy expected value-based goal programing model for product planning using quality function deployment. Engineering Optimization 37(6), 633–647 (2005)

[37] Gao, J., Liu, B.: Fuzzy multilevel programming with a hybrid intelligent algorithm. Computers & Mathematics with Applications 49, 1539–1548 (2005)

[38] Gao, J., Lu, M.: Fuzzy quadratic minimum spanning tree problem. Applied Mathematics and Computation 164(3), 773–788 (2005)

[39] Gao, J.: Credibilistic game with fuzzy information. Journal of Uncertain Systems 1(1), 74–80 (2007)

[40] Gao, J., Gao, X.: A new stock model for credibilistic option pricing. Journal of Uncertain Systems 2(4), 243–247 (2008)

[41] Gao, X.: Some properties of continuous uncertain measure, International Journal of Uncertainty. Fuzziness & Knowledge-Based Systems 17(3), 419–426 (2009)

[42] Gao, X., Gao, Y., Ralescu, D.A.: On Liu's inference rule for uncertain systems. International Journal of Uncertainty, Fuzziness and Knowledge-Based Systems 18(1), 1–11 (2010)

[43] Gao, Y.: Uncertain inference control for balancing inverted pendulum, http://orsc.edu.cn/online/100128.pdf

[44] González, A.: A study of the ranking function approach through mean values. Fuzzy Sets and Systems 35, 29–41 (1990)

[45] Guan, J., Bell, D.A.: Evidence Theory and its Applications. North-Holland, Amsterdam (1991)

[46] Guo, R., Zhao, R., Guo, D., Dunne, T.: Random fuzzy variable modeling on repairable system. Journal of Uncertain Systems 1(3), 222–234 (2007)

[47] Ha, M.H., Wang, X.Z., Yang, L.Z.: Sequences of (S) fuzzy integrable functions. Fuzzy Sets and Systems 138(3), 507–522 (2003)

[48] Ha, M.H., Li, Y., Wang, X.F.: Fuzzy knowledge representation and reasoning using a generalized fuzzy petri net and a similarity measure. Soft Computing 11(4), 323–327 (2007)

[49] Hansen, E.: Global Optimization Using Interval Analysis. Marcel Dekker, New York (1992)

[50] He, Y., Xu, J.: A class of random fuzzy programming model and its application to vehicle routing problem. World Journal of Modelling and Simulation 1(1), 3–11 (2005)

[51] Heilpern, S.: The expected value of a fuzzy number. Fuzzy Sets and Systems 47, 81–86 (1992)

[52] Higashi, M., Klir, G.J.: On measures of fuzziness and fuzzy complements. International Journal of General Systems 8, 169–180 (1982)

[53] Higashi, M., Klir, G.J.: Measures of uncertainty and information based on possibility distributions. International Journal of General Systems 9, 43–58 (1983)

[54] Hisdal, E.: Conditional possibilities independence and noninteraction. Fuzzy Sets and Systems 1, 283–297 (1978)

[55] Hisdal, E.: Logical Structures for Representation of Knowledge and Uncertainty. Physica-Verlag, Heidelberg (1998)

[56] Hong, D.H.: Renewal process with T-related fuzzy inter-arrival times and fuzzy rewards. Information Sciences 176(16), 2386–2395 (2006)

[57] Hu, B.G., Mann, G.K.I., Gosine, R.G.: New methodology for analytical and optimal design of fuzzy PID controllers. IEEE Transactions on Fuzzy Systems 7(5), 521–539 (1999)

[58] Hu, L.J., Wu, R., Shao, S.: Analysis of dynamical systems whose inputs are fuzzy stochastic processes. Fuzzy Sets and Systems 129(1), 111–118 (2002)

[59] Inuiguchi, M., Ramík, J.: Possibilistic linear programming: A brief review of fuzzy mathematical programming and a comparison with stochastic programming in portfolio selection problem. Fuzzy Sets and Systems 111(1), 3–28 (2000)

[60] Ishibuchi, H., Tanaka, H.: Multiobjective programming in optimization of the interval objective function. European Journal of Operational Research 48, 219–225 (1990)

[61] Jaynes, E.T.: Information theory and statistical mechanics. Physical Reviews 106(4), 620–630 (1957)

[62] Ji, X.Y., Shao, Z.: Model and algorithm for bilevel Newsboy problem with fuzzy demands and discounts. Applied Mathematics and Computation 172(1), 163–174 (2006)

[63] Ji, X.Y., Iwamura, K.: New models for shortest path problem with fuzzy arc lengths. Applied Mathematical Modelling 31, 259–269 (2007)

[64] John, R.I.: Type 2 fuzzy sets: An appraisal of theory and applications. International Journal of Uncertainty, Fuzziness & Knowledge-Based Systems 6(6), 563–576 (1998)

[65] Kacprzyk, J., Esogbue, A.O.: Fuzzy dynamic programming: Main developments and applications. Fuzzy Sets and Systems 81, 31–45 (1996)

[66] Kacprzyk, J.: Multistage Fuzzy Control. Wiley, Chichester (1997)

[67] Karnik, N.N., Mendel, J.M., Liang, Q.: Centroid of a type-2 fuzzy set. Information Sciences 132, 195–220 (2001)

[68] Kaufmann, A.: Introduction to the Theory of Fuzzy Subsets, vol. I. Academic Press, New York (1975)

[69] Kaufmann, A., Gupta, M.M.: Fuzzy Mathematical Models in Engineering and Management Science, 2nd edn. North-Holland, Amsterdam (1991)

[70] Ke, H., Liu, B.: Project scheduling problem with stochastic activity duration times. Applied Mathematics and Computation 168(1), 342–353 (2005)

[71] Ke, H., Liu, B.: Project scheduling problem with mixed uncertainty of randomness and fuzziness. European Journal of Operational Research 183(1), 135–147 (2007)

[72] Ke, H., Liu, B.: Fuzzy project scheduling problem and its hybrid intelligent algorithm. Applied Mathematical Modelling 34(2), 301–308 (2010)

[73] Klement, E.P., Puri, M.L., Ralescu, D.A.: Limit theorems for fuzzy random variables. Proceedings of the Royal Society of London Series A 407, 171–182 (1986)

[74] Klir, G.J., Folger, T.A.: Fuzzy Sets, Uncertainty, and Information. Prentice-Hall, Englewood Cliffs (1980)

[75] Klir, G.J., Yuan, B.: Fuzzy Sets and Fuzzy Logic: Theory and Applications. Prentice-Hall, New Jersey (1995)

[76] Knopfmacher, J.: On measures of fuzziness. Journal of Mathematical Analysis and Applications 49, 529–534 (1975)

[77] Kolmogorov, A.N.: Grundbegriffe der Wahrscheinlichkeitsrechnung. Julius Springer, Berlin (1993)

[78] Kosko, B.: Fuzzy entropy and conditioning. Information Sciences 40, 165–174 (1986)

[79] Kruse, R., Meyer, K.D.: Statistics with Vague Data. D. Reidel Publishing Company, Dordrecht (1987)

[80] Kwakernaak, H.: Fuzzy random variables–I: Definitions and theorems. Information Sciences 15, 1–29 (1978)

[81] Kwakernaak, H.: Fuzzy random variables–II: Algorithms and examples for the discrete case. Information Sciences 17, 253–278 (1979)

[82] Lai, Y.J., Hwang, C.L.: Fuzzy Multiple Objective Decision Making: Methods and Applications. Springer, New York (1994)

[83] Lee, E.S.: Fuzzy multiple level programming. Applied Mathematics and Computation 120, 79–90 (2001)

[84] Lee, K.H.: First Course on Fuzzy Theory and Applications. Springer, Berlin (2005)

[85] Lertworasirkul, S., Fang, S.C., Joines, J.A., Nuttle, H.L.W.: Fuzzy data envelopment analysis (DEA): a possibility approach. Fuzzy Sets and Systems 139(2), 379–394 (2003)

[86] Li, D.Y., Cheung, D.W., Shi, X.M.: Uncertainty reasoning based on cloud models in controllers. Computers and Mathematics with Appications 35(3), 99–123 (1998)

[87] Li, D.Y., Du, Y.: Artificial Intelligence with Uncertainty. National Defense Industry Press, Beijing (2005)

[88] Li, J., Xu, J., Gen, M.: A class of multiobjective linear programming model with fuzzy random coefficients. Mathematical and Computer Modelling, Vol 44(11-12), 1097–1113 (2006)

[89] Li, P., Liu, B.: Entropy of credibility distributions for fuzzy variables. IEEE Transactions on Fuzzy Systems 16(1), 123–129 (2008)

[90] Li, S.M., Ogura, Y., Kreinovich, V.: Limit Theorems and Applications of Set-Valued and Fuzzy Set-Valued Random Variables. Kluwer, Boston (2002)

[91] Li, X., Liu, B.: A sufficient and necessary condition for credibility measures. International Journal of Uncertainty, Fuzziness & Knowledge-Based Systems 14(5), 527–535 (2006)

[92] Li, X., Liu, B.: Maximum entropy principle for fuzzy variables. International Journal of Uncertainty. Fuzziness & Knowledge-Based Systems 15(supp.2), 43–52 (2007)

[93] Li, X., Liu, B.: On distance between fuzzy variables. Journal of Intelligent & Fuzzy Systems 19(3), 197–204 (2008)

[94] Li, X., Liu, B.: Chance measure for hybrid events with fuzziness and randomness. Soft Computing 13(2), 105–115 (2009)

[95] Li, X., Liu, B.: Foundation of credibilistic logic. Fuzzy Optimization and Decision Making 8(1), 91–102 (2009)

[96] Li, X., Liu, B.: Hybrid logic and uncertain logic. Journal of Uncertain Systems 3(2), 83–94 (2009)

[97] Li, X., Liu, B.: Conditional chance measure for hybrid events, Technical Report (2009)

[98] Li, Y.M., Li, S.J.: A fuzzy sets theoretic approach to approximate spatial reasoning. IEEE Transactions on Fuzzy Systems 12(6), 745–754 (2004)

[99] Liu, B.: Dependent-chance goal programming and its genetic algorithm based approach. Mathematical and Computer Modelling 24(7), 43–52 (1996)

[100] Liu, B., Esogbue, A.O.: Fuzzy criterion set and fuzzy criterion dynamic programming. Journal of Mathematical Analysis and Applications 199(1), 293–311 (1996)

[101] Liu, B.: Dependent-chance programming: A class of stochastic optimization. Computers & Mathematics with Applications 34(12), 89–104 (1997)

[102] Liu, B., Iwamura, K.: Chance constrained programming with fuzzy parameters. Fuzzy Sets and Systems 94(2), 227–237 (1998)

[103] Liu, B., Iwamura, K.: A note on chance constrained programming with fuzzy coefficients. Fuzzy Sets and Systems 100(1-3), 229–233 (1998)

[104] Liu, B.: Minimax chance constrained programming models for fuzzy decision systems. Information Sciences 112(1-4), 25–38 (1998)

[105] Liu, B.: Dependent-chance programming with fuzzy decisions. IEEE Transactions on Fuzzy Systems 7(3), 354–360 (1999)

[106] Liu, B., Esogbue, A.O.: Decision Criteria and Optimal Inventory Processes. Kluwer, Boston (1999)

[107] Liu, B.: Uncertain Programming. Wiley, New York (1999)

[108] Liu, B.: Dependent-chance programming in fuzzy environments. Fuzzy Sets and Systems 109(1), 97–106 (2000)

[109] Liu, B., Iwamura, K.: Fuzzy programming with fuzzy decisions and fuzzy simulation-based genetic algorithm. Fuzzy Sets and Systems 122(2), 253–262 (2001)

[110] Liu, B.: Fuzzy random chance-constrained programming. IEEE Transactions on Fuzzy Systems 9(5), 713–720 (2001)

[111] Liu, B.: Fuzzy random dependent-chance programming. IEEE Transactions on Fuzzy Systems 9(5), 721–726 (2001)

[112] Liu, B.: Theory and Practice of Uncertain Programming. Physica-Verlag, Heidelberg (2002)

[113] Liu, B.: Toward fuzzy optimization without mathematical ambiguity. Fuzzy Optimization and Decision Making 1(1), 43–63 (2002)

[114] Liu, B., Liu, Y.K.: Expected value of fuzzy variable and fuzzy expected value models. IEEE Transactions on Fuzzy Systems 10(4), 445–450 (2002)

[115] Liu, B.: Random fuzzy dependent-chance programming and its hybrid intelligent algorithm. Information Sciences 141(3-4), 259–271 (2002)

[116] Liu, B.: Inequalities and convergence concepts of fuzzy and rough variables. Fuzzy Optimization and Decision Making 2(2), 87–100 (2003)

[117] Liu, B.: Uncertainty Theory: An Introduction to its Axiomatic Foundations. Springer, Berlin (2004)

[118] Liu, B.: A survey of credibility theory. Fuzzy Optimization and Decision Making 5(4), 387–408 (2006)

[119] Liu, B.: A survey of entropy of fuzzy variables. Journal of Uncertain Systems 1(1), 4–13 (2007)

[120] Liu, B.: Uncertainty Theory, 2nd edn. Springer, Berlin (2007)

[121] Liu, B.: Fuzzy process, hybrid process and uncertain process. Journal of Uncertain Systems 2(1), 3–16 (2008)

[122] Liu, B.: Theory and Practice of Uncertain Programming, 2nd edn. Springer, Berlin (2009)

[123] Liu, B.: Some research problems in uncertainty theory. Journal of Uncertain Systems 3(1), 3–10 (2009)

[124] Liu, B.: Uncertain entailment and modus ponens in the framework of uncertain logic. Journal of Uncertain Systems 3(4), 243–251 (2009)

[125] Liu, B.: Uncertain set theory and uncertain inference rule with application to uncertain control. Journal of Uncertain Systems 4(2), 83–98 (2010)

[126] Liu, B.: Uncertain risk analysis and uncertain reliability analysis. Journal of Uncertain Systems 4(3), 163–170 (2010)

[127] Liu, L.Z., Li, Y.Z.: The fuzzy quadratic assignment problem with penalty: New models and genetic algorithm. Applied Mathematics and Computation 174(2), 1229–1244 (2006)

[128] Liu, S.F., Lin, Y.: Grey Information: Theory and Practical Applications. Springer, London (2006)

[129] Liu, W., Xu, J.: Some properties on expected value operator for uncertain variables. Information: An International Interdisciplinary Journal 13 (2010)

[130] Liu, X.C.: Entropy, distance measure and similarity measure of fuzzy sets and their relations. Fuzzy Sets and Systems 52, 305–318 (1992)

[131] Liu, X.W.: Measuring the satisfaction of constraints in fuzzy linear programming. Fuzzy Sets and Systems 122(2), 263–275 (2001)

[132] Liu, Y.H., Ha, M.H.: Expected value of function of uncertain variables. Journal of Uncertain Systems 4(3) (2010)

[133] Liu, Y.H.: Currency models with uncertain exchange rate, http://orsc.edu.cn/online/091010.pdf

[134] Liu, Y.H.: Uncertain stock model with periodic dividends, http://orsc.edu.cn/online/091011.pdf

[135] Liu, Y.K., Liu, B.: Random fuzzy programming with chance measures defined by fuzzy integrals. Mathematical and Computer Modelling 36(4-5), 509–524 (2002)

[136] Liu, Y.K., Liu, B.: Fuzzy random variables: A scalar expected value operator. Fuzzy Optimization and Decision Making 2(2), 143–160 (2003)

[137] Liu, Y.K., Liu, B.: Expected value operator of random fuzzy variable and random fuzzy expected value models, International Journal of Uncertainty. Fuzziness & Knowledge-Based Systems 11(2), 195–215 (2003)

[138] Liu, Y.K., Liu, B.: A class of fuzzy random optimization: Expected value models. Information Sciences 155(1-2), 89–102 (2003)

[139] Liu, Y.K., Liu, B.: Fuzzy random programming with equilibrium chance constraints. Information Sciences 170, 363–395 (2005)

[140] Liu, Y.K.: Fuzzy programming with recourse. International Journal of Uncertainty. Fuzziness & Knowledge-Based Systems 13(4), 381–413 (2005)

[141] Liu, Y.K.: Convergent results about the use of fuzzy simulation in fuzzy optimization problems. IEEE Transactions on Fuzzy Systems 14(2), 295–304 (2006)

[142] Liu, Y.K., Gao, J.: The independence of fuzzy variables with applications to fuzzy random optimization. International Journal of Uncertainty, Fuzziness & Knowledge-Based Systems 2(supp.2), 1–20 (2007)

[143] Loo, S.G.: Measures of fuzziness. Cybernetica 20, 201–210 (1977)

[144] Lopez-Diaz, M., Ralescu, D.A.: Tools for fuzzy random variables: Embeddings and measurabilities. Computational Statistics & Data Analysis 51(1), 109–114 (2006)

[145] Lu, M.: On crisp equivalents and solutions of fuzzy programming with different chance measures. Information: An International Journal 6(2), 125–133 (2003)

[146] Lucas, C., Araabi, B.N.: Generalization of the Dempster-Shafer Theory: A fuzzy-valued measure. IEEE Transactions on Fuzzy Systems 7(3), 255–270 (1999)

[147] Luhandjula, M.K., Gupta, M.M.: On fuzzy stochastic optimization. Fuzzy Sets and Systems 81, 47–55 (1996)

[148] Luhandjula, M.K.: Fuzzy stochastic linear programming: Survey and future research directions. European Journal of Operational Research 174(3), 1353–1367 (2006)

[149] Maiti, M.K., Maiti, M.A.: Fuzzy inventory model with two warehouses under possibility constraints. Fuzzy Sets and Systems 157(1), 52–73 (2006)

[150] Maleki, H.R., Tata, M., Mashinchi, M.: Linear programming with fuzzy variables. Fuzzy Sets and Systems 109(1), 21–33 (2000)

[151] Merton, R.C.: Theory of rational option pricing. Bell Journal of Economics and Management Science 4, 141–183 (1973)

[152] Mizumoto, M., Tanaka, K.: Some properties of fuzzy sets of type 2. Information and Control 31, 312–340 (1976)

[153] Molchanov, I.S.: Limit Theorems for Unions of Random Closed Sets. Springer, Berlin (1993)

[154] Möller, B., Beer, M.: Engineering computation under uncertainty. Computers and Structures 86, 1024–1041 (2008)

[155] Nahmias, S.: Fuzzy variables. Fuzzy Sets and Systems 1, 97–110 (1978)

[156] Negoita, C.V., Ralescu, D.A.: Representation theorems for fuzzy concepts. Kybernetes 4, 169–174 (1975)

[157] Negoita, C.V., Ralescu, D.A.: Simulation, Knowledge-based Computing, and Fuzzy Statistics. Van Nostrand Reinhold, New York (1987)

[158] Neumaier, A.: Interval Methods for Systems of Equations. Cambridge University Press, New York (1990)

[159] Nguyen, H.T.: On conditional possibility distributions. Fuzzy Sets and Systems 1, 299–309 (1978)

[160] Nguyen, H.T., Nguyen, N.T., Wang, T.H.: On capacity functionals in interval probabilities. International Journal of Uncertainty, Fuzziness & Knowledge-Based Systems 5, 359–377 (1997)

[161] Nguyen, H.T., Kreinovich, V., Wu, B.L.: Fuzzy/probability similar to fractal/smooth. International Journal of Uncertainty, Fuzziness & Knowledge-Based Systems 7, 363–370 (1999)

[162] Nguyen, V.H.: Fuzzy stochastic goal programming problems. European Journal of Operational Research 176(1), 77–86 (2007)

[163] Nilsson, N.J.: Probabilistic logic. Artificial Intelligence 28, 71–87 (1986)

[164] Øksendal, B.: Stochastic Differential Equations, 6th edn. Springer, Berlin (2005)

[165] Pal, N.R., Pal, S.K.: Object background segmentation using a new definition of entropy. IEE Proc. E 136, 284–295 (1989)

[166] Pal, N.R., Pal, S.K.: Higher order fuzzy entropy and hybrid entropy of a set. Information Sciences 61, 211–231 (1992)

[167] Pal, N.R., Bezdek, J.C.: Measuring fuzzy uncertainty. IEEE Transactions on Fuzzy Systems 2, 107–118 (1994)

[168] Pawlak, Z.: Rough sets. International Journal of Information and Computer Sciences 11(5), 341–356 (1982)

[169] Pawlak, Z.: Rough Sets: Theoretical Aspects of Reasoning about Data. Kluwer, Dordrecht (1991)

[170] Pedrycz, W.: Optimization schemes for decomposition of fuzzy relations. Fuzzy Sets and Systems 100, 301–325 (1998)

[171] Peng, J., Liu, B.: Parallel machine scheduling models with fuzzy processing times. Information Sciences 166(1-4), 49–66 (2004)

[172] Peng, J., Zhao, X.D.: Some theoretical aspects of the critical values of biran-dom variable. Journal of Information and Computing Science 1(4), 225–234 (2006)

[173] Peng, J., Liu, B.: Birandom variables and birandom programming. Comput-ers & Industrial Engineering 53(3), 433–453 (2007)

[174] Peng, J.: A general stock model for fuzzy markets. Journal of Uncertain Systems 2(4), 248–254 (2008)

[175] Peng, J.: A stock model for uncertain markets, http://orsc.edu.cn/online/090209.pdf

[176] Peng, Z.X.: Some properties of product uncertain measure, http://orsc.edu.cn/online/081228.pdf

[177] Peng, Z.X., Iwamura, K.: A sufficient and necessary condition of uncertainty distribution, http://orsc.edu.cn/online/090305.pdf

[178] Peng, Z.X.: Uncertain systems are universal approximators, http://orsc.edu.cn/online/100110.pdf

[179] Puri, M.L., Ralescu, D.: Fuzzy random variables. Journal of Mathematical Analysis and Applications 114, 409–422 (1986)

[180] Qin, Z.F., Li, X.: Option pricing formula for fuzzy financial market. Journal of Uncertain Systems 2(1), 17–21 (2008)

[181] Qin, Z.F., Gao, X.: Fractional Liu process with application to finance. Math-ematical and Computer Modelling 50(9-10), 1538–1543 (2009)

[182] Ralescu, D.A.: A generalization of representation theorem. Fuzzy Sets and Systems 51, 309–311 (1992)

[183] Ralescu, D.A., Sugeno, M.: Fuzzy integral representation. Fuzzy Sets and Systems 84(2), 127–133 (1996)

[184] Ralescu, A.L., Ralescu, D.A.: Extensions of fuzzy aggregation. Fuzzy Sets and Systems 86(3), 321–330 (1997)

[185] Ramer, A.: Conditional possibility measures. International Journal of Cyber-netics and Systems 20, 233–247 (1989)

[186] Ramík, J.: Extension principle in fuzzy optimization. Fuzzy Sets and Sys-tems 19, 29–35 (1986)

[187] Robbins, H.E.: On the measure of a random set. Annals of Mathematical Statistics 15(1), 70–74 (1994)

[188] Saade, J.J.: Maximization of a function over a fuzzy domain. Fuzzy Sets and Systems 62, 55–70 (1994)

[189] Sakawa, M., Nishizaki, I., Uemura, Y.: Interactive fuzzy programming for two-level linear fractional programming problems with fuzzy parameters. Fuzzy Sets and Systems 115, 93–103 (2000)

[190] Shafer, G.: A Mathematical Theory of Evidence. Princeton University Press, Princeton (1976)

[191] Shannon, C.E.: The Mathematical Theory of Communication. The University of Illinois Press, Urbana (1949)

[192] Shao, Z., Ji, X.Y.: Fuzzy multi-product constraint newsboy problem. Applied Mathematics and Computation 180(1), 7–15 (2006)

[193] Shih, H.S., Lai, Y.J., Lee, E.S.: Fuzzy approach for multilevel programming problems. Computers and Operations Research 23, 73–91 (1996)

[194] Shreve, S.E.: Stochastic Calculus for Finance II: Continuous-Time Models. Springer, Berlin (2004)

[195] Slowinski, R., Teghem, J.: Fuzzy versus stochastic approaches to multicriteria linear programming under uncertainty. Naval Research Logistics 35, 673–695 (1988)

[196] Slowinski, R., Vanderpooten, D.: A generalized definition of rough approximations based on similarity. IEEE Transactions on Knowledge and Data Engineering 12(2), 331–336 (2000)

[197] Steuer, R.E.: Algorithm for linear programming problems with interval objective function coefficients. Mathematics of Operational Research 6, 333–348 (1981)

[198] Sugeno, M.: Theory of Fuzzy Integrals and its Applications. Ph.D. Dissertation, Tokyo Institute of Technology (1974)

[199] Szmidt, E., Kacprzyk, J.: Entropy for intuitionistic fuzzy sets. Fuzzy Sets and Systems 118, 467–477 (2001)

[200] Taleizadeh, A.A., Niaki, S.T.A., Aryanezhad, M.B.: A hybrid method of Pareto, TOPSIS and genetic algorithm to optimize multi-product multi-constraint inventory control systems with random fuzzy replenishments. Mathematical and Computer Modelling 49(5-6), 1044–1057 (2009)

[201] Tanaka, H., Asai, K.: Fuzzy solutions in fuzzy linear programming problems. IEEE Transactions on Systems, Man and Cybernetics 14, 325–328 (1984)

[202] Tanaka, H., Guo, P.: Possibilistic Data Analysis for Operations Research. Physica-Verlag, Heidelberg (1999)

[203] Tian, D.Z., Wang, L., Wu, J., Ha, M.H.: Rough set model based on uncertain measure. Journal of Uncertain Systems 3(4), 252–256 (2009)

[204] Torabi, H., Davvaz, B., Behboodian, J.: Fuzzy random events in incomplete probability models. Journal of Intelligent & Fuzzy Systems 17(2), 183–188 (2006)

[205] Wang, G.J., Wang, H.: Non-fuzzy versions of fuzzy reasoning in classical logics. Information Sciences 138(1-4), 211–236 (2001)

[206] Wang, G.J.: Formalized theory of general fuzzy reasoning. Information Sciences 160(1-4), 251–266 (2004)

[207] Wang, X.S., Yang, W.G.: Method of moments for estimating uncertainty distributions, http://orsc.edu.cn/online/100408.pdf

[208] Wang, X.Z., Wang, Y.D., Xu, X.F.: A new approach to fuzzy rule generation: fuzzy extension matrix. Fuzzy Sets and Systems 123(3), 291–306 (2001)

[209] Wang, Z.Y., Klir, G.J.: Fuzzy Measure Theory. Plenum Press, New York (1992)

[210] Yager, R.R.: On measures of fuzziness and negation, Part I: Membership in the unit interval. International Journal of General Systems 5, 221–229 (1979)

[211] Yager, R.R.: Entropy and specificity in a mathematical theory of evidence. International Journal of General Systems 9, 249–260 (1983)

[212] Yager, R.R.: On the entropy of fuzzy measures. IEEE Transactions on Fuzzy Systems 8, 453–461 (2000)

[213] Yager, R.R.: On the evaluation of uncertain courses of action. Fuzzy Optimization and Decision Making 1, 13–41 (2002)

[214] Yang, L.X., Liu, B.: On inequalities and critical values of fuzzy random variable. International Journal of Uncertainty, Fuzziness & Knowledge-Based Systems 13(2), 163–175 (2005)

[215] Yang, L.X., Liu, B.: On continuity theorem for characteristic function of fuzzy variable. Journal of Intelligent & Fuzzy Systems 17(3), 325–332 (2006)

[216] Yang, L.X., Li, K.: B-valued fuzzy variable. Soft Computing 12(11), 1081–1088 (2008)

[217] Yang, N., Wen, F.S.: A chance constrained programming approach to transmission system expansion planning. Electric Power Systems Research 75(2-3), 171–177 (2005)

[218] Yang, X.H.: Moments and tails inequality within the framework of uncertainty theory. Information: An International Interdisciplinary Journal 13 (2010)

[219] Yazenin, A.V.: On the problem of possibilistic optimization. Fuzzy Sets and Systems 81, 133–140 (1996)

[220] You, C., Wen, M.: The entropy of fuzzy vectors. Computers & Mathematics with Applications 56(6), 1626–1633 (2008)

[221] You, C.: Some convergence theorems of uncertain sequences. Mathematical and Computer Modelling 49(3-4), 482–487 (2009)

[222] Zadeh, L.A.: Fuzzy sets. Information and Control 8, 338–353 (1965)

[223] Zadeh, L.A.: Outline of a new approach to the analysis of complex systems and decision processes. IEEE Transactions on Systems, Man and Cybernetics 3, 28–44 (1973)

[224] Zadeh, L.A.: The concept of a linguistic variable and its application to approximate reasoning. Information Sciences 8, 199–251 (1975)

[225] Zadeh, L.A.: Fuzzy sets as a basis for a theory of possibility. Fuzzy Sets and Systems 1, 3–28 (1978)

[226] Zadeh, L.A.: A theory of approximate reasoning. In: Hayes, J., Michie, D., Thrall, R.M. (eds.) Mathematical Frontiers of the Social and Policy Sciences, pp. 69–129. Westview Press, Boulder (1979)

[227] Zhang, X.F., Peng, Z.X.: Uncertain predicate logic based on uncertainty theory, http://orsc.edu.cn/online/091204.pdf

[228] Zhao, R., Liu, B.: Stochastic programming models for general redundancy optimization problems. IEEE Transactions on Reliability 52(2), 181–191 (2003)

[229] Zhao, R., Liu, B.: Renewal process with fuzzy interarrival times and rewards. International Journal of Uncertainty, Fuzziness & Knowledge-Based Systems 11(5), 573–586 (2003)

[230] Zhao, R., Liu, B.: Redundancy optimization problems with uncertainty of combining randomness and fuzziness. European Journal of Operational Research 157(3), 716–735 (2004)

[231] Zhao, R., Liu, B.: Standby redundancy optimization problems with fuzzy lifetimes. Computers & Industrial Engineering 49(2), 318–338 (2005)

[232] Zhao, R., Tang, W.S., Yun, H.L.: Random fuzzy renewal process. European Journal of Operational Research 169(1), 189–201 (2006)

[233] Zhao, R., Tang, W.S.: Some properties of fuzzy random renewal process. IEEE Transactions on Fuzzy Systems 14(2), 173–179 (2006)

[234] Zheng, Y., Liu, B.: Fuzzy vehicle routing model with credibility measure and its hybrid intelligent algorithm. Applied Mathematics and Computation 176(2), 673–683 (2006)

[235] Zhou, J., Liu, B.: New stochastic models for capacitated location-allocation problem. Computers & Industrial Engineering 45(1), 111–125 (2003)

[236] Zhou, J., Liu, B.: Analysis and algorithms of bifuzzy systems. International Journal of Uncertainty, Fuzziness & Knowledge-Based Systems 12(3), 357–376 (2004)

[237] Zhou, J., Liu, B.: Modeling capacitated location-allocation problem with fuzzy demands. Computers & Industrial Engineering 53(3), 454–468 (2007)

[238] Zhu, Y., Liu, B.: Continuity theorems and chance distribution of random fuzzy variable. Proceedings of the Royal Society of London Series A 460, 2505–2519 (2004)

[239] Zhu, Y., Liu, B.: Some inequalities of random fuzzy variables with application to moment convergence. Computers & Mathematics with Applications 50(5-6), 719–727 (2005)

[240] Zhu, Y., Ji, X.Y.: Expected values of functions of fuzzy variables. Journal of Intelligent & Fuzzy Systems 17(5), 471–478 (2006)

[241] Zhu, Y., Liu, B.: Fourier spectrum of credibility distribution for fuzzy variables. International Journal of General Systems 36(1), 111–123 (2007)

[242] Zhu, Y., Liu, B.: A sufficient and necessary condition for chance distribution of random fuzzy variables. International Journal of Uncertainty, Fuzziness & Knowledge-Based Systems 2(supp.2), 21–28 (2007)

[243] Zhu, Y.: A fuzzy optimal control model. Journal of Uncertain Systems 3(4), 270–279 (2009)

[244] Zhu, Y.: Uncertain optimal control with application to a portfolio selection model, http://orsc.edu.cn/online/090524.pdf

[245] Zimmermann, H.J.: Fuzzy Set Theory and its Applications. Kluwer Academic Publishers, Boston (1985)

[246] Zuo, Y., Ji, X.Y.: Theoretical foundation of uncertain dominance. In: Proceedings of the Eighth International Conference on Information and Management Sciences, Kunming, China, July 20-28, pp. 827–832 (2009)

[247] Zuo, Y.: Critical values of function of uncertain variables, http://orsc.edu.cn/online/090807.pdf

List of Frequently Used Symbols

$\mathcal{M}$	uncertain measure
$(\Gamma, \mathcal{L}, \mathcal{M})$	uncertainty space
ξ, η, τ	uncertain variables
Φ, Ψ, Υ	uncertainty distributions
μ, ν, λ	membership functions
$\mathcal{L}(a, b)$	linear uncertain variable
$\mathcal{Z}(a, b, c)$	zigzag uncertain variable
$\mathcal{N}(e, \sigma)$	normal uncertain variable
$\mathcal{LOGN}(e, \sigma)$	lognormal uncertain variable
(a, b)	rectangular uncertain set
(a, b, c)	triangular uncertain set
(a, b, c, d)	trapezoidal uncertain set
E	expected value
V	variance
H	entropy
$\xi_{\sup}(\alpha)$	α-optimistic value to ξ
$\xi_{\inf}(\alpha)$	α-pessimistic value to ξ
μ_α	α-cut of membership function μ
X_t, Y_t, Z_t	uncertain processes
C_t	canonical process
N_t	renewal process
$\emptyset$	empty set
$\Re$	set of real numbers
$\vee$	maximum operator
$\wedge$	minimum operator
$\triangleright$	imaginary inclusion
$a_i \uparrow a$	$a_1 \leq a_2 \leq \cdots$ and $a_i \to a$
$a_i \downarrow a$	$a_1 \geq a_2 \geq \cdots$ and $a_i \to a$
$A_i \uparrow A$	$A_1 \subset A_2 \subset \cdots$ and $A = A_1 \cup A_2 \cup \cdots$
$A_i \downarrow A$	$A_1 \supset A_2 \supset \cdots$ and $A = A_1 \cap A_2 \cap \cdots$

Five Plus One

Uncertainty theory is a branch of mathematics based the following five axioms plus one principle:

Axiom 1. (Normality Axiom) $\mathcal{M}\{\Gamma\} = 1$ for the universal set Γ.

Axiom 2. (Monotonicity Axiom) $\mathcal{M}\{\Lambda_1\} \leq \mathcal{M}\{\Lambda_2\}$ whenever $\Lambda_1 \subset \Lambda_2$.

Axiom 3. (Self-Duality Axiom) $\mathcal{M}\{\Lambda\} + \mathcal{M}\{\Lambda^c\} = 1$ for any event Λ.

Axiom 4. (Countable Subadditivity Axiom) For every countable sequence of events $\Lambda_1, \Lambda_2, \cdots$, we have

$$\mathcal{M}\left\{\bigcup_{i=1}^{\infty} \Lambda_i\right\} \leq \sum_{i=1}^{\infty} \mathcal{M}\{\Lambda_i\}.$$

Axiom 5. (Product Measure Axiom) Let $(\Gamma_k, \mathcal{L}_k, \mathcal{M}_k)$ be uncertainty spaces for $k = 1, 2, \cdots, n$. Then the product uncertain measure $\mathcal{M}$ is an uncertain measure on the product σ-algebra $\mathcal{L}_1 \times \mathcal{L}_2 \times \cdots \times \mathcal{L}_n$ satisfying

$$\mathcal{M}\left\{\prod_{k=1}^{n} \Lambda_k\right\} = \min_{1 \leq k \leq n} \mathcal{M}_k\{\Lambda_k\}.$$

That is, for each event $\Lambda \in \mathcal{L}_1 \times \mathcal{L}_2 \times \cdots \times \mathcal{L}_n$, we have

$$\mathcal{M}\{\Lambda\} = \begin{cases} \sup\limits_{\Lambda_1 \times \Lambda_2 \times \cdots \times \Lambda_n \subset \Lambda} \min\limits_{1 \leq k \leq n} \mathcal{M}_k\{\Lambda_k\}, \\ \qquad \text{if} \quad \sup\limits_{\Lambda_1 \times \Lambda_2 \times \cdots \times \Lambda_n \subset \Lambda} \min\limits_{1 \leq k \leq n} \mathcal{M}_k\{\Lambda_k\} > 0.5 \\ 1 - \sup\limits_{\Lambda_1 \times \Lambda_2 \times \cdots \times \Lambda_n \subset \Lambda^c} \min\limits_{1 \leq k \leq n} \mathcal{M}_k\{\Lambda_k\}, \\ \qquad \text{if} \quad \sup\limits_{\Lambda_1 \times \Lambda_2 \times \cdots \times \Lambda_n \subset \Lambda^c} \min\limits_{1 \leq k \leq n} \mathcal{M}_k\{\Lambda_k\} > 0.5 \\ 0.5, \qquad \text{otherwise.} \end{cases}$$

Principle. (Maximum Uncertainty Principle) For any event, if there are multiple reasonable values that an uncertain measure may take, then the value as close to 0.5 as possible is assigned to the event.

Index